The New Europe
Into the 1990s

G.N. Minshull, MA, MSc(ECS).
*Formerly Senior Lecturer in Geography
Worcester College of Higher Education*

Hodder & Stoughton
LONDON SYDNEY AUCKLAND

Acknowledgements

Adam Opel A G (p. 54); Aerofilms Ltd (pp. 53, 158, 160); Articapress (p.121); Bart Hofmeester (pp. 33, 166, 251); Belgian Institute of Information and Documentation (pp. 11, 55, 90, 231, 236); Belgian National Tourist Office (p. 48); British Steel Corporation (p. 85); The Danish Tourist Board (pp. 133, 258, 261); The Embassy of the Federal Government of West Germany (pp. 29, 42, 208); The French Government Tourist Office (pp. 79, 311); The German National Tourist Office (pp. 103, 213, 224); Horizon Holidays Ltd (p. 321); Imperial Chemical Industries (p. 116); The Italian Institute (pp. 94, 274, 277, 279); The Italian State Tourist Office (p. 272); Kommunalverband Ruhrgebiet (p. 212); National Tourist Organisation of Greece (pp. 299, 300, 302); Netherlands National Tourist Office (p. 127); Portuguese National Tourist Office (pp. 320, 325); The Royal Danish Ministry for Foreign Affairs (p. 261); The Royal Netherlands Embassy (pp. 164, 243, 244); Schiphol Airport Authority (pp. 248); The Shannon Free Airport Development Authority (p. 297).

The photograph on the front cover shows Frankfurt city centre. Source: Stadt Frankfurt – am – Main.

British Library Cataloguing in Publication Data

Minshull, G.N. (Gordon Neil), 1933–1989
  The new Europe: an economic geography of Europe in the 1990s. – 4th ed.
  1. European Community countries. Economic conditions. Geographical aspects
  I. Title
  330.94

  ISBN 0 340 505125

First printed 1978
Second edition 1980
Third edition 1985
Fourth edition 1990
Impression number   10 9 8 7 6 5 4
Year   1998 1997 1996 1995 1994 1993

© 1990 G. N. Minshull

All rights reserved. No part of this publication may be reproduced or transmitted in any form or by any means, electronic or mechanical, including photocopy, recording, or any information storage and retrieval system, without permission in writing from the publisher or under licence from the Copyright Licensing Agency Limited. Further details of such licences (for reprographic reproduction) may be obtained from the Copyright Licensing Agency Limited, of 90 Tottenham Court Road, London W1P 9HE.

Typeset by Gecko Limited, Bicester.
Printed in Great Britain for the educational publishing division of Hodder and Stoughton Limited, Mill Road, Dunton Green, Sevenoaks, Kent by St Edmundsbury Press Limited, Bury St Edmunds, Suffolk.

Contents

Acknowledgements ii

Preface iv

1 The development of Western Europe: the European Community 1
2 Energy: a variety of sources 24
3 Employment: Industrial change and the Tertiary sector 47
4 The iron and steel industry: integration and rationalisation 81
5 Other industries: automobiles, textiles and chemicals 99
6 Agriculture: the Common Agricultural Policy 118
7 Trade: the world's largest trading group 144
8 Transport: movement on a continental scale 153
9 Population: the axis of city development 171
10 Regional disparities: core and periphery 187
11 North-Rhine Westphalia: crossroads of Europe 205
12 The Middle Rhinelands: three city regions 217
13 Belgium: a study in regional contrasts 227
14 Randstad Holland: the Ring City 240
15 Denmark: dairy specialist and gateway to northern Europe 255
16 Piedmont, Lombardy and Liguria: the economic core of Italy 264
17 The Mezzogiorno: Italy's problem region 276
18 Greece and Ireland: the periphery 291
19 The Paris region: a problem of definition 304
20 Spain and Portugal 317
21 The integration process: The Single European Act and progress towards the Single European Market 329
Appendix: key dates in European integration 338
Glossary 342
References 343
Index 346

Preface to the Fourth Edition

One of the most significant developments in Western Europe since 1945 has been the growth of integration. This book deals with the emergence of the European Community, formerly known as the EEC or Common Market, as a geographically significant, economically powerful and politically durable unit. The book gives an overview of the origins and evolution of the European Community since 1951, integration policies, the dramatic growth in material wealth during the 1960s, the resource base and patterns of economic and regional geography. Since the first oil price shock of 1973, the industrialised world has experienced major changes in its terms of trade. Conditions of rapid inflation have been followed by the depression of 1979–82, industrial decline, unemployment, and severe regional problems. Nevertheless, the European Community is a very important economic unit and a major trading bloc. Its institutions have become more democratic, and its external relationships, particularly with the Third World, have become much more significant.

The difficult economic period of the late 1970s and early 1980s was accompanied by a slowing down of the pace of integration. This has been referred to as 'Eurosclerosis'. However, the accession to the Community of Greece, Spain and Portugal, has added a new geographical dimension and has assisted in a 'great leap forward' in integration. The European Single Act of 1987 is accompanied by the removal of all remaining barriers to trade and economic harmonisation which culminates in the European Single Market by 1992. Moves towards economic and monetary union (EMU) were initiated at the Madrid summit in 1989. In a global context, as the relative decline of the superpowers takes place, the European Community is experiencing a major enhancement of its economic strength and political status. It remains to be seen what impact recent developments in Eastern Europe have.

The fourth edition of this book deals with all twelve members of the European Community, with particular reference to the four largest member states, West Germany, France, the United Kingdom and Italy, who are all members of the 'Group of Seven' most powerful world industrial states. The economic geography of the book stresses the key concept of core and periphery and is organised around a balance between systematic and regional geography. Major revisions to all chapters have been made, with statistics updated to the late 1980s, and there is a new chapter 21 which deals with the Single European Market and its potential for continued progress in the 1990s.

This experiment in European Integration is of interest to geographers in demonstrating the indivisibility of knowledge. It is a case study of the developed world which illustrates the physical character and resource-base of Western Europe, the historical and cultural diversity of nation states, regional contrasts between core and periphery, the shrinkage of distance, and the restructuring of economic and social space by means of political cooperation.

G N Minshull

1

The development of Western Europe: the European Community

Introduction

The theme of this book is the economic geography of Western Europe and the internal and external pressures which are altering its whole economic and political structure. Essentially it relates to the growth and enlargement of the European Community, often known as 'The Common Market' or EEC (fig. 1.1).

The principal characteristics of Western Europe may be summarised thus: countries with a relatively small land area and great physical diversity; many nation states separated by historical precedent and language barriers; wealthy countries with relatively dense populations; an industrialised and urbanised way of life; a well-watered environment with adequate fertile land and a temperate climate which supports a commercial system of agriculture.

Physical configuration

Western Europe is a triangular-shaped sub-continent with its base to the east merging into the Eurasian land-mass. It is a rim land tapering towards the west and having the Atlantic Ocean on its north-western side and the Mediterranean Basin on the south. The overriding characteristics are a relatively small land-area and an irregular configuration consisting of diverse peninsulas, islands and shallow continental seas.

There are four broad physiographic regions (fig. 1.2) and these underline the diversity and emphasise an east-west alignment. On the north-western rim lie the Atlantic uplands and Scandinavian mountains. The single most significant physical region is the north-west European Plain, an extensive lowland stretching from the Bay of Biscay across France into the Rhinelands. A complex central region of mountains, valleys and plateaux (Hercynian) lies from the Massif Central across into the Rhine Uplands. This is succeeded to the south by the Alpine system, the highest in Western Europe with peaks exceeding 4000 m, which stretches across the Mediterranean coastlands from Spain to Italy and Greece. Contained within the Alpine system are further examples of hercynian plateaux such as the Spanish Meseta, basins of tertiary deposits such as the Ebro valley in Spain and the Plain of Lombardy in Northern Italy, and the fringing lowlands and islands of the Mediterranean Basin.

2 THE NEW EUROPE

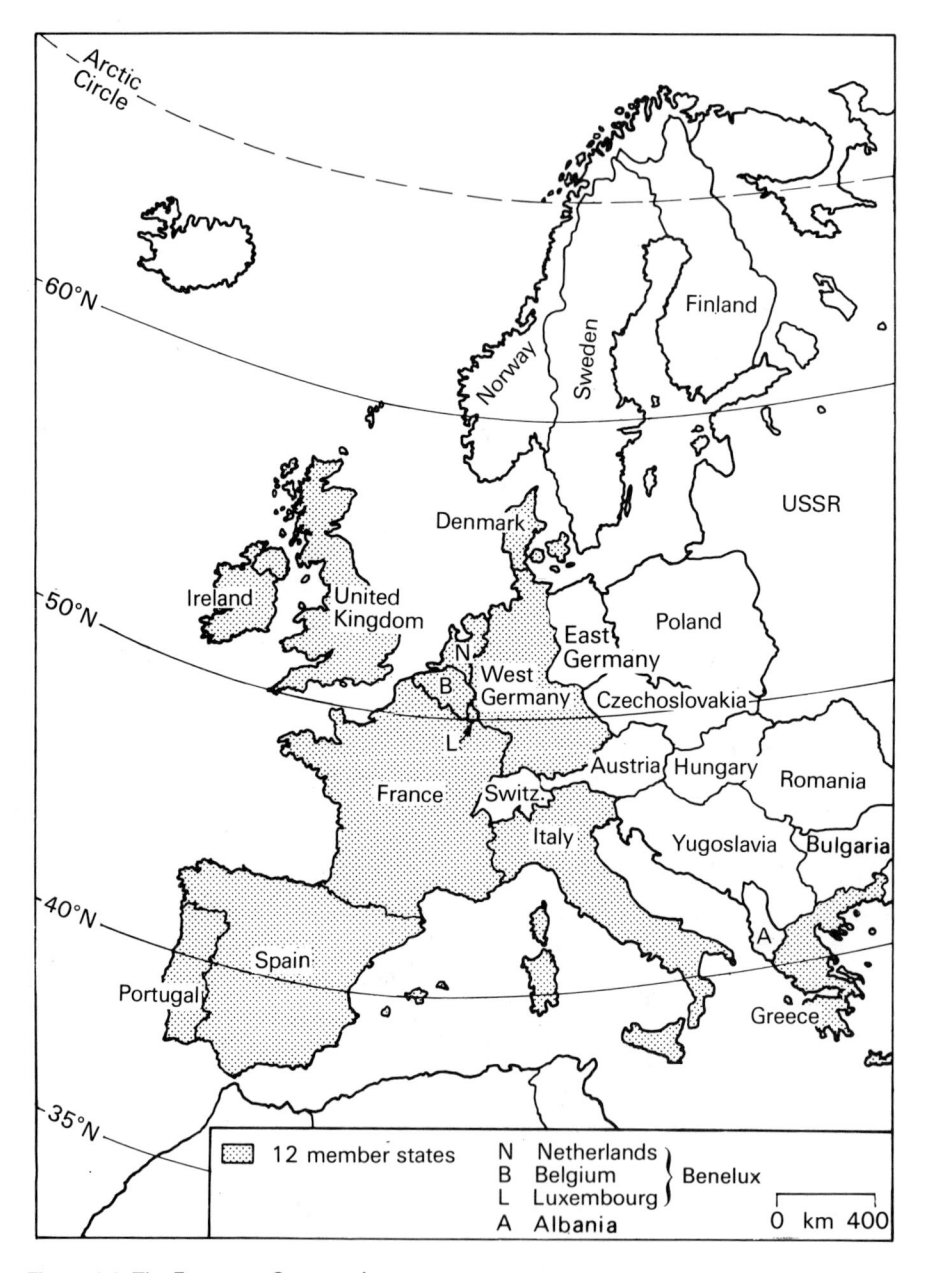

Figure 1.1 The European Community

There are *three* significant implications for the human and economic geography of Western Europe:

(a) The most extensive areas of fertile productive lowlands lie in the north and west, whilst the Mediterranean regions are mountainous, with small, discontinuous lowlands, and generally adverse for development. The

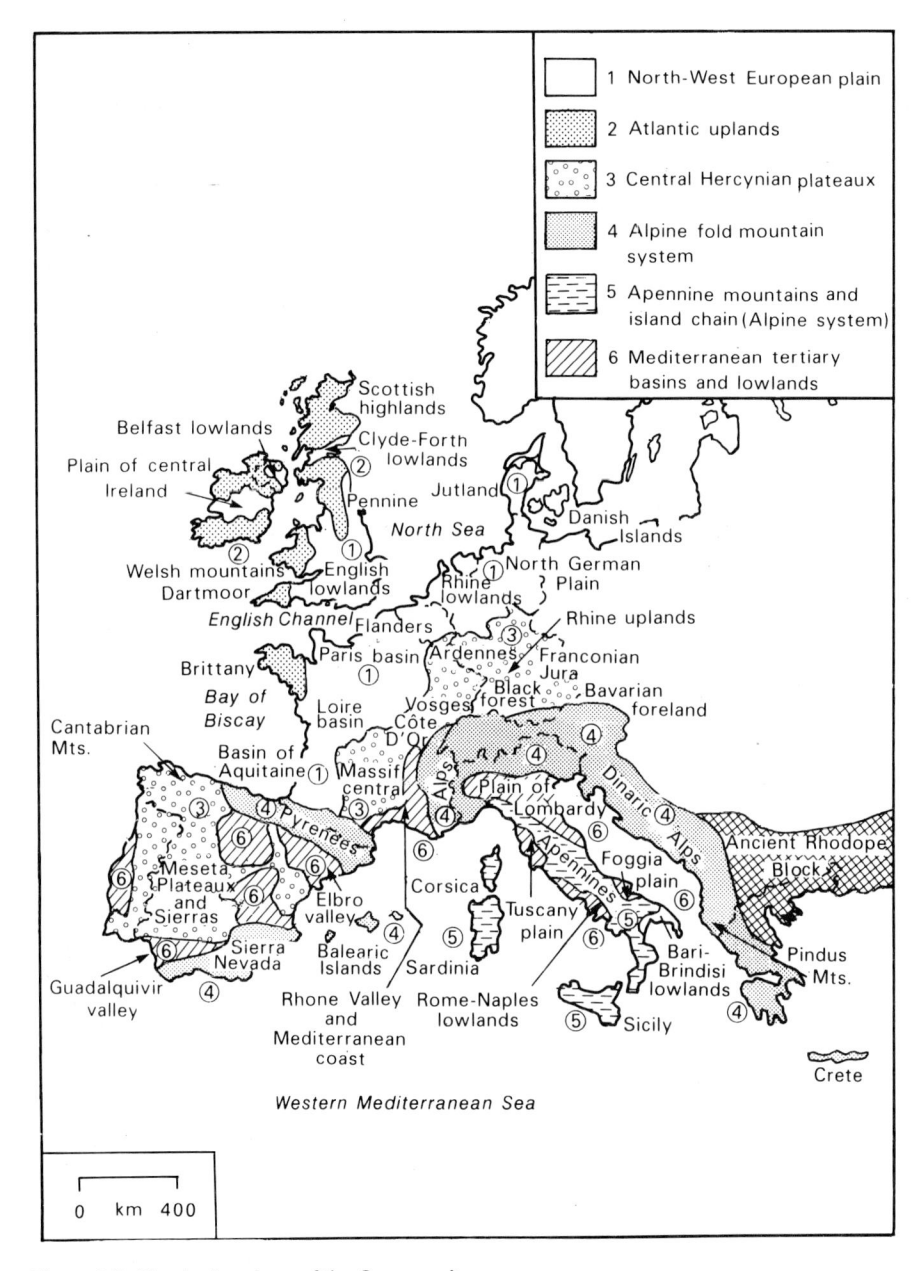

Figure 1.2 Physical regions of the Community

agricultural productivity, mineral resources and population density of north-west Europe have enabled it to become the 'core' region, whilst the Mediterranean regions have in many ways become the 'peripheral' zones.
(b) There is a contrast to be found between largely continental states like France and Germany, and traditionally maritime states such as the

4 THE NEW EUROPE

United Kingdom and the Netherlands. The configuration of the coastline has helped to create a basic diversity of interests.

(c) The east-west trend of the major physical features is pierced by several very important north-south routeways based on river valleys. The two most important of these are the rivers Rhine and Rhône and their tributaries, and their economic effect has been immense. They form major core areas, particularly in the case of the Rhine valley, and they have been instrumental since ancient times in creating routeways and links between the Mediterranean and the North Sea. With the additional element of Alpine passes such as the St. Gotthard and Mt. Cenis, the mountain systems have never been the barrier they might have been. Trading links and interdependence between north and south have been considerable since Roman times.

Political fragmentation

Western Europe is politically most diverse with many different groups of race, culture, language and religion. It was here that the nation-state originated. During the Roman Empire, and later under Charlemagne, considerable parts of the sub-continent lay under one rule, but this has never been the normal pattern. It has only occurred for short periods since then, during war, and imposed by force, as in 1939–45.

The earliest nation states included France and England. France is a relatively compact nation state based around its core area, the Paris Basin. It is a key country forming a 'bridgeland' and having a nodal position between the Atlantic Ocean, Mediterranean Sea, English Channel and Rhinelands. In England's case, her expansion to become the United Kingdom was assisted by the unifying and protective effects of the sea. The sea has also had a stimulating effect upon her outward looking maritime character and was a major factor in the establishment of her empire.

Other nation states illustrate the changing patterns of territory and the strains of nationalism. The Netherlands originated in the sixteenth century after a war of independence from Spain, and Belgium, created as a buffer state in 1830, is an uneasy coalition of Flemings and Walloons. Small territories such as Luxembourg are a reminder of the medieval importance of the Grand Duchy as a fortified city-state. Italy and Germany were both merely 'geographical expressions' until 1866 and 1871 respectively. Since then, particularly in Germany's case, the impact of nationalism and militarism has been responsible for many territorial upheavals. The Rhineland economic centre of Western Europe was an area of political instability and divided between five countries. Alsace, Lorraine and the Saarland have changed hands between France and Germany on several occasions.

Wealth from trade

Western Europe produced great trading nations from very early times. Rome was a great trading empire as well as a military one, and the city states of Venice and Genoa were founded upon trade with the Mediterranean and the East. As early as the eleventh and twelfth centuries the Netherlands and Belgium were the centres for the trading activities of the Hanseatic League, and Bruges, Ghent and Amsterdam were famous trading ports. From 1492 Spain and Portugal led the European thrust into the 'New World' and built up great empires in South and Central America, Africa and the Indies. They were followed by France, The Netherlands and in particular by Great Britain whose overseas trade and empire was based upon sea power. The accrued wealth from commerce helped build up capital and laid the foundations for the Industrial Revolution. The supply of raw materials from overseas has thus become an essential part of life in Western Europe and the area's dependence upon world trade has become a dominant characteristic. An example is the supply and price of oil from the Middle East, which has now become a major cause of economic concern and indicates the interdependence of the industrialised world with the Third World.

Western Europe: the site of the 'Industrial Revolution'

Industrial growth began in Western Europe from the late eighteenth century onwards. The use of coal and steam power together with iron and steel products forged the basic industrial sinews which enabled these countries to act as world powers. In Great Britain the canals and railways enhanced the movement of raw materials and industrial products. A powerful navy provided maritime supremacy and the consolidation of imperial possessions over a quarter of the world's surface. Raw materials from the colonies and the demands of their market served industries and increased the export trade. Great Britain in particular, but also France, the Netherlands and eventually Germany, become world powers, with large colonial empires. The phrase 'workshop of the world' was used to describe Great Britain in the midnineteenth century. This was no myth: Great Britain was the first in the field and had little or no competition. In 1870 she produced seventy per cent of the world's ships and exported coal, steel and textiles to all parts of the world. Many of the world's railway systems such as those of Argentina and India were built with British capital, skill and manufactured products. At the end of the nineteenth century Western Europe as a whole accounted for approximately ninety per cent of world industrial production.

The diversity was now compounded. Industrial growth was centred on Great Britain, Belgium and Germany, those countries which had major resources of coal and iron-ore. Industry and dense populations developed in the heavy industrial triangle and along the Rhinelands. Great Britain at this time was more concerned with her imperial possessions than with Europe. This was her period of 'splendid isolation'. Germany from 1871 onwards was powerful, nationalistic, industrial and urbanised. By contrast, France, once the most powerful nation in Western Europe, had few coal resources, remained under-industrialised, and went into a long period of decline. Italy

6 THE NEW EUROPE

and the Mediterranean states remained largely backward with peasant-based economies. Spain and Portugal, losing most of their colonial empires during the nineteenth century, and lacking in mineral resources, failed to industrialise and became economically peripheral to Western Europe. They entered a period of political instability and eventually dictatorship. The diversity of interests and the different stages of development were the basis of the growing French fear of German domination. This developed into war on three occasions, 1870, 1914 and 1939, and the major scene of contention was the Rhineland, now the key economic area of Western Europe.

The loss of world power status: elements of weakness

By the mid-twentieth century, however, it was apparent that the economic supremacy of Western Europe could not last. The small nation states of Europe were caught up in the exhaustive process of fighting two world wars,

	USA	USSR	UK	EC 12
Area (1000 km^2)	9363	22402	244	2261
Population (millions)	239	279	57	322
Cars per 1000 inhabitants	550	30	312	340
Steel production (million tonnes)	81	154	16	136
Steel consumption per head (kilogrammes)	377	571	254	346
Energy consumption per head (kilogrammes)	5200	3900	3665	3232
Grain production (million tonnes)	320	179	24	162
Meat production (million tonnes)	26	17	3	28
Motor car production (millions)	8.2	1.3	1.0	10.9
Exports (million ECU)	279319	116154	131621	849923
Crude oil refining capacity (million tonnes)	760	612	90	598
Coal production (million tonnes)	464	360	104	227
Gross Domestic Product (1000 m ECU)	5172	na	595	3231
Merchant shipping (million tonnes gross)	19.5	24.7	14.3	88.2

na – not available
ECU – EC unit of account

Figure 1.3 Comparative economic resources, 1985/86

as a result of which their infrastructure of housing, railways, port facilities and industries was badly damaged. The cost of fighting the Second World War left the UK in debt to the USA, and Western Europe had to accept Marshall Aid from the USA. The French railway system had to be completely rebuilt. The German Ruhr was devastated. Production of basic industries throughout Europe was dislocated until 1948, after which the flow of economic aid through the Organisation for European Economic Co-operation (OEEC) began a slow recovery. The wartime disruption of trade had forced the primary producing countries to look elsewhere for their imports of manufactured goods and after the war these markets were no longer assured to the Western European nations. The change in attitudes and political awareness throughout the world forced the rapid dissolution of the colonial empires so that from 1945 to the mid-1960s practically all the former European colonies became independent. This meant that the sources of raw materials and the colonial markets were no longer under the political control of Europe. Great Britain and France were reduced effectively to nations of approximately 50 million people. Germany was reduced in size and partitioned. The new state of West Germany, though larger and more powerful than East Germany, was in no position to exert control over central Europe, (Mitteleuropa) as she had done previously. Great Britain and West Germany were industrial and urbanised, and depended heavily upon imports of food and raw materials. Many raw materials were exhausted. Coal was now more expensive to mine in Western Europe than in other parts of the world. This reduction of economic, imperial and political influence in the world after 1945 has been called 'the dwarfing of Europe'. It marked a geographical, economic and political watershed.

The emergence of the 'super powers'

The declining advantage of Western Europe has contrasted strongly with the growth of the 'super powers', the USA and USSR. By the twentieth century the land powers based upon huge continental interiors had overtaken the smaller Western European nation states. These continental powers have certain basic advantages. Their huge size was once a disadvantage, but now they have efficient internal lines of communication, brought by the transcontinental railway and airlines. They gain from their large land space, huge populations and vast reserves of mineral wealth, which makes them less dependent on world trade. In the USA, San Francisco and New York, over three thousand miles apart, are in fact only hours distant from each other. The USSR is ninety times as large as the United Kingdom and nine times larger than Western Europe. Its vast land extent was referred to by Sir Halford Mackinder in his 'heartland' theory. Apart from their great size and resources the two super powers also dwarf the individual Western European countries in terms of their industrial production. Reference to the steel industry shows the weakness in size of the individual European countries. In 1986 even West Germany's steel output of 37 million tonnes does not bear comparison with that of the USSR (154 million tonnes), or the USA (81 million tonnes). Figure 1.3 shows that the contrast is particularly dramatic in the case of Great Britain which was certainly a world power politically, economically and militarily during the nineteenth and early twentieth centuries. Only in terms of export trade, and per capita wealth and consumption is there any valid comparison with the USA and the USSR. The continental powers are practically self-sufficient. Their latitudinal extent and extensive agricultural interiors mean that many of their food supplies can be produced within their own frontiers. The comparisons of food production show why the USA, although the richest country in the world, is not, pro rata, the greatest trading nation. She does not need to trade as do the Western European nations.

New elements: economies of scale and global competition

By the 1950s, Western Europe was a rapidly shrinking sub-continent in distance terms. Modern communications, including air transport, motorways, the telephone and television, created speed and flexibility and opened up new contacts. Tourism was becoming a major Europeanising force. A rapid increase in intra-European trade took place in the immediate post-war years, by contrast with the tariff barriers which had severely curtailed trade in the 1930s. By 1953 trade was over three times the value of twenty years previously. Food production rose rapidly as recovery got under way, with the large population stimulating demand, so that it was possible to see a potential self-sufficiency level for most foodstuffs.

There is also a very important economic factor. In modern technological civilisation the product of a large economic unit is more efficient than the sum of its parts. In a continental unit such as the USA there is unhindered movement of raw materials, labour, capital and the finished product, to all parts of the country, allowing high productivity and specialisation in favourable localities. By contrast the national frontiers of Europe have hitherto hindered the most economical use of resources. A large population provides a basic and substantial home market. Without this assured home market it is very difficult for industry to sustain large-scale production. It is not worth investing huge amounts of money in plant and equipment unless a profit is guaranteed, and this profit comes primarily from the home market. Mass production and assembly line methods work most efficiently under economies of scale, and when they serve a large market. The American car industry is so large that one single company, General Motors, produces more than the entire British car industry. The more that is produced, the less proportionately each unit costs. Thus the Americans can sustain superior technology, large-scale production, efficiency and relatively low costs. Also the cost of industrial technology, research and specialised equipment is now so great that smaller countries have difficulty in finding the money needed for research and development. An example here is the space programme and aero-engine production. Rolls Royce in 1971 had developed the RB211 engine. So much research and development money was required before any payment was forthcoming from US aircraft companies. Costs escalated far beyond the estimates, the company could not find the money and went bankrupt. On the other hand, British and French firms have combined to make Concorde, the most advanced civil aircraft in the world.

A study of Great Britain during the mid-twentieth century reveals a prolonged fight for her export markets against increasing, efficient, and often superior competition. There has been such a low growth rate because of the near static home market, a steady invasion of American industry, and lack of investment in new equipment and industrial technology. Japan, which began to industrialise in the 1930s, and which rapidly recovered after World War II, began a sustained, efficient onslaught on European and American markets and by the 1960s had become a world economic power, second only in wealth to the USA. First in heavy industry such as shipbuilding, then in the car industry and electronics, Japan has successfully secured a large segment of world markets and is an economic power of world dimensions. In Western Europe only West Germany has successfully competed on a broad front.

The threshold for economies of scale has been raised by new technology and changing raw material sources. The steel and petroleum industries are now sited at waterside locations dependent upon imports of raw materials in large bulk carriers. Steel works are usually now vertically integrated with all processes carried out in the same plant. The vehicle industry is dominated by a few large companies and is horizontally integrated. High technology industry requires large-scale financial backing and research capability. The establishment of branch plants outside the national territory for marketing purposes has led to the development of the multinational company.

Conclusions may be drawn from these facts that the consolidation of a single European market in this rapidly changing scene appeared logical and economically justifiable. In countries like France and the UK fifty-five million people is not a sufficiently large economic base in the rigorous conditions of modern international competition. The optimum size for the modern industrial state might well appear to be 250 million people. Reference to figure 1.3 illustrates how, when Western Europe is taken as a whole, its productive capacity bears comparison with that of the super powers. The integration of production in a highly developed sub-continent such as Western Europe compares favourably with the American model. The replanning of economic operations from national scales to a European scale was the justifiable basis upon which the ideas of a Common Market in Western Europe germinated in the post 1945 period.

Post-war integration in Europe: the initial stages

A change of the greatest significance has been taking place in Western Europe since 1945. In its economic context it is a reflection of the relative decline in status of these nation states since the nineteenth century. The replacement of the fifteen or so nations of Western Europe by a unified state has never really been an accomplished fact since the days of the Roman

Empire, or Charlemagne in the ninth century. All attempts since then by individual nations to impose unity upon Europe have resulted in wars. The revolution in political co-operation which has occurred stems from the desire for creating unity, maintaining peace and increasing security. The radical difference is that the movement towards unity is now accomplished by agreement between the nations concerned.

A large number of attempts at increased co-operation emerged out of the economic chaos of 1945. Recovery required a joint programme of economic aid and monetary co-operation which was given form and direction by OEEC, the Organisation for European Economic Co-operation. This involved very substantial American aid which was received by eighteen nations in all. In the defence sphere, the Western European Union, comprising the United Kingdom, France and the Benelux countries, was supplanted in April 1949 by that most important body, the North Atlantic Treaty Organisation (NATO). Aimed at the collective defence of North America, the North Atlantic and Western Europe, this has been one of the most significant and successful steps towards continental security. Political co-operation began in May 1949 with the Council of Europe. This is a consultative assembly in Strasbourg with deals with cultural, administrative, environmental and social matters of general interest. It has no executive powers, but is a useful sphere of co-operation.

Benelux

In 1947 the Benelux Union was formed by Belgium, the Netherlands and Luxembourg. Although these are three of the smaller nations in Western Europe, it was an initial move of the greatest significance. The union allowed for the free movement of capital, persons, services and goods across frontiers, for the co-ordination of economic policy, and for a common trade policy towards countries external to the union. These criteria of economic integration were a pointer to the future.

The European Free Trade Association

The European Free Trade Association (EFTA), formed in November 1959, was a loose association of seven nations with a limited object of 'the abolition of all tariffs on industrial goods between the seven member states'. The member states were the United Kingdom, Norway, Sweden, Denmark, Portugal, Switzerland and Austria. EFTA was not a customs union, but was a means of explanding trade and ensuring that a broad alignment in tariff reductions was kept in line with that of the EEC, which was already in existence. The United Kingdom certainly saw it as a 'bridge', or bargaining counter, for the time when she could negotiate full membership of the EEC. Although the free trade area still functions, it is at a reduced level since many of its members are now members of the EEC.

The European Economic Community (the Common Market)

The first community 1951 – 72: 'The Six'

The machinery of economic co-operation was established by six countries initially, all of whom were strongly motivated in different ways. France required a practical means of controlling the powerful German coal and steel industry in a supra-national way. She also saw the great advantages to be gained from expanding her agriculture in a European market. West Germany was anxious to be rehabilitated into the European family of nations, and also needed markets for her rapidly expanding industries. The smaller Benelux countries were great trading nations and keen on European unity for both commercial and security reasons. Italy, not part of the trunk of Europe, had always felt isolated and needed the links which would help expand her economy. The Common Market, though primarily economic in its objectives, was also seen as a great reconciliation between France and West Germany in particular.

One market for coal and steel

The European Coal and Steel Community (ECSC) was set up on 18 April 1951 by France, West Germany, Italy and the three Benelux Union countries. A supra-national authority was formed to administer the coal and steel industries of the six countries, and to 'abolish and export duties, subsidies and

Brussels: the Berlaymont building, headquarters of the European Commission

restrictive practices, and to establish free and unrestricted movement of coal, iron-ore, scrap, pig-iron and steel between the member countries'. As an experiment in co-operation this was a crucial pointer to the future. Six European countries had placed a basic sector of their economy into a common pool administered by a supra-national authority. This psychological success was matched by dramatic practical results. The production of crude steel and rolled products by 'The Six' rose dramatically from 1952 (42 million tonnes) to 1971 (103 million tonnes). The Heavy Industrial Triangle (chapters 2 and 3) found a new impetus with the cross-frontier integration of its resources. The ECSC also found a successful social and regional role. From 1957 the run-down of the coal industry necessitated plans for the retraining of workers, industrial development loans to overcome unemployment, and the adaptation of the industrial structure of declining areas (chapter 2.)

The Common Market (EEC)

The Treaty of Rome was signed on 25 March 1957. This established the basic tenet of economic integration by means of four progressive harmonisation processes. These were:

1. **Internal tariffs**: Barriers to the free flow of trade between 'The Six' were to be dismantled and customs duties on goods bought from each other were to disappear in stages.
2. **Customs union**: All six member states were to apply to external countries a common external tariff. This created a customs union with the six states progressively becoming one trading unit.
3. **Internal mobility**: There was to be free movement of labour, goods, services and capital between the six member countries.
4. **Economic integration**: Common policies throughout 'The Six' were to be applied to harmonise transport, industry, energy and agriculture.

Euratom

The European Atomic Energy Authority (Euratom) was set up for the co-ordination of nuclear research and to provide the conditions necessary for the ultimate production of nuclear energy on a large scale.

Integration policies and Economic growth during the 1960s

The customs union and abolition of internal tariffs was completed by 1968, ahead of schedule, and during 1965 the three executives, ECSC, EEC and Euratom had been merged into a single executive, known as the EEC Commission. Meanwhile the essential policies of economic integration were being created. In 1961 anti-monopoly regulations and steps to create free movement of labour, capital and services were devised. Implementation of the Common Agricultural Policy began on 14 January 1962. In 1963 the Yaoundé Convention was signed with eighteen African States. This was essentially a preferential trading relationship for the supply of tropical food and raw materials to the EEC in return for aid and a guaranteed market. It marked the emergence of the EEC as a trading unit of world dimensions.

The most startling measure of success during the 1960s was the rate of economic growth of 'The Six'. Comparisons between 'The Six' and the United Kingdom (figs. 1.4, 1.5) reveal many disparities in terms of gross domestic product and exports. The steel industry (fig. 4.1) is a significant pointer to national growth, and this too shows rapid growth on the continent in contrast to the UK. There are many and complex reasons for this pattern and the

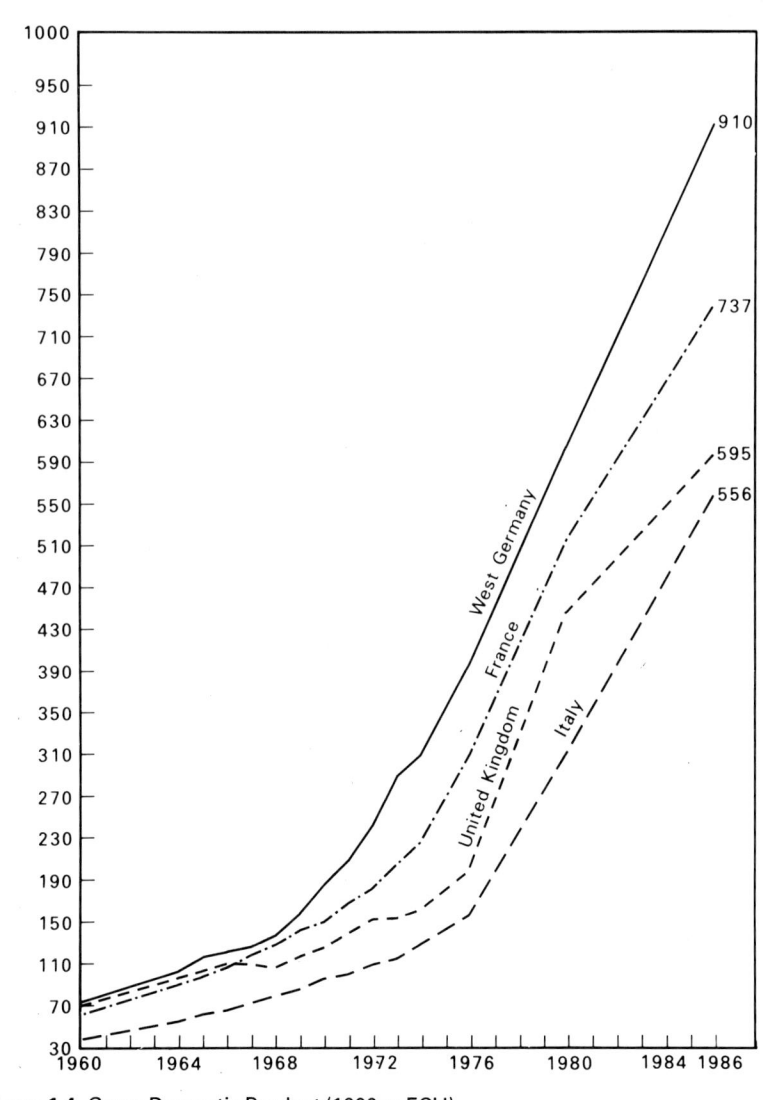

Figure 1.4 Gross Domestic Product (1000 m ECU)

stimulus given by the Common Market may be only partially responsible for the rapid growth. West Germany in particular gained a great deal from Marshall Aid, and her industrial structure was completely rebuilt, gaining the benefits of new technology. Her defence costs were much lower and labour troubles much less of a problem than those in Great Britain. However, the greatest single stimulus on the continent was probably the change which was taking place in the composition of the labour force (fig. 6.10). The movement off the land caused by agricultural improvements created an enormous reservoir of labour for industry in Italy, West Germany and France. This allowed considerable industrialisation without labour shortages and consequent rises in labour costs, such as those which had occurred in Great Britain. West Germany had the benefit of large numbers of refugees from the east many of whom were skilled workers, and an influx of migrant workers from Southern European and Mediterranean countries such as Turkey and Yugoslavia. West Germany rapidly became the largest single industrial power in the EEC.

Other notable wealth expansion however was in France and Italy. France now gained the benefits of expanded markets for her food production and became the 'granary of Europe'. French and Italian industry rapidly expanded on the basis of favourable terms of trade with the Third World raw materials producers, new energy sources based on oil and the development of consumer industry based on an expanding industrial-urban labour force. The industrial revolution which had been only partial in France and even less in Italy now completed its development, and these two countries became major industrial powers in a short space of time.

The six Common Market countries as a whole enjoyed remarkable growth and prosperity during the period up to 1973. They developed into a large and powerful economic group, capable of comparison with the major world powers, the USA and the USSR (fig.1.3). Furthermore, the EEC as a whole is potentially self-sufficient in food. Across Western Europe there are large stretches of agriculturally productive lowlands (fig. 1.2). There is a considerable latitudinal extent and a climatic range which gives a wide spectrum of agricultural types, and enables the EEC to be a major food producer. The advantages of scale and latitudinal extent are perhaps the two most significant criteria upon the success of the Common Market is based.

Geographical enlargement: The Nine(1973)

During the 1960s there was a realisation in the United Kingdom that a significant and momentous change had occurred on the continent. The United Kingdom was a small, highly developed and populated island with a static home market, an economy based upon trade and without the resource base of an empire, and was offshore to a developing continental unit of powerful proportions. The United Kingdom exports one third of its car production, two fifths of its engineering products and one third of its chemical manufactures. During the 1960s, the "Wind of change" speech by Prime Minister

Harold Macmillan could be applied not only to the decolonisation process in Africa and elsewhere, but also to the closer real and perceived links (psychological, physical and economic) which the United Kingdom now had with her neighbours in Western Europe. On 8 November 1961 the United Kingdom, Ireland, Denmark and Norway began negotiations for membership

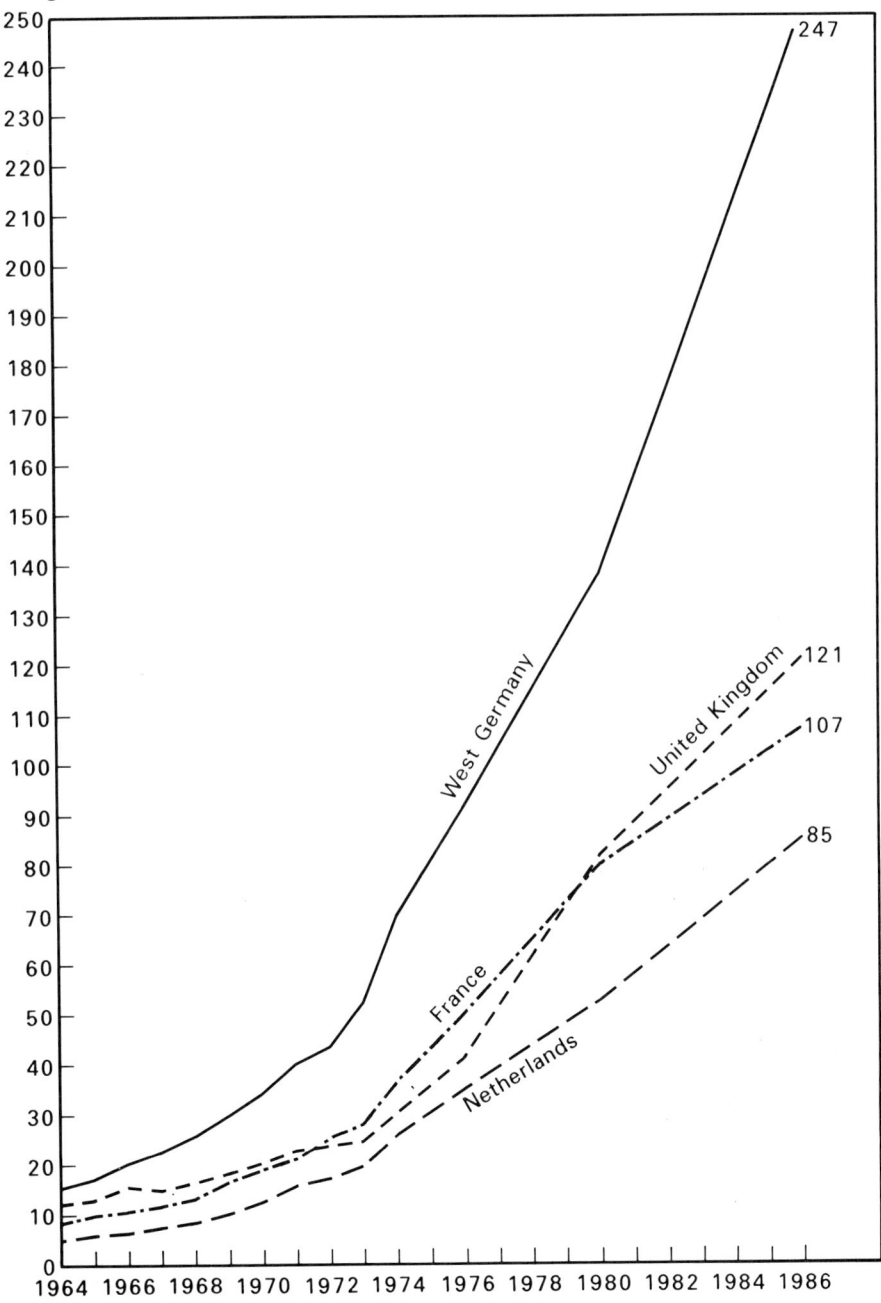

Figure 1.5 Total export trade (1000 m ECU)

		Area (1000 km^2)	Population (millions)
SIX 1951	France	547	55.2
	West Germany	249	61.0
	Italy	301	57.1
	Netherlands	41	14.5
	Belgium	31	9.8
	Luxembourg	3	0.4
NINE 1973	United Kingdom	244	56.6
	Ireland	70	3.5
	Denmark	43	5.1
TWELVE 1981–6	Greece	132	9.9
	Spain	505	38.5
	Portugal	92	10.1
Total		2258	321.7

Figure 1.6 Area and population figures

of the EEC. These lasted for nearly a decade, largely because of the intransigence of the French President, Charles de Gaulle, who insisted that the UK was not ready for Community membership. Immediately after the resignation of de Gaulle in April 1969, negotiations were resumed and succeeded very quickly. By this stage the original Six had the will to enlarge and strengthen the Community, and on 1 January 1973 the United Kingdom, Ireland, and Denmark acceded to the EEC, now enlarged to nine member countries (fig. 1.6).

The Mediterranean enlargement, 1981 – 86

When Greece, Spain and Portugal emerged as democratic states in the 1970s after varying periods of dictatorship, they moved rapidly to seek admission to the European Community. The accession of Greece in 1981 and Spain and Portugal in 1986 involved a whole range of strategic, economic and geographical implications for the community.

Greece had had an association agreement with the European Community since 1962. Between 1967 and 1974 the monarchy was replaced by a military dictatorship. During this period relations with the EEC were suspended and not resumed until 1975 when Greece, having restored a democratic government, applied to join the EEC as a full member. She acceded to memebership in January 1981. Having been one of the pillars of western civilisation in classical times, with a great empire and the home of great philosophers and architects, Greece fell into a long decline, and for several centuries was mis-ruled as part of the Ottoman Empire in the Balkans. She emerged in the twentieth century as an under-industrialised, semi-developed country with the bulk of its population working on the land. Greece has a population of 9.6 million, and in area is almost the same size as England and Wales. The enviroment is extremely difficult, being mountainous and desiccated with a semi-arid mediterranean climate. Mount Olympus, 3200 m, is the highest mountain. There are three physical regions, the rugged Pindus mountain-chain which covers eighty per cent of the country, and several small areas of lowland, principally around Athens, Thessaloniki and in Thessaly. Finally there are the Greek islands in the Aegean Sea, a complex group which includes Crete and Rhodes and the Cyclades archipelago.

Portugal was one of the first European colonial powers, but since the seventeenth century has suffered a long decline for several reasons. It is almost totally lacking in mineral resources, and hydro-electric power has been retarded by irregular river regimes and industrialisation is limited. The African colonial empire proved a costly drain on scarce resources during the long guerilla war in the 1960s and the 1970s, after which the last two major colonies, Angola and Mozambique, became independent. There was a long period of dictatorship under which economic development was neglected. Portugal emerged from this with a democratic government in 1976. The single most significant source of income has been the revenue from the tourist industry in Lisbon and the Algarve.

Spain, with 37 million people, is by far the largest economy, next to Italy, in the Mediterranean region. It is a country of great potential and with significant resources, but one in which industrialisation started late. On the north coast are coal reserves. Bilbao mines high grade iron-ore and is the centre of a major steel-producing area. The Sierra Morena is a substancial producer of copper, lead, silver, zinc, manganese and mercury. Major rivers such as the Guadalquivir, Tagus and Guadiana are being harnessed for hydro-electric power. As well as irrigation control, and increased agricultural productifity, this is significant for industrial development. Spain now closely follows Norway and Italy in total HEP production. Though semi-developed, it is thus experiencing transformation from an agricultural into an industrial country. The change has been relatively rapid during the last two decades, for several reasons. These include the opening up of the country to western influence from American air bases; the enormous influx of capital from the tourist trade since the 1960s which has given Spain finance for development projects; the end of the dictatorship on General Franco's death and the subsequent change to a constitutional monarchy. The rapid transformation in the economy is reflected in vast differences between underdeveloped regions, and core regions in which development has been concentrated.

Problems

There are considerable differences in levels of development between the three new Mediterranean members and the rest of the Community. The convergence of economies is basic to Community policy, but this has been made more difficult by the economic recession, and the advent of the three countries adds to the problems. Spain has a high growth potential, and in many ways is comparable to Italy. Greece has a per capita income rather lower than Ireland. Portugal has a lower per capita income than any other member of the Community, and will need considerable aid and investment assistance. The proportion of peripheral regions and problem economic sectars in the Community has increased substantially.

There is a further tilt towards agricultural interests, with an increase of over fifty per cent in the number of people working in agriculture, and a twenty four per cent increase in agricultural production. These Mediterranean products add to the imbalances of production already existing in the Community. Wine, olive oil and certain types of fruit and vegetables are already close to a structural surplus, and this tendency is exacerbated. French wine producers are very concerned about Spanish competition in the area.

Greece, Spain and portugal depend on imports for over 80 per cent of the energy, compared to 54 per cent for the other EEC countries. With development in prospect, energy consumption is expected to rise steeply, thus placing increased burdens on the energy supply situation. Thus a fresh impetus to reduce dependence upon imported energy is vital.

The budgetary cost to the Community will be substantial, involving in particular the Regional Development Policy, Social Fund, agricultural support, and the EIB. The Mediterranean states will face severe adjustment problems. Spanish industry in particular has developed behind strong projective barriers, and the removal of these will cause pressure, unemployment, and closure for vulnerable industries. As a result, a long transitional period is proving necessary as the Spanish economy gradually reduces its tariffs and harmonises its competition rules, integrates its trading patterns, restructures industry and diversifies agriculture to face competition in a common market.

Finally there are political and strategic considerations. The accession of the three new members may eventually cause a geographical shift in emphasis southwards to the Mediterranean basin. Three new languages may cause a slowing-up of integration, and there is always the danger that a community of twelve may have to be organised into a loose confederation of two tiers of developed and semi-developed economies; a core and a periphery. It is to combat that possibility that the European Commission in 1985 produced the Integrated Mediterranean Programmes (IMP) designed to raise income levels and improve employment possibilities in the Mediterranean regions of the Community. However any possible eastward expansion of the Community following political changes in Eastern Europe would have inevitable consequences for IMP.

Geographical and strategic advantages

Geographically, Spain and portugal will help to create a coherent European framework. They are physically contiguous with Western Europe, whereas

Greece lies in the eastern Mediterranean basin and is physically detached. All three Mediterranean states help to complete the natural southern limits of Europe; the European Community's role on both sides of the Mediterranean will be strengthened and its strategic links with north Africa extended. Economically, there are 55 million people in the three Mediterranean countries, with Greece as a major force in world shipping, and Spain as a substancial industrial power with rich mineral resources. Already the world's largest trading unit, the Community's weight in international trade will increase still further. In view of the historical, cultural and linguistic links of Spain and Portugal with the 500 million people of Latin America there should be a probability of considerable bridgebuilding with that continent.

The Community institutions

The main distinguishing feature of the Community is its decision-making process (fig. 1.7).

(a) It is essentially a confederation of national states, each with its own government. The Heads of Government meet periodically at 'Summit Meetings', known as 'The European Council'.
(b) The governments are represented by one minister each on the 'Council of Ministers'. This body is the effective link between the national governments and the Commission, and it receives policy proposals from the Commission and takes decisions which are then passed to the Commission for implementation.
(c) The European Commission is the executive, permanent civil service of workhorse of the Community, with its headquarters in the Berlaymont Building in Brussels. It is directed by seventeen members, with France, West Germany, Spain, the United Kingdom and Italy having two members each, and the Netherlands, Belgium, Ireland, Luxembourg, Greece, Portugal and Denmark one each. The Commission initiates policies, drafts legislation, and administers the day-to-day mechanics of a customs union of 320 million people.
(d) A large body of some 9 000 people is the working secretariat for the Commission and Council of Ministers.
(e) The Court of Justice, which sits in Luxembourg, administers Community law and arbitrates in disputes involving the Community treaties.
(f) The European Parliament at Luxembourg and Strasbourg is the developing political assembly. It works on a consultative basis, with specialised committees which review policy-making and the compostion of the budget. Since June 1979 its members have been elected directly by the electorates of the member countries.
(g) At the present time the EEC consists essentially of a confederation of countries which are cumulatively delegating broad areas of economic policy to a common machinery, so that decision-making lies between national governments (represented by the Council of Ministers), the

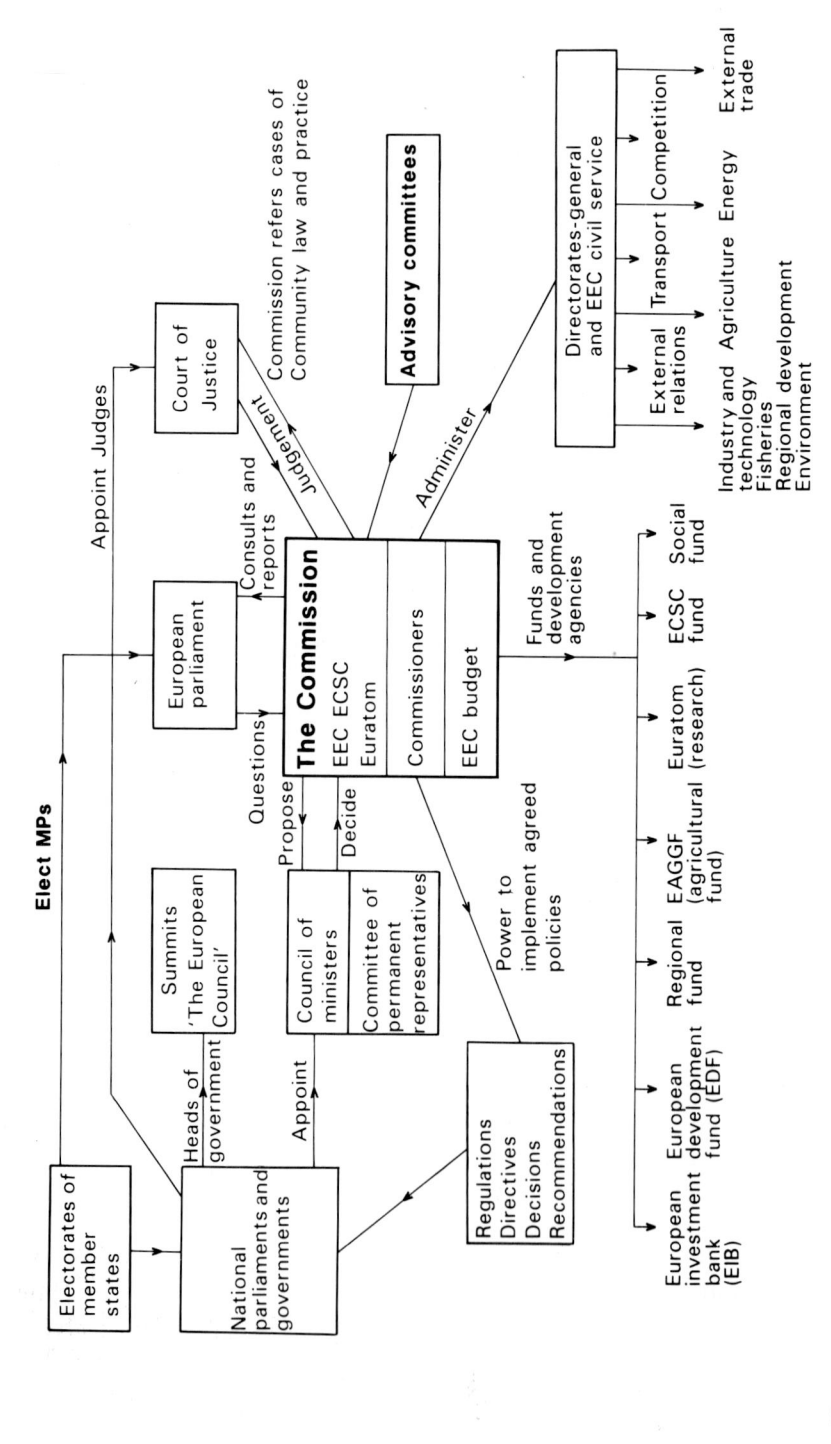

Figure 1.7 The structure and activities of the European Community

European Commission and the European Parliament, in a tripartite system of control.

The Community budget, funds and policies

Since the phasing-out of national contributions, from 1 January 1979, the Commission has had three direct sources of income. These are its 'own resources' based upon agricultural levies and customs duties, plus a percentage of Value Added Tax proceeds. The budget is one of the most important single elememts which allows the European Community to function as an effective supra-national organisation and to administer policies on a European scale.

The Community agencies are the practical means by which the policies of economic integration are implemented (fig. 1.7). One of the earliest successes was the ECSC (chapter 2). The European Social Fund has become the principal agency in the fight against unemployment with industrial retraining and social welfare schemes. The Common Agricultural Policy (chapter 6) has been concerned primarily with funds to stabilise farmers' incomes by means of the EAGGF (European Agricultural Guidance and Guarantee Fund). The EDF (European Development Fund) finances economic development schemes in countries which have association and trade agreements with the Community. Most significant is the EIB (European Investment Bank) (chapter 10). This works closely with the Commission and provides loans for development and modernisation projects. Examples are projects like the Val d'Aosta motorway, an essential link in the early integration of Italy and the rest of the Community. A significant development was in 1974 with the inception of the Regional Policy (chapter 10). The financial aid available under this agency underlines the important principle that economic harmonisation between regions is as important as that between national economies.

The Contemporary Economic Situation: Towards the 1990s

Since the oil price rise of 1973 which triggered inflation and recesion, there has been a major reduction in economic activity during the 1970s and particularly from 1979 to 1982. As an example of this, energy consumption fell by thirteen per cent in Western Europe between 1975 and 1985. Unemployment levels rose rapidly and have remained stubbornly high and there has been a lessening of the pace of economic integration during which the process of convergence in the European Community has been slowed considerably. The danger of protectionism has been revived but the peripheral countries of the Community have in the end been worst hit by the effects of the recession.

Other international factors have become more significant. One of the most important is the process of Third World industrialisation based upon cheap labour and lower costs in which countries such as Taiwan, Hong Kong and Brazil have become major industrial competitors with more established older industrial countries. This is the price of the international redistribution of labour which has become very significant in the last fifteen or twenty years. The need for fewer raw materials and the use of replacement materials has led to the decline of many basic industries.

High labour costs in Western Europe and robotisation of many basic processes have led to the shedding of labour, the so called 'shake-out' or the restructuring of the labour force in traditional industries such as steel, ship-building, textiles, and heavy engineering. There is a greater emphasis on skills, services, high value added products such as electronics and consumer goods, and tertiary and quaternary sector growth which includes tourism, financial and legal services, ect. The mis-match of employment during these major changes from the secondary to the teritary and quaternary sector caused severe unemployment during the late 1970s and early 1980s. The regional impact has been severe for heavy industrial districts, particularly those inland such as Lorraine, the West Midlands and even Ruhr. Equally the dispersal of industry and the movement of population into suburban areas, and the decline of innercities with associated unemployment and social tensions have become major problems. Major revival policies are necessary for many of the older industrial conurbations in the European Community. Easier access to the low wage economies of Eastern Europe could make difficulties for such policies.

A major rural crisis has also developed. The over production of foodstuffs in the European Community has now reached crisis levels and the budgetary cost levels of the Community Agricultural Policy have become impossible to support. The CAP takes some 70 per cent of the European Community budget and it is necessary for some redirection of the use of these resources, hence the Community Agricultural Policy has been under major political pressure for some time. Renegotiations of levels of price support quotas and production levels are at present being implemented. At the same time as this agricultural crisis is continuing, levels of enviromental concern have increased quite dramatically. The farmer needs to be the custodian of the countryside and a major redirection of land use into forestry, conservation and leisure pursuits is very necessary. The Green or Ecology parties have become a major political force in many West European countries, particularly in West Germany. Resources need to be redirected from the countryside into declining urban and industrial areas.

In spite of the slowdown in economic activity during the recession, the absoption of the three Mediterranean countries and a lessening of the pace of economic integration, nevertheless the European Community has made a great achievement during the last fifteen or twenty years. This has been the progressive lowering of barriers (economic, psychological, political) between the member states. Internal trade patterns between the member states are emerging so that nearly 60 per cent of their total trade is within the Community. Even the United Kingdom now does 50 per cent of its trade with its partners and this percentage is rising each year.

The Customs Union has gone some way towards economic integration but a great deal remains to be done. The principal of changes required are reform of the budget, the transference of funds from agriculture, and the reform and expansion of the other major structural funds. Industrial, urban, social, regional and enviromental problems are now the major considerations. Further movement towards high level policy making is necessary, including monetary union and a European currency, common taxation and fiscal policies and a common budget.

Contemporary political developments

The greatest degree of integration occured in the European Community during the period 1951 to the late 1960s and there was a pause in the rate of integration after that. From the mid 1970s onwards some major political changes took place. Cumulative causation, illustrated by the co-operation, harmonisation and political processes which had worked since the mid 1960s had become a habit. The change of name from the EEC to the European Community signalled a change from an economic bureaucracy to a political unit. The Community was beginning to co-operate in external affairs, having a single spokesman for the member states in major world conferences.

In 1979 two major steps forward were taken. The European Monetary System, the EMS, was established on 13 March 1979 with the European Currency Unit, the ECU, and the exchange rate intervention mechanism creating a zone of monetary stability throughout the Community. This was only a partial success since the United Kingdom did not fully join the system. In 1979 the first direct elections to the European Parliament in Strasbourg took place. This meant that MEPs are directly elected by constituents and are responsible to them. The European Parliament was remarkable for the early establishment of major groups by political tendency rather than nationality so that a normal political spectrum from left to right soon emerged as a major integrating force.

In the European Summit Meetings of 1984 and 1985 action on two fronts to lead to a new dimension in Europe integration was achieved. This was agreement on constitutional reform and an approach to a 'People's Europe'.

In December 1985 in Luxembourg the Single European Act was drawn up by the European Council for the creation of a frontierless internal market by the end of 1992. The development of this Single European Market is the principal activity in which the Community is now engaged. (This is dealt with more fully in chapter 21).

2
Energy: a variety of sources

The Western European economy is based upon manufacturing industry and therefore has a fundamental requirement for large quantities of energy. Furthermore, energy needs to be available on a long-term, substantial and low-cost basis if industry is to plan and compete effectively. This chapter seeks to illustrate the diversity of energy resources which are now available to the EC nations, their changing geographical location, and the political economy of the supply and demand of energy.

There are five main primary sources of energy: coal, oil, natural gas, nuclear energy and water power. These are used for the direct production of heat; for industrial purposes, this is frequently converted immediately into electrical energy, which may be regarded as secondary energy. Electricity is perhaps the most significant of all twentieth-century forms of energy because of its easily distributable nature and its liberalising effects upon industrial location.

Coal

Coal provided the energy for the industrial revolution which began in Western Europe in the eighteenth and nineteenth centuries. It occurred in thick seams at, or relatively near, the earth's surface and many coalfields such as the Ruhr, or Northumberland and Durham, were by rivers or the sea, facilitating easy movement in the days before railway networks were established. More important in the days of the nineteenth-century technology was its versatility. As a source of direct heat it was used as a home fuel; as coke it could smelt iron-ore, and could be used in the manufacture of pottery, glass and in most basic industrial processes. It was the energy source which encouraged the development of steam-driven machinery and the locomotive. Coal powered the railway network of Europe and the navies which maintained British, French, Dutch and German colonies, trade and military supremacy. Its usefulness as an important source of energy has been prolonged into the twentieth century by its availability as a generator of electricity in power stations. In addition to contributing to the early industrialisation of the Continent, coal also dominated the original location of large-scale European manufacturing industry. Coal was low in value in relation to its bulk, being expensive to transport and inefficiently used, so that factories tended to concentrate close to the pit-head. Most nineteenth-century industrial regions developed on or near coalfields. The most industrialised nations - West Germany, the United Kingdom and Belgium – were those with the major coalfields. France, Spain and Italy, in particular, lagged behind the other nations because of their relatively poor endowment of coal. Italy had scarcely any coal at all, France only one major coalfield, the Nord/Pas de Calais, and Spain the Oviedo coalfield on the north coast.

The major coalfields

There are two major coal provinces within Western Europe (fig. 2.1). In the United Kingdom the large coalfields are found generally north of a line from

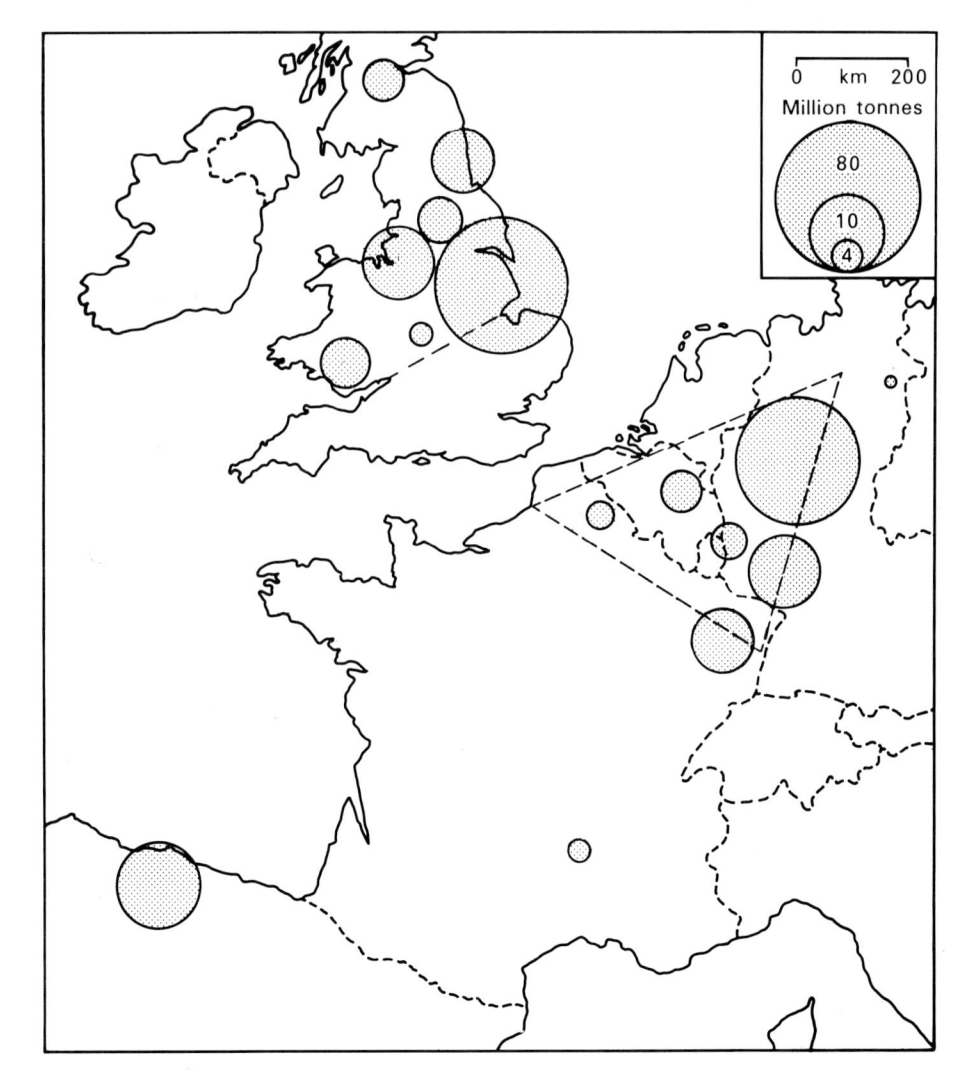

Figure 2.1 European Community coal producing regions 1986 (million tonnes) Problem: Identify each coalfield or producing region. What significant groupings are there?

Bristol to the Wash, and these gave rise to the concentration of British manufacturing in the Midlands and the North. Each coalfield specialised in a major branch of manufacturing: for example steel and non-ferrous metallurgy in South Wales, ship-building on the Clyde, and textiles in Lancashire and West Yorkshire. Such manufacturing zones developed their coal-based industrial activities during the nineteenth century and these dominated the United Kingdom economy until the inter-war period when coal was beginning to give way to the pressure of new energy sources.

The second major coal-based industrial area stretches from Lille in France across to Dortmund in West Germany. This contains the Nord/Pas de Calais field of France, stretching to the Sambre/Meuse coalfield of Belgium; the Kempenland and Limburg fields, extending eastwards to Aachen and the Ruhr in West Germany. At the southern apex of the triangle lies the Saarland coal-basin and the coal and minette iron-ores of Lorraine and Luxembourg. This is the 'Heavy Industrial Triangle' which as late as 1952 was responsible for 95 per cent of the coal, and 86 per cent of the steel produced in the six EEC countries. It is still the most significant single area of heavy industry on the Continent.

Falling demand for coal

Technological change and the discovery of new energy sources have caused a dramatic reduction in the role of coal since the early 1950s. The exhaustion of the best and most easily accessible coal seams, the labour-intensive character of coal-mining and the expense of transporting it compared to alternative, competitive fuels, has meant that coal has become a very much more expensive source of energy. This process has been accompanied by the demise of the open coal-fire, together with the change in energy consumption from coal to oil and electricity by major users such as the railways, the shipping fleets and the steel industry. Air pollution has created a need for cleaner forms of energy. Primary hydro-carbon fuels such as oil and natural gas, the new technology fuel uranium, hydro-electric power, and the secondary energy source, electricity, have all combined to produce a highly varied energy supply (fig. 2.2). In addition, they are much more easily distributed and transferable by pipeline or cable, and are less bulky and heavy to transport. Coal, therefore, has become a high-cost and inconvenient fuel despite research into increasing its efficiency. As late as 1951, at the beginning of the EC's existence, coal provided two-thirds of its primary energy and in the United Kingdom the proportion was as high as 90 per cent. Since then, however, although total energy requirements have risen, the share of coal has fallen dramatically, both relatively and absolutely. Coal production has fallen from 447 million tonnes in 1962 to 227 million tonnes in 1986.

The smaller, modern coal industry

It has been necessary to contract and modernise the coal industries of Western Europe considerably (fig. 2.4). The National Coal Board (now

	EEC 6		UK		France	EC		
	1951	1972	1951	1986	1986	1982	1985 Commission Medium-term Objectives	1986
Coal	67	21.5	90	31.9	10.5	20.5	16	22.2
Lignite		2.8				3.8		
Oil		59.5		37.1	42.8	48.7	40	45.5
Natural Gas	33	11.6	10	23.2	12.3	17.8	25	17.9
Hydro/electric and nuclear energy		4.5		7.8	32.7	9.2	19	14.1
Total energy consumption (mtoe)	247	675	158	193	175	961		1042

mtoe = million tonnes of oil equivalent

Figure 2.2 Percentages of primary energy consumption

British Coal) of the United Kingdom is an example of relative success in restructuring an old industry into a more competitive form. Several hundreds of the small and less productive pits have been closed and production concentrated in modern efficient pits. Productivity has been increased considerably by long-wall coal-face machinery, by power loading, and by the construction of power stations for electricity generation adjacent to the pitheads, thus reducing the transport costs of coal.

The labour force of over a million men in 1914 has thus been dramatically reduced (fig. 2.5). Mechanised coal-cutting and loading has increased from two per cent of total production in 1947 to over 90 per cent now. The East Midlands coalfield (Derbyshire, Nottingham, Leicester and Yorkshire) has now emerged as the most important region, producing half the United Kingdom total – the pace of contraction has slowed considerably now that British Coal is producing 105 million tonnes of coal per year and the industry has slimmed to a more efficient size.

The other major coal province in the EC has experienced similar problems. Although there has been considerable retrenchment in total production, the pattern varies considerably. Productivity has been constantly low in the small isolated coal basins of the Centre/Midi of France and the South Belgian coalfield. The Nord/Pas de Calais has always suffered from thin, disturbed seams and, with reserves practically exhausted and mining increasingly expensive, output is being cut back by some two million tonnes per year; in 1986 production was 1.9 million tonnes (fig. 2.4). By contrast, the most important single coalfield in Western Europe, the Ruhr, is far more productive. Its output per man is the highest in the EC. Lorraine is another efficient coalfield and has become relatively much more important to France. As in the United Kingdom, modernisation and rationalisation in these areas has been undertaken by large-scale authorities. The Charbonnages de France is a nationalised body created in 1946, and the Société Générale of Belgium controls forty per cent of the coal industry. Ruhrkole AG now controls production in the Ruhr coal basin.

The Spanish coal industry has followed a significantly different process. Production from the small Oviedo coalfield has been the basis for the heavy industrial districts centred upon Bilbao and the northern coastal region. As the Spanish economy has rapidly developed since the 1960s the production of coal has been raised to nearly sixteen million tonnes. The industry is, however, overmanned and under-capitalised with far too many small collieries and faces considerable rationalisation (fig. 2.3).

Co-operation in the coal industry

The importance of coal to the economies of Western Europe was recognised by the formation in 1951 of the first agency of co-operation, the ECSC (European Coal and Steel Community). Although it has since been integrated into the EC treaties, it was the first significant area of economic co-operation and success in the post-war period. By the Paris Treaty of 1951, the original Six decided to remove internal price barriers and transport discrimination, customs duties and quotas on coal, coke, pig-iron, scrap and steel, thereby establishing a single market for these products. For the first time in history the great resource area of the 'Heavy Industrial Triangle' could be considered as one efficient geographical unit. Instead of four groups of national coalfields, separated by national frontiers, there was now a single resource of some 230 million tonnes of coal per annum on the market at a common price. The ECSC played a major role in co-ordinating coal production and reducing the internal costs of energy. More significantly, this co-operation had a major psychological value in pooling the resources of an area which had been a major bone of contention between France and Germany for over 150 years.

ECSC policy experienced a marked change in emphasis due to the changing role of coal. In 1951 coal played the largest part in energy supplies, providing 67 per cent of the total energy of the original Six. This was a period of rapid post-war industrial growth and with an acute shortage of fuel, priority was given to the maximum output of coal. The rapid growth of thermal electricity production from coal-fired power stations was a major growth factor. As an example, French production rose to a high point of 59 million tonnes in 1958.

During the 1960s this pattern changed dramatically. The days of energy self-sufficiency based upon coal had gone. The ECSC controlled and guided the decline in coal production caused by competition from oil and natural gas, and cheaper imported non-Community coal. There was a phased reduction of output, and closure and amalgamation of collieries to create fewer but more

Bergkamen 'A' in the Ruhr: a coal-fired power station

	Coal production by country (million tonnes)				Significant Collieries	Underground workers		
	1962	1971	1981	1986	1986	1971	1982	1986
West Germany	147.1	117.1	95.5	87.1	33	135 000	122 000	107 000
France	52.3	33.0	18.6	14.4	15	60 000	28 100	18 500
Netherlands	11.8	3.7	nil	nil	nil	6 000	nil	nil
Spain	12.6	10.6	14.6	15.9	272	–	33 200	33 600
United Kingdom	200	147.1	125.3	104.6	110	221 000	170 000	108 400
Belgium	21.2	10.9	6.1	5.6	5	24 000	16 000	13 300
Total EC	**447.1**	**312.4**	**245.6**	**227.9**				

Luxembourg, Italy, Ireland, Denmark, Greece and Portugal – insignificant production

Figure 2.3 EC coal production 1961-86. Which countries have the most significant volume of production? Which EC countries have the most efficient coal industry? In which countries has coal declined most, and why? How does Spain compare with the rest of the EC?

efficient units. Annual demand for coal in the original Six dropped by over two-thirds up to 1972 (fig. 2.2), and coal now provides just over one-fifth of total energy consumption. In the United Kingdom, although the picture of decline is similar, the degree of dependence upon coal is much greater than on the Continent, with coal still providing thirty two per cent of total energy requirements. The improvement in the efficiency of the coal industry has been remarkable, with modernisation and mechanisation of collieries, Community grants were provided for retraining redundant miners, housing grants, improved safety techniques and research into new uses for coal and its by-products. Particularly in the Franco-Belgian coalfield, there has been a wave of development of new industrial estates in the decaying mining areas. Altogether, about half a million workers have been retrained. The basic achievement of the ECSC has been to adapt its policies to periods of expansion and contraction and to have aided the restructuring of an ailing industry.

The coal industry remains problematic. After the actions of the OPEC (Organisation of Petroleum Exporting Countries) oil control from 1974 onwards, it was envisaged that the role of coal would grow again, with a particular increase in cheap imports from the USA and Poland. Domestic coal output has continued to decline however since 1973. The Netherlands has terminated production. In France the original intention to close the Nord/Pas de Calais coalfield has been reappraised in view of the political consequences in a declining industrial area. In the UK and West Germany, the best endowed countries, there has been closure of uneconomic pits and major investment in new deep mines. There are regional, political and social implications, however, and the 1984 labour unrest in the UK was a symptom of this. The indigenous production of coal remains subject to internal labour problems, environmental problems, cost and efficiency considerations and to external pressures from the fluctuating cost of oil and cheap imports of coal from the USA and Poland. It is caught up in the global political economy.

Oil

The consumption of oil in Western Europe increased enormously during the 1950s and 1960s, at an average rate of ten per cent per year. This was due partly to the relatively low cost of production of the refined product, and partly to the advantage of oil in servicing large, expanding and non-competitive markets such as aviation fuels. In addition, oil was competitive in the production of electricity in thermal power stations, in home heating, and in the petrochemical industry with products such as artificial fibres and plastics.

Consumption of oil by the enlarged EC was over 500 million tonnes in 1986 (fig. 2.6). This represented approximately one-third of the international movement of oil with most imports coming from the Middle East. The oil proportion in the total fuel bill has trebled during the period from 1951. This

Coalfield	1961	1971	1972	1974	1981	1986	Comment
Belgium							
Campine (Kempenland)	9.6	7.3	7.3	6.3	5.8	5.6	Now the only Belgian coalfield. 5 collieries in 1986
South Belgium	11.9	3.6	3.1	2.6	0.3	nil	Sambre-Meuse valley. Nil production
West Germany							
Ruhr	120.3	96.4	88.9	83.0	76.7	67.7	1986 production very efficient from 26 large collieries
Ville (Cologne) (lignite field)	–	–	110.0	–	122.9	117.0	Made into briquettes as fuel
Aachen	8.7	6.8	6.5	6.2	5.2	4.9	
Saar	16.0	10.6	10.4	8.9	11.4	11.5	
Lower Saxony	2.0	2.9	2.6	2.2	2.6	2.4	
France							
Nord/Pas de Calais	26.9	14.5	12.6	9.0	3.9	1.9	Production reduced massively and to be phased out
Lorraine	14.0	11.5	10.9	9.1	10.9	9.4	The most efficient French coalfield with reserves of 400 million tonnes, (5 collieries in 1981)
Centre/Midi	11.2	6.9	6.2	4.8	3.7	2.6	Small isolated coal basins to be phased out
Netherlands							
Limburg	12.9	3.7	2.9	0.8	nil	nil	Production phased out completely in 1976
Spain							
Oviedo	12.6	10.6	–	–	14.6	15.9	Asturias province
United Kingdom							
Derby, Nottinghamshire and Yorkshire	82.7	68.5	55.8	55.0	60.3	65.5	The largest and most efficient British coalfield. It contains the newly discovered Selby coalfield and the Vale of Belvoir, with estimated reserves of over 700 million tonnes
Northumberland and Durham	33.7	19.2	15.2	12.9	13.7	10.8	Considerable rationalisation
West Midlands	16.1	8.6	6.5	5.4	8.6	4.9	Production now confined to Cannock and Warwickshire
Wales	18.2	12.1	9.5	7.4	7.6	6.9	Considerable exhaustion. Only 7 mines open
North West	12.8	13.5	10.6	9.9	11.4	10.5	Main productive area is the North Staffordshire coalfield
Scottish	17.5	12.5	9.8	8.6	7.4	5.6	
Kent	1.6	1.0	0.7	0.6	0.6	0.4	
Opencast mines	8.6	10.6	10.4	9.8	14.4	14.0	Low cost, high productivity

Figure 2.4 EC Coalfields production (million tonnes)

32 THE NEW EUROPE

	Number of miners	Collieries	Production (tonnes)
1947	900 000	908	195
1956	704 700	840	209
1967	409 700	483	184
1971	285 000	292	147
1973	250 000	285	118
1977	240 000	231	121
1982	288 000	176	121
1986	140 000	110	105

Figure 2.5 Changes in the United Kingdom coal industry

dependence on imported oil is therefore one of the most vulnerable areas of the European economy both from a strategic and a financial point of view.

In spite of this the EC relies upon oil for a high proportion of its energy requirements. The low cost oilfields exploited in many parts of the world are supplemented by ocean-going supertankers of one hundred thousand tonnes and over dead weight. These carry crude oil to the principal estuaries of Western Europe, at which point the oil is refined and then carried by pipeline,

	Production	Imports	
West Germany	4.1	114.8	
United Kingdom	129.7	−49.7	(net exporter)
France	3.5	80.8	
Italy	2.6	82.1	
Netherlands	5.0	29.3	
Belgium	0	22.9	
Luxembourg	0	1.1	
Denmark	3.7	7.5	
Ireland	0	5.0	
Greece	1.3	12.1	
Spain	2.2	38.5	
Portugal	0	9.6	
Total	152.1	354.2	

Figure 2.6 EC consumption of crude oil 1986 (million tonnes)

Botlek oil terminal at Rotterdam

road or railway. This is a well-integrated production - supply pattern, largely under the control of the major Western oil companies during the 1960s, which encouraged a huge expansion of low-cost industrial production. The pattern of oil refining in the EC is almost entirely of coastal, river-side or estuarine refineries with ocean terminals for crude oil input (fig. 2.7). As consumption increased, refining capacity enlarged dramatically reaching 598 million tonnes by 1986. Italy has the largest refining capacity (119 million tonnes), reflecting her almost total dependence upon imported oil for energy, followed by France, West Germany and the United Kingdom. The refining capacity of the Netherlands is high, and dominated by the Rotterdam-Europoort agglomeration which is at the centre of the transnational pattern of pipelines (fig. 2.7). As well as the North Sea oil terminals, the Mediterranean Sea has become a major focus with oil pipelines reaching north to the Rhinelands from Lavera, Genoa and Trieste. However, since 1973/4 a range of new and complex factors have transformed the position of oil. The rising pace of energy consumption until then was changed by the external situation. The actions of the OPEC cartel to control operations and quadruple prices ended the low-cost advantage of oil and sent the industrial world into recession. Since then the price of oil has fluctuated depending upon the political effectiveness of the cartel in controlling production, and during the 1980s the cartel's power has diminished and the price of oil has fallen markedly again. The EC is now a net consumer of energy, heavily dependent upon imports, and is locked into the global political economy.

Consumption of oil has stabilised from the high point of 1973, and there have been refinery closures and a drop in capacity of twenty five per cent to its 1986 levels of 598 million tonnes. Another aspect of this complex supply situation is the attempt to diversify sources of supply so reducing the strategic risk of over-dependence upon any one area. In 1958, the Middle East supplied 77 per cent of the original six EC countries' oil, whereas by 1986 this had fallen to 39 per cent. Today, Africa, including Libya, Algeria and Nigeria supply 34 per cent of the EC's oil requirements and increasing amounts are coming from the Far East, Venezuela, Mexico and the USSR. Exploration within the Community has been intensified. Land deposits of oil have been disappointingly small and there are only small producing fields in the Lacq and Parentis areas of Aquitaine–Pyrenees; at Emsland and Lower Saxony in the North European basin; in the Rhine Valley and in the Paris Basin. In the UK there are small fields in the Trent Valley and Dorset, but these account for a small proportion of total UK consumption.

The Commission's medium-term guidelines for 1985 (fig. 2.2), although not fully realised, reflect some success in reducing the role of oil in relation to other energy sources.

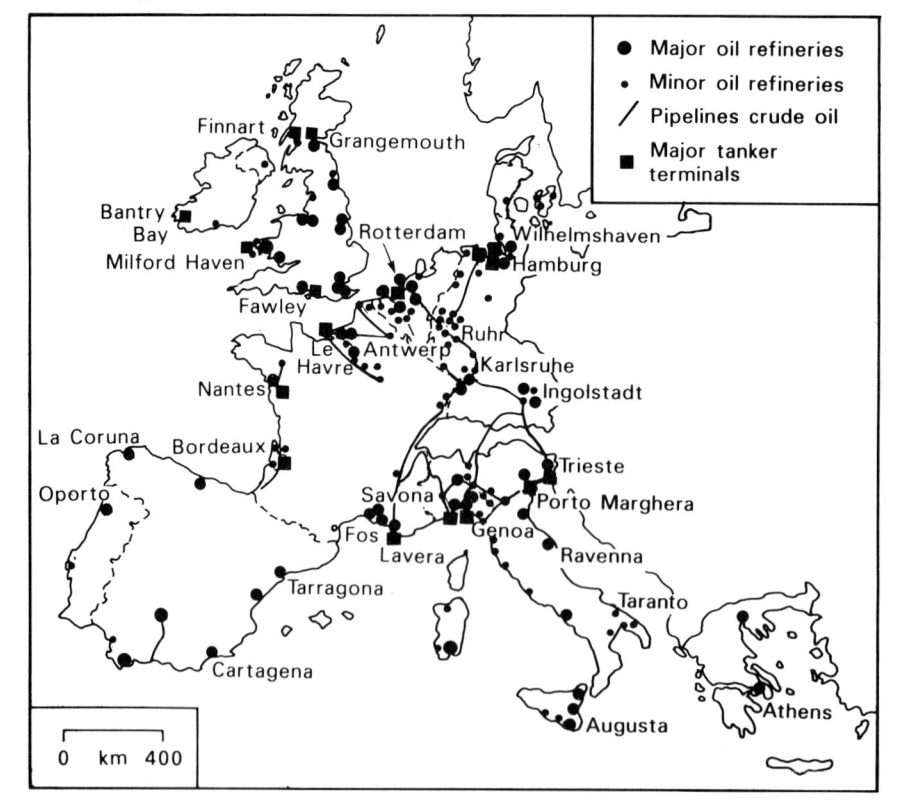

Figure 2.7 EC oil refining capacity

North Sea gas and oil

A significant new energy province, the North Sea, emerged during the late 1960s. Combined with the effects of the accession of the United Kingdom to the EC, it has had the effect of altering the location of European energy supplies. The centre of indigenous energy production is now in and around the North Sea, the so-called 'Home Energy Zone', (Parker 1979). The economic value of the North Sea is in supplementing EC energy supplies by up to one-third, in reducing oil imports substantially, and in moderating the power of OPEC. If they are taken in their wider context, the continental shelf, the energy reserves could last well into the next century, although opinions vary considerably.

The source rocks

In the early part of the Permian period, (circa 250 million years ago) the Rotliegendes sandstone was deposited in the North Sea area over coal deposits which were the source of natural gas. In later Permian times the Zechstein Sea, in which large deposits of salts and carbonates were laid down, covered much of the area of north west Europe and the North Sea. Later the North Sea was a sedimentary basin for hundreds of millions of years, throughout the Triassic, Jurassic, Cretaceous and Tertiary periods, and so has become a vast area of marine deposits, including hydrocarbons, interruptions of which could only have been on a minor scale. The main natural gas concentrations are trapped in porous sections of the Rotliegendes sandstone and below the impermeable salt layers. The oil occurs mainly in the younger rocks where they are thickest in the centre of the North Sea.

The discovery of North Sea wealth

Exploration on the continent of Europe and in the United Kingdom, from the 1930s onwards, provided evidence of small oil and gas fields, but there was little encouragement to undertake the expense of searching for and developing resources under the North Sea on this evidence alone. In addition, marine technology in this field has had to develop very rapidly during the last few years to be capable of constructing and maintaining the large semi-submersible platforms necessary for drilling in up to 200 metres of stormy seas.

At Slochteren, a village in the Groningen province of Northern Holland, Shell and Esso struck natural gas in enormous quantities on 14 August 1959. Further drillings confirmed the vast size of the find - the estimated reserves of gas in this field are 2400 million tonnes of coal-equivalent. It may seem strange that a major gas field like this one in Groningen province had remained undiscovered for so long. One reason was the great depth of the gas, over 3 km below the surface.

Studies of the rock structures beneath the North Sea began with an airborne magnetometer survey of the entire area from southern Norway to the Straits of Dover, which indicated broad areas within which hydrocarbons were likely to be found. These are normally dome-like zones in which the oil and gas are trapped beneath impermeable rock layers. The first seismic survey of the United Kingdom part of the North Sea began in 1962 as a joint enterprise of Shell, Esso and British Petroleum. In July 1964, the Continental Shelf Act defined the territorial areas of the North Sea for the seven nations bordering it. The boundaries between the areas of the states concerned were defined by reference to a 'median line' (fig. 2.8). Concessions were granted to 23 consortia to drill for natural gas in specified 'blocks' in the United Kingdom area, each about 250 km^2 in area.

Natural gas

The first major strike of natural gas was made by British Petroleum late in 1965 in the Rotliegendes sandstone, some 70 km off the Humber estuary. An 80 km pipeline under the sea brought the first experimental quantities of gas from the under-sea wells to Easington on Humberside early in March 1967. A 60 cm feeder main, 120 km long, from Easington across the Humber feeds into the existing Canvey–Leeds methane pipeline. By May 1967, out of some 50 odd wells, sixteen had shown significant amounts of gas – a phenomenal score by all past drilling experience. So far, seven gas fields have been located, Rough, West Sole, Vulcan, Ann, Viking, Leman Bank and Hewett (fig. 2.8). The possibility of gas fields extending to the mainland has also emerged with the discovery of Home Oil of a substantial find in Yorkshire. A second pipeline brings gas to shore at Bacton in Norfolk, from where it is fed into the Canvey–Leeds pipeline near Rugby, and a third pipeline extends to Theddlethorp in Lincolnshire.

Natural gas has become a substantial and rapidly-growing element in the EC energy pattern during the last decade. The United Kingdom and the Netherlands have between them approximately 80 per cent of EC reserves of natural gas. In the Netherlands natural gas now provides 51 per cent of the total primary energy consumption. Dutch natural gas production has now been trimmed back to minimum levels to conserve the remaining reserves. In the EC as a whole it provided 18 per cent of primary energy consumption in 1986. The Shetland Basin, with its large natural gas reserves in fields such as Frigg and Sleipner, will be of great significance in prolonging these supplies and in making natural gas a long-term economic factor.

The oil discoveries

The oil companies discovered significant evidence of oil deposits as well as natural gas in the North Sea. Whilst natural gas is present in the sandstones found to the south of the Humber estuary, oil-bearing layers are to be found in the younger Cretaceous, Jurassic and Tertiary rocks in the Scottish and East Shetland Basin. The northern part of the North Sea is extremely hazardous, especially in winter. Nevertheless, the prospect of large indigenous supplies of oil in Western Europe was such a great attraction that during the latter part of the 1960s exploration drilling rigs began to move north. The

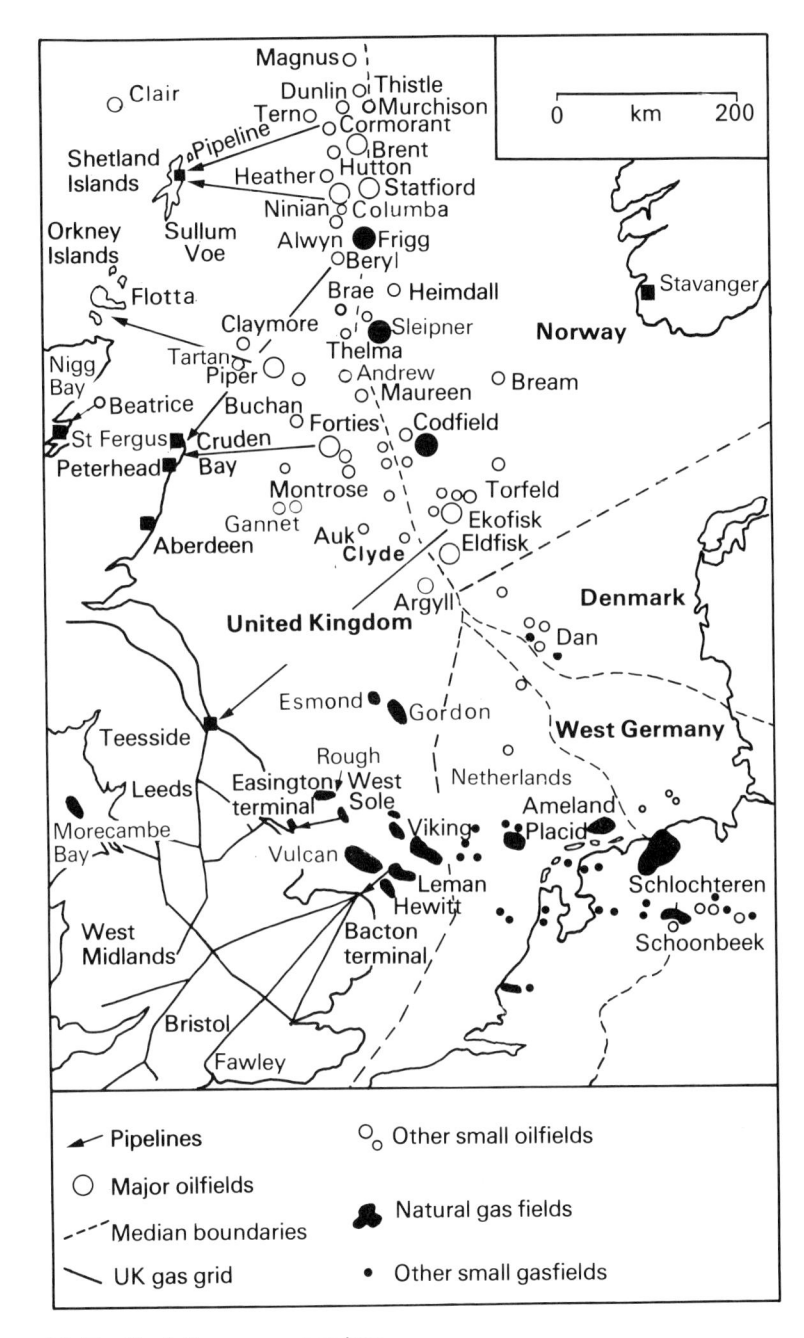

Figure 2.8 The North Sea energy province

major world oil companies have been heavily involved, with thirty consortia working in the area and creating a new marine industrial technology in the shape of the offshore drilling platform. British Coal is involved with the major oil companies in oil exploration as a natural extension of its offshore drilling for coal. The British National Oil Corporation (BNOC or Britoil) was set up to secure Government participation but has now been sold to BP as government involvement has been reduced.

Norway and the United Kingdom gained the major share of this new natural resource as most deposits of oil lie along the median line dividing British from Norwegian waters. There are smaller deposits in Danish waters. Norway, having a much smaller population, benefits in per capita terms more than the United Kingdom, although in any case the whole of Western Europe benefits from having a politically stable source of oil.

Assessing the importance of the oil and natural gas

There is a complex series of physical, economic, geographical and political considerations involved in assessing the significance of the resources.

1. Major economic development has come to the Shetland Islands, Aberdeen and Stavanger. Rotterdam is particularly well-placed to refine North Sea oil.
2. There are over thirty oilfields with recoverable reserves already in production. Some of these such as the Brent field are large by world standards.
3. The UK produces approximately 120 million tonnes of oil per year, is one of the top ten world oil producers, and has become a large scale oil exporting country. Production peaked at 123 million tonnes in 1985/6. Self sufficiency will last well into the 1990s on the basis of present operating oilfields. According to Department of Energy figures in 1989, up to forty new developments are anticipated by 2000 AD. Thus recoverable reserves continue to promise self-sufficiency until well into the twenty first century.
4. Production will begin to decline slowly unless further extensive exploration is carried out. BP estimate that total recoverable reserves, under present technological conditions, are 3000 million tonnes. This gives thirty years' production at an extraction rate of 100 million tonnes per year.
5. The world oil market and price are the key factors in determining the value of North Sea oil. Many of the present fields would not be exploited now had not OPEC increased the price of oil so dramatically in 1973.
6. British government taxation policy in some of the smaller, marginal oilfields is now sufficiently flexible to allow for them to be profitably developed. These include the Columba field, close to Ninian, and the Andrew field close to Forties. These marginal oilfields are crucial to the extension of North Sea production. Improving technology should allow for the proportion of recoverable reserves to increase.
7. Estimates made at Rotterdam University by Professor P. R. Odell in 1975 suggest that, providing exploration continues under favourable political

and economic terms, oil-bearing strata in other parts of the continental shelf such as the Bay of Biscay and the coastal waters of Spain, Greece and Italy could allow for recoverable reserves to be available until the year 2020 AD. Recent reviews have shown that there is twenty-five per cent more oil in the UK sector than originally envisaged.
8. Policies of oil exploitation illustrate basic geographical differences. Norway, having a small population and major HEP endowment has taken a much slower approach to her oil wealth. In the UK with its large energy requirement, very rapid exploitation from 1975 onwards has benefited the country principally during the period of high global oil prices 1973 to 1986 and the money has been used in the reconstruction of the UK economy.

The principal oilfields in 1986 (fig. 2.8) are:

(a) **The Norwegian Group.** Ekofisk is an oilfield in the Norwegian sector discovered by Phillips Oil consortium. Its position is 290 km west of Stavanger and its probable capacity gives reserves of oil half as large as Alaska's Prudhoe Bay. Production is more than half a million barrels per day and is expected to last thirty years. The problems of landing the oil are, however, very difficult. The Norwegian trench, 250 metres deep, prevents a pipeline from being built to Norway and has been the deciding factor in the construction of a pipeline to Teesmouth in the UK. Also, the UK oil and gas market is much larger so that the prospects for piping the oil to Teesmouth are much more viable economically. Around Ekofisk are a number of other fields, the Torfeld and West Ekofisk, whilst much farther to the north are four fields of importance, Codfield, Heimdall, Statfiord (which straddles the median boundary and is therefore of value to both the UK and Norway) and the Anglo-Norwegian Frigg gas and oil field. The Sleipner gas and oil complex is a major new development.

(b) **The Forties Group off East Scotland.** The Forties field was discovered by British Petroleum 160 km east of Aberdeen, and is capable of producing 400 000 barrels of oil per day (one sixth of current UK consumption) although production has already peaked. This group of oilfields stretches, however, from the smaller Auk and Argyll fields northwards to the Piper field. Pipelines run from Forties to Cruden Bay and from the Piper group to Scapa Flow in the Orkneys. Nigg Bay is another major petro-chemical complex.

(c) **The Shetland Group.** This is the richest series of strikes so far. At least six large fields have been discovered, the best known being the giant Brent field. This is linked, with other fields such as the Cormorant and Thistle, to the Brent pipeline. Further south, the Ninian and Heather fields are linked by a second pipeline, running to Sullum Voe in the Shetlands, the largest oil terminal in Europe.

Prospecting activity is already moving north towards Rockall and Iceland, and into the Irish Sea. Three areas are of the greatest significance. The first lies to the north-west of Scotland and is known as the West Shetland Basin. Of twenty-five wells drilled here, three have contained oil, one natural gas, and seven more have given oil shows. In the Irish (Celtic) Sea a sizeable gas field has been discovered off Morecambe, and oil exists at Kinsale Head, just south of Cork. Oil has been discovered west of the Shannon estuary. Perhaps the greatest potential lies in the Western Approaches between Britain and France. The two countries succeeded, during 1978, in defining the median line between them, and three favourable hydrocarbon areas straddle this line. One lies off-shore of Dorset and the Isle of Wight, one south of Plymouth, and one between the Scilly Isles and Brittany. France, which is heavily deficient in energy, has few oil reserves, and has taken a considerable interest in the possibilities for oil exploration in the English Channel.

Nuclear energy

Nuclear energy is seen by many as the long term energy source of the future, although there are enormous problems of pollution and waste-disposal still to be overcome. Its economics are related to the availability and efficiency of competing fossil fuels, and it appears to have the greatest potential for

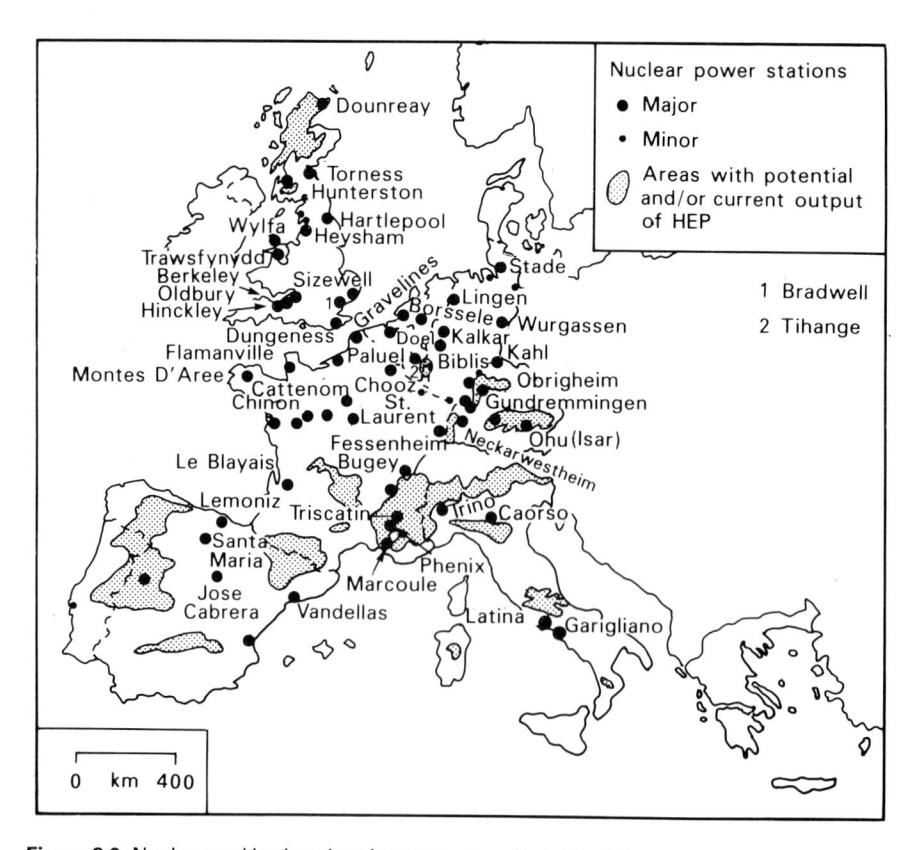

Figure 2.9 Nuclear and hydro-electric power capacity in the EC

technological improvements and cost reduction. Its great advantages are that its fuel, uranium, is needed in relatively small quantities and is therefore easily transportable. Its main use is the generation of electricity, that most flexible form of secondary energy. The development of the fast-breeder reactor gives the prospect of vastly increased energy from each tonne of uranium and its by-product, plutonium, more of which can be produced than is used. Although construction costs of nuclear power stations are higher than those of a conventional thermal power station, generating costs are now lower. The drawbacks of nuclear energy include safety factors, disposal of nuclear waste and the uncertain research factor. Plutonium, in particular, remains radio-active for periods up to 200000 years.

Geographical locations for nuclear power stations are dominated by the need for cooling facilities, remoteness from major population centres, and the provision of electricity in energy deficient areas. Coastal areas and large rivers are therefore ideal. When the pressures of development lead to sites being close to centres of population, such as Stade near Hamburg, then social, political and environmental objections arise.

The UK led the world with its installation of nuclear power stations (fig. 2.9). As a highly industrialised country with an independent military deterrent, major research capability, declining coal production and large oil imports, it has had a long-term nuclear programme since the 1950s. Magnox Mark I reactors were followed by the Mark II, or advanced gas-cooled reactor (AGR), built at Dungeness 'B', Hinkley 'B', Hunterston 'B' and Hartlepool. Generating costs are lower than the latest thermal power stations such as Drax, in Yorkshire, and Pembroke. The second generation AGRs are much larger than the original power stations, but these have been subject to long delays because of development problems, and the UK is now undertaking the construction of PWRs (Pressurised Water Reactors). From an average 400 megawatts in the 1950s the projection for Heysham power station is 2500 megawatts. In 1989 there were twenty-five major nuclear power stations in the United kingdom, with Dungeness, Hartlepool, Heysham and Torness the latest additions. The nuclear capacity in 1986 contributed 18 per cent of the national electricity supply.

On the Continent, Euratom was set up as one of the original institutions of the Common Market, with its aim being the co-ordination of nuclear energy production and research. During the 1960s, however, its work was severely curtailed by lack of funds and by the lack of urgency because of cheap and abundant oil. In 1970 installed nuclear capacity on the Continent was minimal compared to that of the United Kingdom.

However, the oil shock of 1973 gave a major boost to nuclear power programmes, and new construction was considered necessary and desirable. Large new installations, particularly in West Germany and France, boosted this capacity to 10000 megawatts by the end of 1977, thus supplying nearly 10 per cent of the total electricity generation. The energy costs of the 1970s, and particularly the rise in the price of oil, stimulated a major programme of research and development.

Biblis nuclear-power plant under construction in West Germany

A further large increase in the mid-1990s is in line with the Commission's medium-term objective (fig. 2.2) of nuclear power supplying 17 per cent of EC primary energy generation. In terms of electrical energy production alone the percentages are vastly increased. Belgium already has 73 per cent of her electricity generated from nuclear power, and West Germany had twenty-one nuclear plants in operation or under construction in 1986. France has an impressive development programme. By 1986 it had over thirty-five nuclear plants in operation producing 69 per cent of French electrical energy. This pattern of development was rudely interrupted, however, by the nuclear accident at Chernobyl in the USSR in 1986. The effects of the disaster triggered a slow retreat from nuclear energy. Controversy is widespread and has made it likely that Italy, West Germany, the Netherlands and Belgium will produce no more nuclear reactor construction programmes for the foreseeable future.

With lower demand for energy since 1973 and cheaper oil post 1986, the further development of nuclear energy is once more in the balance in financial as well as in environmental terms.

Hydro-electric energy

Hydro-electric energy is one of the less important sources of primary energy in the Community. The best sites have already been harnessed, and its relative contribution to primary energy consumption has declined to two per cent (fig. 2.2). The high relief areas of the Alps, Apennines, Massif Central and Central South Germany, and major rivers such as the Rhine and Rhône (fig. 2.10) are the principal areas of HEP generation on the continent. For electricity generation, HEP is locally very significant, particularly in the case of France and Italy, and its overall contribution to the EC electrical energy balance is thirteen per cent (fig. 2.11). Spain's development programme since the 1950s has included a major HEP programme in the basins of the Tagus, Ebro, Guadiana and Guadalquivir rivers. HEP contributes twenty per cent to her electrical energy generation. In Portugal HEP supplies some forty per cent of electricity generation from the Tejo and Douro rivers.

The Shannon basin project in Ireland harnesses the energy of the tides. A similar scheme is in operation in the Rance estuary in Brittany, and there are

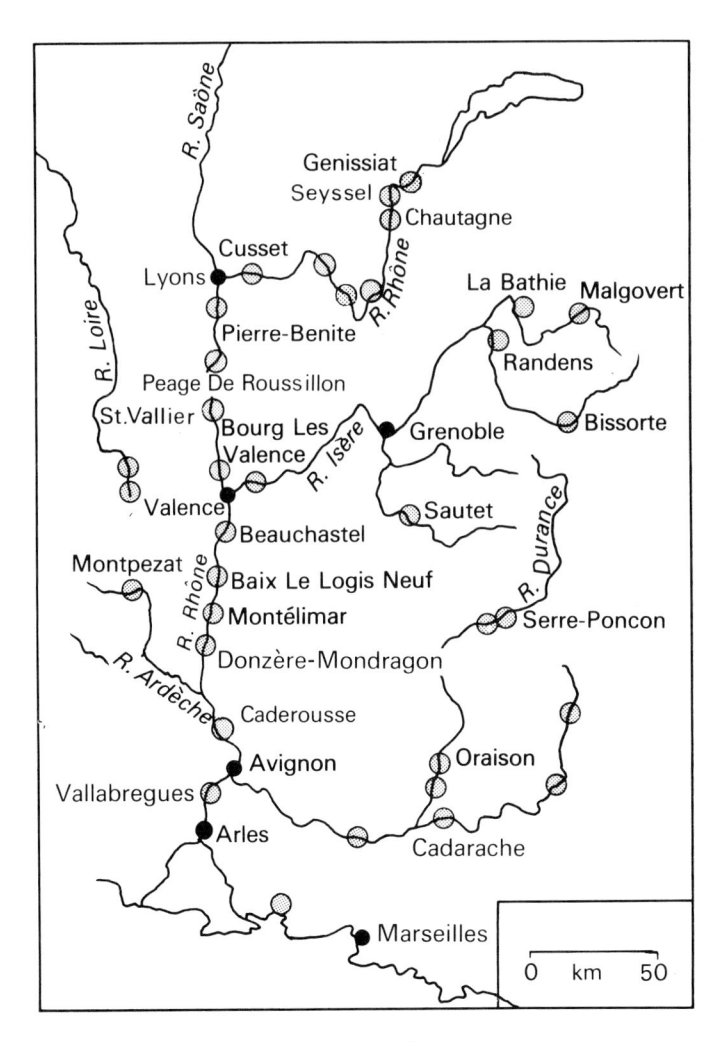

Figure 2.10 Hydro-electric power in the Rhone valley

possible schemes suggested for the UK on the river Severn, at Morecambe Bay, and at the Wash. The number of sites is limited, and this source of energy is still in the development stage.

Electricity

The transformation of the primary fuels described in this chapter into the secondary source electrical energy is one of the most important single aspects of the energy revolution. The generation of electricity depends upon the use of coal, oil and natural gas in thermal power stations, together with nuclear power stations and hydro-electric power stations (fig. 2.11). There are considerable variations between the member states. In the United Kingdom coal provides 67 per cent of electricity generation, and nuclear power 18 per cent. In France some 19 per cent and in Italy some 25 per cent of electricity generation is from HEP. Italy is heavily dependent upon petroleum, the Netherlands upon natural gas, and in France the dominant contribution is by nuclear power. The generation of electricity illustrates a range of contrasts in development, natural endowments, localised resources and government policies.

Electricity demand in the EC has doubled every ten years since the 1950s. This has been accompanied by a much greater efficiency in production and transmission. The same amount of primary fuel as used in 1958 will now generate 45 per cent more electricity and the super-grid at 400 kilovolts will now transfer it over large areas with a minimal loss of energy.

Electricity has enormous advantages. It is a clean form of energy, and in the long term may well be generated by nuclear power, probably completely replacing the fossil fuels. It is the most suitable energy form for automated industrial processes, including robotisation and data processing. Most important of all, the distribution of electricity via the grid over large areas of Western Europe has helped to break down the old pattern of heavy industrial regions restricted to the coalfields in favour of a widespread and flexible distribution of industry.

Towards a Community Energy Policy

The reduction in the importance of coal, matched by a rapid rise in oil consumption and the growing use of natural gas and nuclear energy, together with the local importance of hydro-electric power, has created a great variety of choice of energy sources in the EC of the 1990s (fig. 2.2). Since the 1950s energy self-sufficiency has become a thing of the past. This reduction in indigenous supplies leads, however, to a dependence upon global supplies where stability of supply is not assured. The difficulties of formulating a common energy policy have been obvious since the Paris summit meeting of 1972. The irony is that the ECSC and Euratom were two of the earliest agencies of European integration. The abundance of oil and the diversification of energy supplies during the 1960s meant that attitudes to energy

	Spain	Portugal	Greece	United Kingdom	West Germany	Belgium	Netherlands	Italy	France	EC
Thermal power stations:										
Total	48.7	56.7	87.2	79.8	65.7	30.3	93.9	69.9	11.7	56.0
Coal	27.3	14.3	3.2	66.5	33.0	19.5	22.6	13.7	8.2	30.1
Lignite	14.2	–	62.0	–	20.2	–	–	0.5	n	7.4
Petroleum	4.6	39.3	21.8	10.7	3.3	4.1	5.1	39.9	1.1	9.8
Natural gas	1.7	–	0.2	0.6	6.2	1.5	61.9	13.8	0.8	6.5
Other	0.9	3.1	–	2.0	3.0	5.1	4.2	2.0	1.5	2.2
HEP/ Geothermal	22.0	43.3	12.8	2.4	4.8	2.5	–	25.5	18.6	11.7
Nuclear Power	29.3	–	–	18.4	29.5	67.2	6.1	4.6	69.7	32.3

n = negligible

Figure 2.11 Electricity production in the EC 1986 (percentage)

planning were unprepared by 1973. The oil crisis and its aftermath provided a great incentive for an energy policy but also illustrated a great divergence of interest amongst the member states. National attitudes and the balance of energy consumption vary according to resource endowment, trading policy, research and investment. Italy and France are energy deficient and keen to secure access to energy at the lowest possible price. The UK, as a major oil producer, would like to see a minimum price for oil. The Netherlands and West Germany would like an open energy market. The UK and West Germany are concerned to maintain their large coal industries, to secure new markets and perhaps to wait until coal becomes an economic substitute for oil in the 1990s. The three newest members, Spain, Greece and Portugal are energy deficient, and with major economic development programmes, their increasing requirement for energy imports may well alter the EC energy balance.

Although the EC still lacks a coherent energy policy, there has been a growing awareness of the likely onset of an 'energy gap' by 1995. The Commission estimates that world oil prices will continue to rise in the long term, with periodic fluctuations in supply and demand making costs unstable. The North Sea and wider continental shelf supplies of oil and natural gas are vital for several decades to come, but increasingly will only partially counteract the deficiency in the EC as a whole. The Bremen European Council in July 1978 stressed the need for co-ordination and a common energy policy.

The Commission's medium-term guidelines for 1985 suggested urgent action to support the Community's coal industry, exploration and development of new indigenous sources of oil, gas and uranium, an increase in the role of natural gas, steady development of nuclear generating capacity, and a drastic reduction in the EC's dependence on imported oil. The Council resolution of 17 December 1974 laid down that by 1985 the Community's dependence on imported energy should be reduced from 61 per cent (1975) to 50 per cent, and possibly to 40 per cent. There should be a more efficient use of energy with a corresponding reduction in overall consumption of 15 per cent, an increase in the share of indigenous natural gas to 25 per cent of the primary energy total, an increase in the role of electricity, and the generation of 75 per cent of electricity from indigenous solid fuels and nuclear power. Fig. 2.2 illustrates the real difficulties of achieving these figures. In 1986 oil and coal are still consumed at levels above those in the 1985 objectives. Since 1984 total energy consumption has risen again and is likely to increase in the 1990s. Policy objectives for 1995 need to be formulated now. The Community has a research and development programme for energy conservation and the production of hydrogen, wind, solar and geothermal energy. The JET project for research into nuclear fusion is sited at Culham, Oxfordshire. It is vital to the Community's future interest to lessen its dependence on external supplies and to build up internal dependable sources of energy. Security of supply is a more significant factor than that of cost.

3
Employment: industrial change and the tertiary sector

The process of employment change is often dramatic, and is related to the initial decline of the primary sector during rapid industrial development and the subsequent transformation of the secondary sector as it is increasingly replaced by the tertiary or quaternary stage. Great Britain was the world's first modern industrial power, followed closely by Germany, and in varying degrees by the other West European powers. A significant characteristic during the nineteenth century was an overwhelming reliance upon heavy and staple industries. In the 1880s the United Kingdom produced about 70 per cent of the world's ships; as late as 1907 coal, iron and steel, and textiles accounted for 46 per cent of her gross domestic product and 70 per cent of her exports. The twentieth century has witnessed the relative decline of these basic industries, as they are the first stages on the road to a mature economy. The growing complexity of industrial structure has meant that they have been supplemented by the assembly-line techniques of the car industry, the technical accuracy of the machine-tool industry, consumer-based light industries, and science-based, specialised 'hi-tech' industries. Associated with this is the increasing dispersal of location, caused largely by the freeing of industry from coalfield locations and its tendency to become 'foot-loose'. This chapter will also examine the rapid growth of employment in the service sector, a phenomenon of recent and variable development in the advanced industrial economies of the EC. There are four principal themes:

(a) Free market factors of location;
(b) Industrial production, organisation and financial factors;
(c) Government intervention and supra-national factors;
(d) The growth of employment in the service sector, including tourism.

Free market factors of location

Industrial location has traditionally been affected most by the availability of energy, capital, raw materials and skilled labour, by market potential, and by transport facilities. It is the changing relative importance of these factors which sets up decline or growth in, and disparities between, regions. The extent of concentration in the nineteenth century on the older coalfield

48 THE NEW EUROPE

The Sambre-Meuse valley, with a ribbon of settlement, coal-tips in the background and factory chimneys adding to the features of an old industrial landscape

industrial areas (figs. 3.1 and 3.2) illustrates the essential importance of coal as the original location factor, due primarily to its bulk and weight, which prohibited extensive movement, but also because it suffered little competition for over a century. This is explained by Weber, (1909), in whose 'least cost location' theory transport costs were usually minimised by location at the source of raw material. Significant areas of 'heavy' industry are still coalfield-located, although they owe their survival and present position partly at least to new factors which have tended to preserve geographical inertia.

In these coalfield areas there is a predominance of heavy or staple industry, and a tendency to a distinctive single industry, based upon the original local raw materials. Such an emphasis is summed up in the term 'monotechnic'. The Ruhr, the Potteries, Liège and Tyneside all have this characteristic dominant industry, although less so than in the past. The age of development of such areas means they also have a large proportion of old and obsolescent housing and factories in densely populated conurbations.

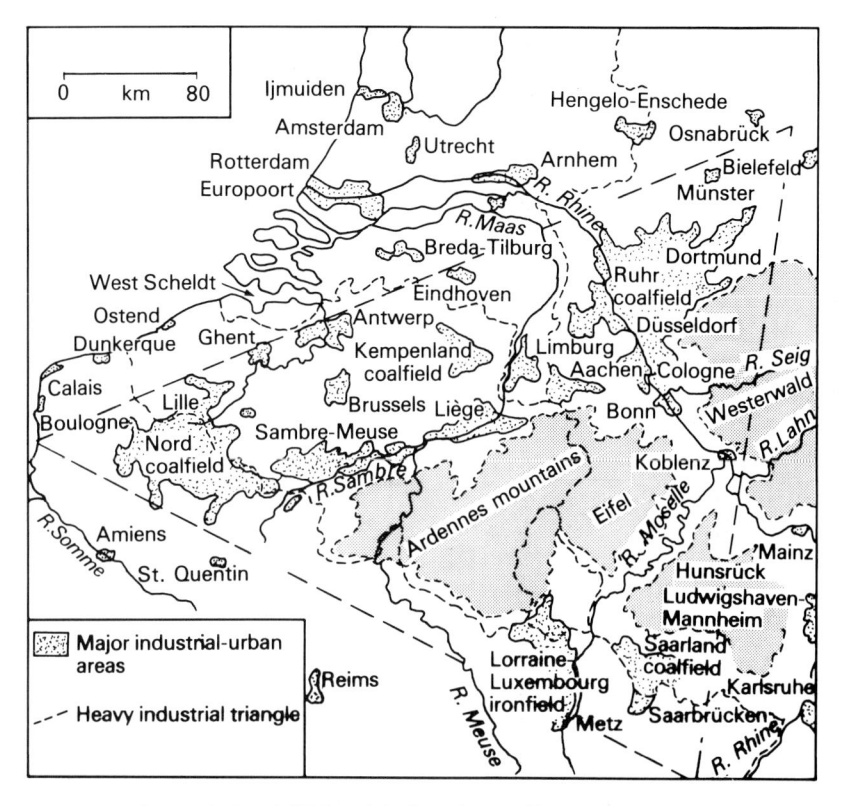

Figure 3.1 The 'Heavy Industrial Triangle' of north-west Europe

The United Kingdom coalfields

On a large-scale map of the EC (fig. 3.2) the areas dominated by long-established heavy industry are dispersed around the margins of the highland zone of the United Kingdom. Individually, however, they are easily definable regions. Dense clusters of industry are found in central Scotland along Clydeside, in the north-east around Newcastle-upon-Tyne, and at Teesside. These are the traditional areas of heavy industry which specialise in iron and steel, shipbuilding, marine engineering, and heavy structural engineering, with chemicals particularly on Teesside. East Lancashire, centred on Manchester, and West Yorkshire are the two areas traditionally associated with the textile industry, but the shipbuilding and chemicals associated with Merseyside, and the Sheffield steel industry are also close by. The West Midlands conurbation was originally dependent upon the steel and metallurgy of the 'Black Country', but is now one of the most varied and complex engineering regions in Europe. One of the most distinctive monotechnic

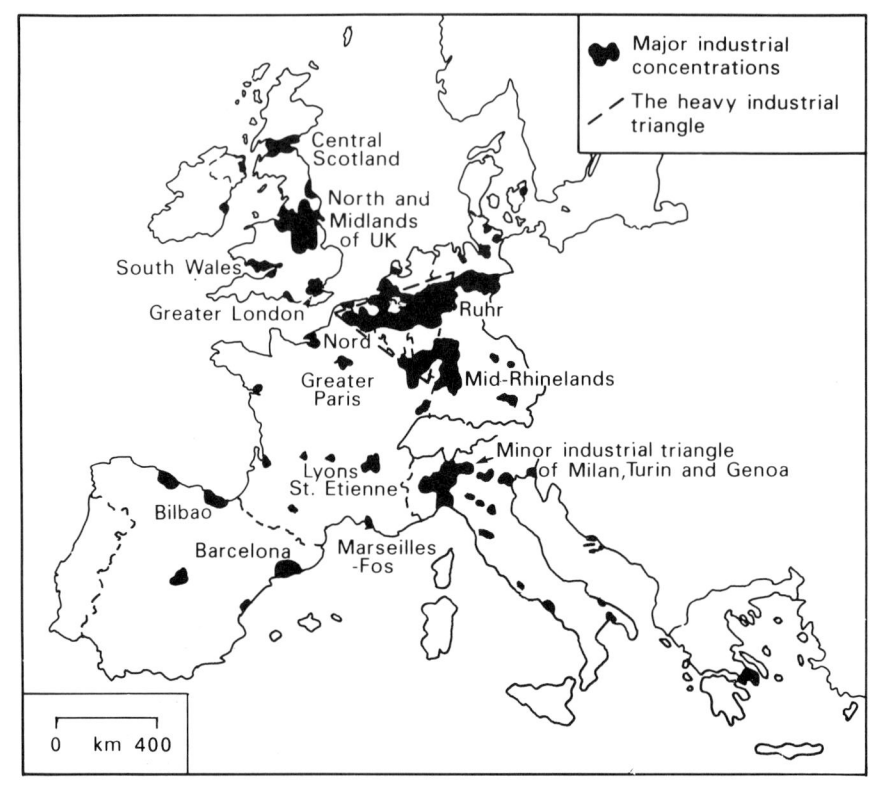

Figure 3.2 Major industrial areas in the EC

industrial areas in the world, the North Staffordshire coalfield, continues to produce high-quality china, pottery and earthenware goods. It deals with a high-value product in great consumer demand, and herein lies its success. The small North Wales coalfield between Wrexham and the Dee developed a small but important steel and chemical industry. The area is relatively isolated and the medium-sized Shotton steel works on Deeside was closed in 1982 as part of the rationalisation programme of the British Steel Corporation. The South Wales coalfield is a classic example of intra-regional migration towards coastal locations such as Port Talbot and Llanelly. The decline of coal production in the confined valleys of the Rhonddha and Taff, combined with the low cost of imported raw materials has led to industrial concentration of the steel, tin-plate and metal-processing industries along the coast.

The 'Heavy Industrial Triangle'

This zone is the principal area of industrial activity in which most of the coal, staple industries, steel-making and heavy engineering of continental Europe are found (fig. 3.1). Bounded at its apices by the Nord coalfield of France, the

Ruhr coalfield of West Germany, and the Lorraine iron-ore field, it contains even now 95 per cent of coal production of the original Six, and about 50 per cent of their steel-making capacity. The French Nord/Pas de Calais region stretches in an arc from Dunkerque to Lille and through into Belgium as the Sambre-Meuse coalfield. Mons, Charleroi and Liège are the central industrial foci in the Sambre-Meuse valley and are important for textiles, heavy metallurgy, chemicals and glass. Extensions of industry are found north-east in the Kempenland, more recently-developed concealed coalfield, where Genk and Hasselt are chemical and metallurgical centres. Across the Dutch border is the small Limburg coalfield, around Maastricht, and in West Germany lies the small Aachen coalfield. The Ruhr is the greatest concentration of all, and extends from its Duisburg-Dortmund axis south to Cologne, and north towards Münster. The southern apex of Lorraine, based upon Minette iron-ores and a modern coalfield, includes the steel and engineering centres of Nancy, Thionville and Metz, and extends into Luxembourg. Slightly to the north-east is the Saarland, an old coal and metallurgy region with Saarbrücken and Volklingen as the principal towns.

The Nord/Pas de Calais coalfield

The Nord/Pas de Calais coalfield provides a useful case-study which illustrates decline and change in these old manufacturing zones in the twentieth century. In 1962 it had over half its total employment (335 000 out of 607 000) in three staple industries: coal, steel and metallurgy, and textiles. The coalfield, stretching from Bruay to Valenciennes (fig. 3.3) has been badly affected, with coal production falling from twenty-seven million tonnes in 1961 to fifteen million tonnes in 1971 and to two million tonnes in 1986. Under the latest rationalisation plan, it is due to cease production altogether, since even with recent changes in the costs of other primary fuels, notably oil, it is a very high-cost coalfield. The seams are thin and disturbed, productivity is low and reserves are practically exhausted. The excessive investment costs which would be required in a declining resource are therefore ruled out. The legacy of the area, in addition, is a cluster of small mining towns known as 'cités minières' with all the environmental disadvantages of obsolescence and a need for urban renewal. The Nord is referred to as the 'Pays Noir'.

The textile industry lies mainly to the north of the coalfield itself at Lille, Roubaix, Tourcoing and Armentieres. It is based on wool and cotton and has experienced severe contraction, with employment reduced from 224 000 in 1931 to 168 000 by 1954, and 40 000 by 1986. Foreign competition, a slowness to adapt and introduce up-to-date machinery, and loss of markets to synthetic fibres are the main causes in this decline, resulting in the closure of many small firms and regrouping of others.

The Douai-Valenciennes area has traditionally been the centre of the steel and heavy metallurgical industries, with specialised steel and machinery to the south in Arras, Cambrai and Maubeuge. Employment is relatively more stable, and production of steel in the whole area has actually increased,

52   THE NEW EUROPE

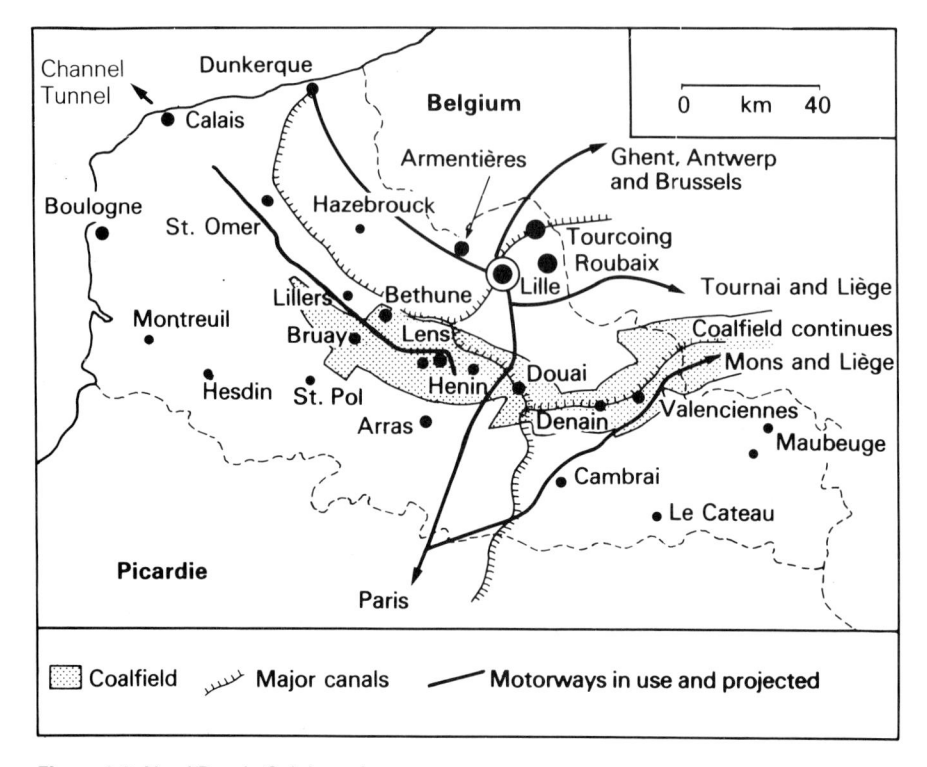

Figure 3.3 Nord/Pas de Calais region

entirely owing to the new integrated Usinor steelworks at Dunkerque which produces over five million tonnes, nearly 70 per cent of the region's steel output. Here is the essence of the Nord's problem. The older parts of the coalfield have an over-dependence upon heavy metallurgy. Factories are often small, in inconvenient situations and use obsolete machinery. There is a lack of light industry and of the specialised high-value products which would help to diversify the industrial structure.

The region needs adjustment and diversification away from basic industries, together with the renewal of its equipment. However, it has great potential advantages, the foremost of which is that there are 60 million people within 300 kilometres, forming a huge consumer market. A large adaptable labour force has the potential to attract capital investment on a large scale and many new industrial estates have been sited in the coalfield towns. New industries include plastics, petrochemicals, cars and components, electronics, and consumer durables. The old canal system has been upgraded with deepwater channels taking 3 000 tonne barges from Dunkerque to Lille and Paris. Finally, there is the new motorway infrastructure and the future link with the channel tunnel. The real value of the motorway network is that it places the Nord athwart the Paris-Lille-Brussels axis.

Lille: crowded factories, railways and canal represent the old smoke-stack industrial landscape

Industrial adjustment

There is a continuing major significance for the coal-based industrial areas in spite of the radical changes now taking place. The process of geographical inertia is based largely upon their skilled labour supplies, traditional industrial linkages, fixed capital investment and large urban consumer market. With the commitment which national governments now have to aiding these old industrial regions, and with the EC's regional aid policy, these areas will continue to attract capital investment on a large scale. The problem of the residual staple and heavy industries, and the resulting imbalances, must be corrected by the lengthy process of diversification into new industries.

Of great significance is the position of each industrial area relative to its national economic core. It is arguable that the Nord/Pas de Calais, Belgian Sambre-Meuse, Ruhr and West Midlands, are in a favourable position, lying as they do close to the core areas and in the mainstream of activity. The

54 THE NEW EUROPE

Bochum in the Ruhr: the Opel car assembly plant

north-east of England, and the Scottish coalfields, both of which lie close to the oil and natural gas resources of the North Sea, may well have a similar advantage. The North Sea littoral is certainly likely in the future to be a major area of concentration for industrial development and it may well have a revitalising effect upon these older coal-based industrial areas. By contrast, industrial regions lying on the periphery of activity within Western Europe face adjustment in a more difficult context. South Wales, the Basque region of Spain and areas around the Massif Central such as Decazeville, Alés and Commentry have no such advantages of centrality and the process of industrial adjustment will take longer, and may not be as successful.

Imports of raw materials: coastal industrial locations

Imports of raw materials are becoming increasingly important. Iron-ore, associated with coal in the blackband deposits of the industrial revolution, is now left in such small uneconomic quantities that the steel industry imports the bulk of its iron-ore from Sweden, Brazil and North Africa. Only French Lorraine and Spain produce any significant quantity in Western Europe. Oil is the single most important element of change, but all forms of raw materials which are heavy, bulky and costly to transport overland are increasingly causing industrial concentrations to develop on coastal, estuarine and major waterway sites. Heavy processing industry, including steel, non-ferrous metals, oil-refining, chemicals and petrochemicals, cement and electrical power generation, need to be sited at low-cost importation points, with good facilities for transhipment inland and large flat sites for construction of installations. Ports are increasingly the cheapest points at which receipt, processing, manufacture and distribution can take place (fig. 3.2). These are break of bulk points. The sea-coast from the Seine to Denmark is the single most important stretch, with Le Havre, Calais, Dunkerque, the Rhine estuary (Rotterdam and Antwerp), Bremen and Hamburg, Kiel and Copenhagen being the main ports. On the Mediterranean coast are Barcelona, Venice, Genoa and the Marseilles-Fos area (figs. 3.4 and 3.5), whilst in the United Kingdom the nodal points for new heavy industry are Thames-side, Severnside, Humberside, and Southampton Water. The importance of river ports is illustrated by Rouen, Duisburg and Mannheim which are centres of major industrial concentrations on waterside sites.

'Footloose' industry: the dispersal process

The liberation of manufacturing industry from the confined zones of the coalfields and the coastal break of bulk points has been achieved by a number of factors acting in conjunction. These are: the widespread cheap transmission of electricity; the development of fast transport links especially motorways so that movement of raw materials is cheaper and more efficient; the growth of light industry, precision engineering and science-based industries in

Antwerp: A dockside industrial zone based upon break-of-bulk with oil storage and refining, and petrochemicals

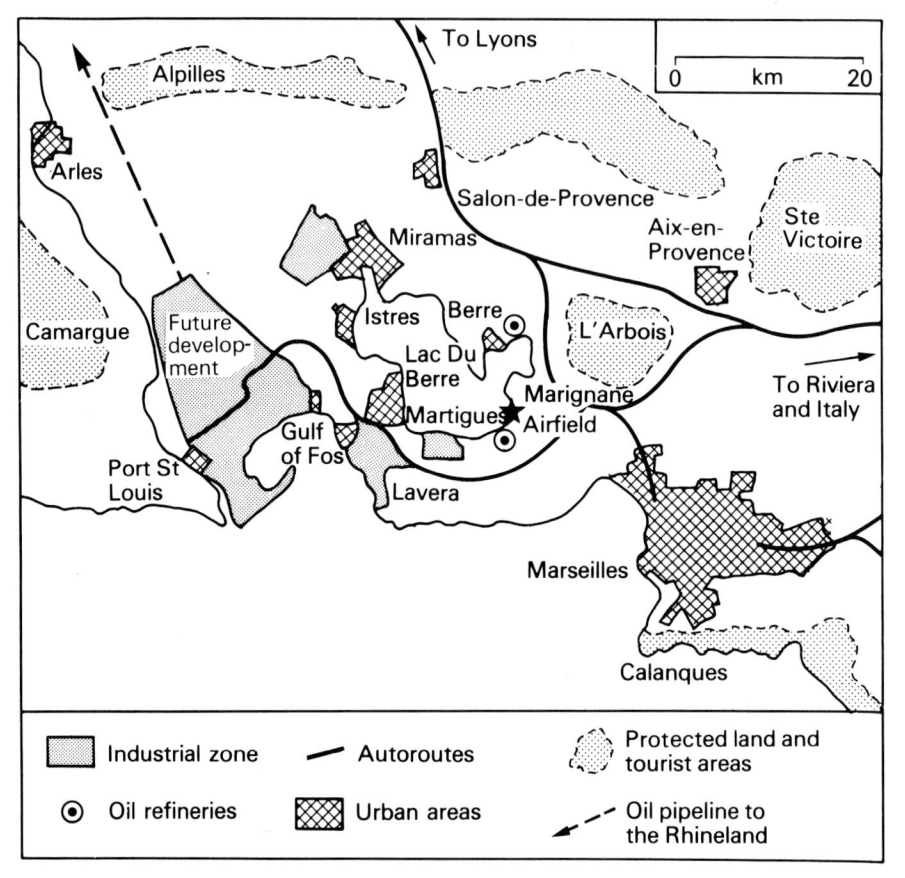

Figure 3.4 Fos-sur-Mer/Lavera industrial zone

which a greater emphasis is placed on skilled labour than on raw materials; improvements in technology in which the product has a higher value-added content.

Modern growth industries, vehicles and components, electrical engineering, synthetic fibres, electronic and optical equipment, all have a flexible choice of location. Industry has become 'footloose' with many factors being important in the location of factories. Labour costs in a wealthy society such as that of Western Europe are often the single largest element in manufacturing overheads, whilst the consumer goods destined for the market need to be adjacent to that market, or within easy transport distance. Industrial estates are found in almost every town and city, usually as part of a planned expansion of employment during the last three decades when population was growing rapidly. They have facilities such as road layout, electrical power, water and sewage supply, and standard factory units close to suburbs and housing estates. They often have commercial and retail facilities such as hypermarkets, and are sited on the periphery of the town in close proximity to major routeways for improved access.

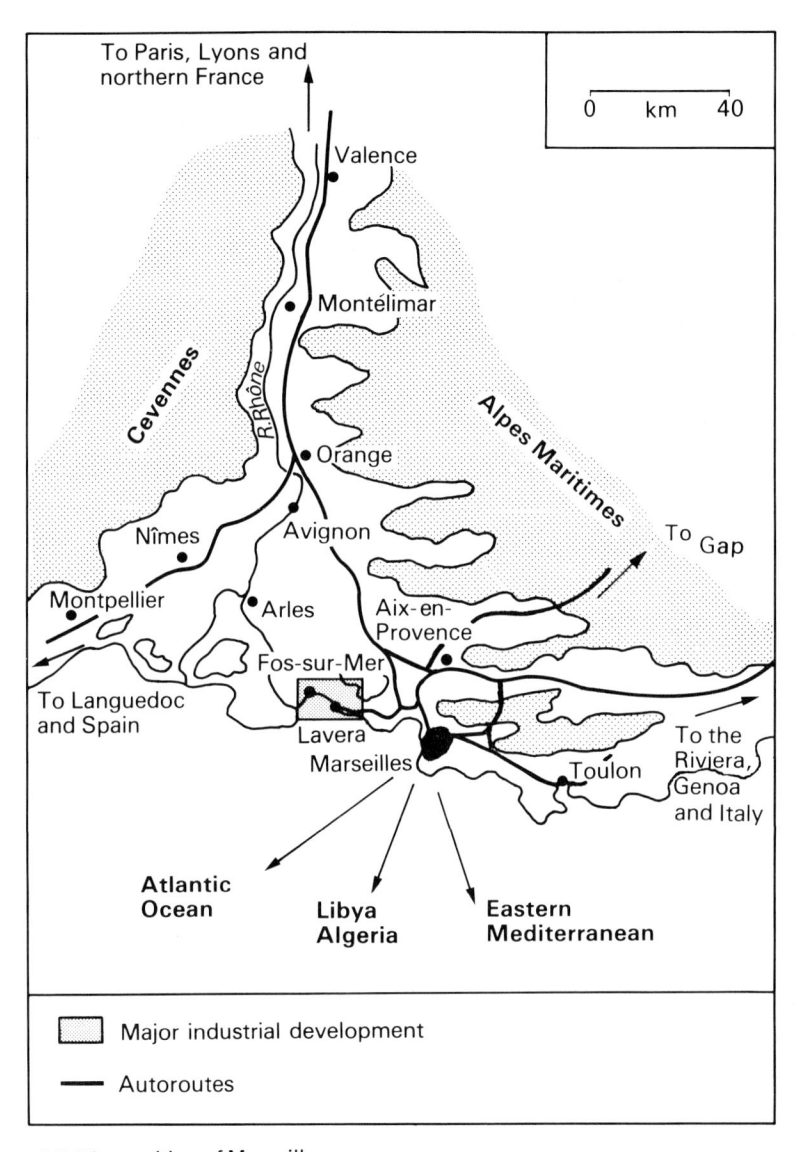

Figure 3.5 The position of Marseilles

Rennes in Brittany illustrates modern industrial estate development as an example of industrial dispersal and mobility. It is the capital city of Brittany and lies in a rich agricultural area, the Rennes basin (fig. 3.6(a)). It is a bridge-point on the Vilaine River and has developed as a market, administrative and cultural centre. Its population thirty years ago was about seventy thousand, but it has experienced dramatic growth since then and now has a population of 250 000. It lies on the main routeways passing into and through Brittany, including the main railway link from Paris to Brest, and north-south road links from St. Malo to Nantes, and the east-west roads, the N24 to

58   THE NEW EUROPE

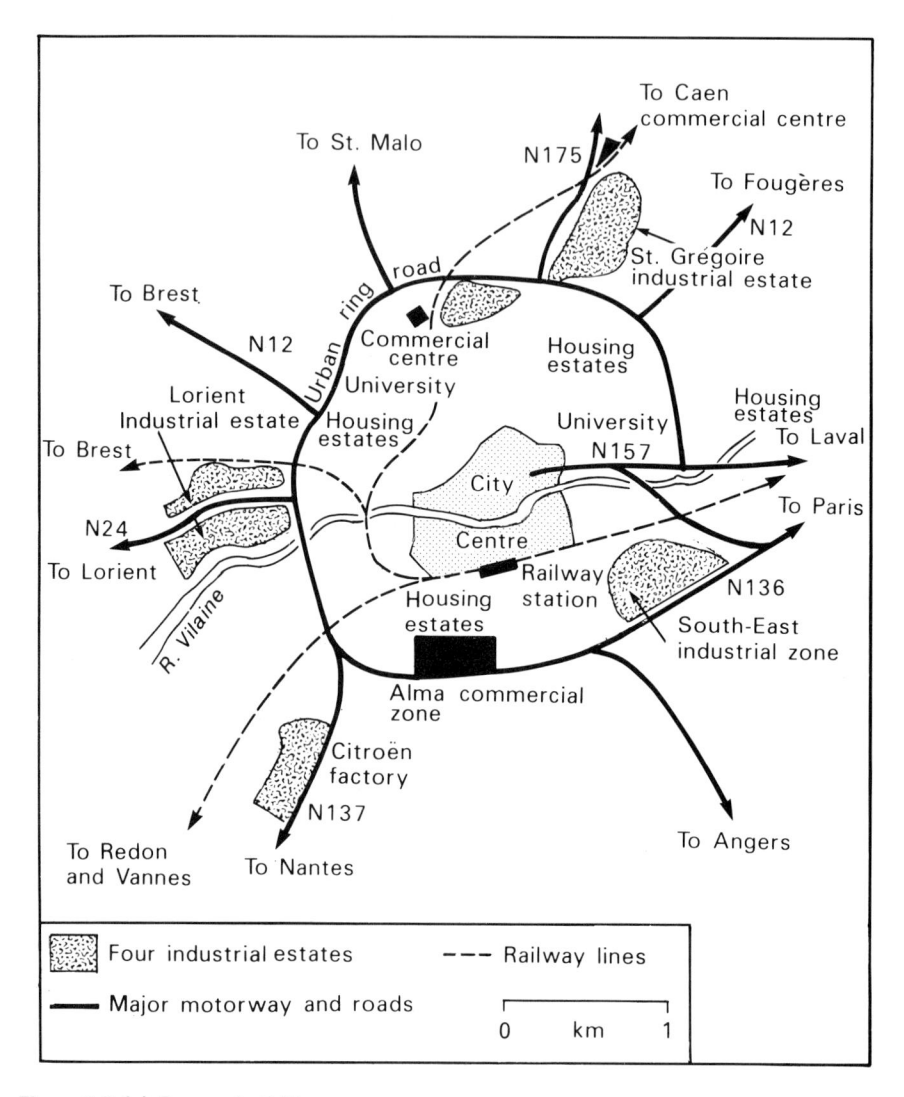

Figure 3.6 (a) Rennes in Brittany

Lorient and the N12 to Brest. It lies closest of all towns in Brittany to Paris and is now linked to Paris by the motorways A11 and A81. It was selected by the French government as a growth point for industrial investment in the 1960s. The old city has been encircled by an urban motorway, the Rocade N130, and large industrial estates have developed at access points (fig. 5.14). St. Gregoire industrial estate lies at the junction of the St. Malo road with the Rocade, the Lorient industrial estate containing the Citroën car works on the western side on the N24, the south-east industrial estate at Chantepie, whilst to the south on the route de Nantes lies the second Citroën car works. The

Lorient industrial estate is the largest with a great variety of vehicle components, engineering, light and consumer industry to be found (fig. 3.6(a)). The two Citroën factories employ over 16 000 workers.

Manufacturing in cities

The most significant positive development relating to market and labour supply factors has been the growth of major manufacturing areas associated with large centres of population, in particular economic core cities and capital cities which are prestige centres acting as the administrative, commercial and financial centres of the country, and the centres of the transport network. Greater Paris, Greater London, Brussels, Frankfurt, Copenhagen and Milan all have one major resource - population. Labour requirements are well-qualified graduates for technical research and managerial staff, and a pool of skilled labour and technicians for the operation of complex machinery.

Mass-market production techniques require areas with a high purchasing power and with an income level above average. Finally, the complexity of modern industry, needing frequent contact and movement by executives, requires good communications and locations near motorways and airports. All these factors have tended to favour the growth of manufacturing in and around the cities of Europe.

The 'minor industrial triangle' of Milan-Turin-Genoa in Northern Italy owes its origins to the accumulation of capital from trade in the medieval city states of the Plain of Lombardy. A range of historical and geographical advantages (see chapter 16) of Northern Italy, has meant that it, rather than the national capital city of Rome, has become the industrial core of the country.

Other major city regions are those in the middle Rhineland, such as Frankfurt, and the southern German cities of Munich, Stuttgart and Nuremberg, noted for their wide range of high-value components and precision engineering. The cities of Saxony, particularly Hanover and Brunswick, are important for vehicles and components. Greater Lyons, second city of France, stands in the centre of the Rhône-Saône corridor. Originally a silk manufacturing centre, it has become important for engineering, metallurgy and chemicals, with an impressively varied industrial structure.

Greater Paris and Greater London, each with about eight million people, are capital cities of world rank, and have an extremely varied manufacturing base. Their political, financial and commercial wealth has been responsible for the creation of a 'metropolitan structure of industry'. This includes a complete range of industrial types over the whole city, ranging from the heavy industry of the port areas to the specialised industries of the central area.

The industrial structure of London and its region

A study of six type zones illustrates the factors involved in the concentration of manufacturing in a major capital city (fig. 3.6(b)).

60   THE NEW EUROPE

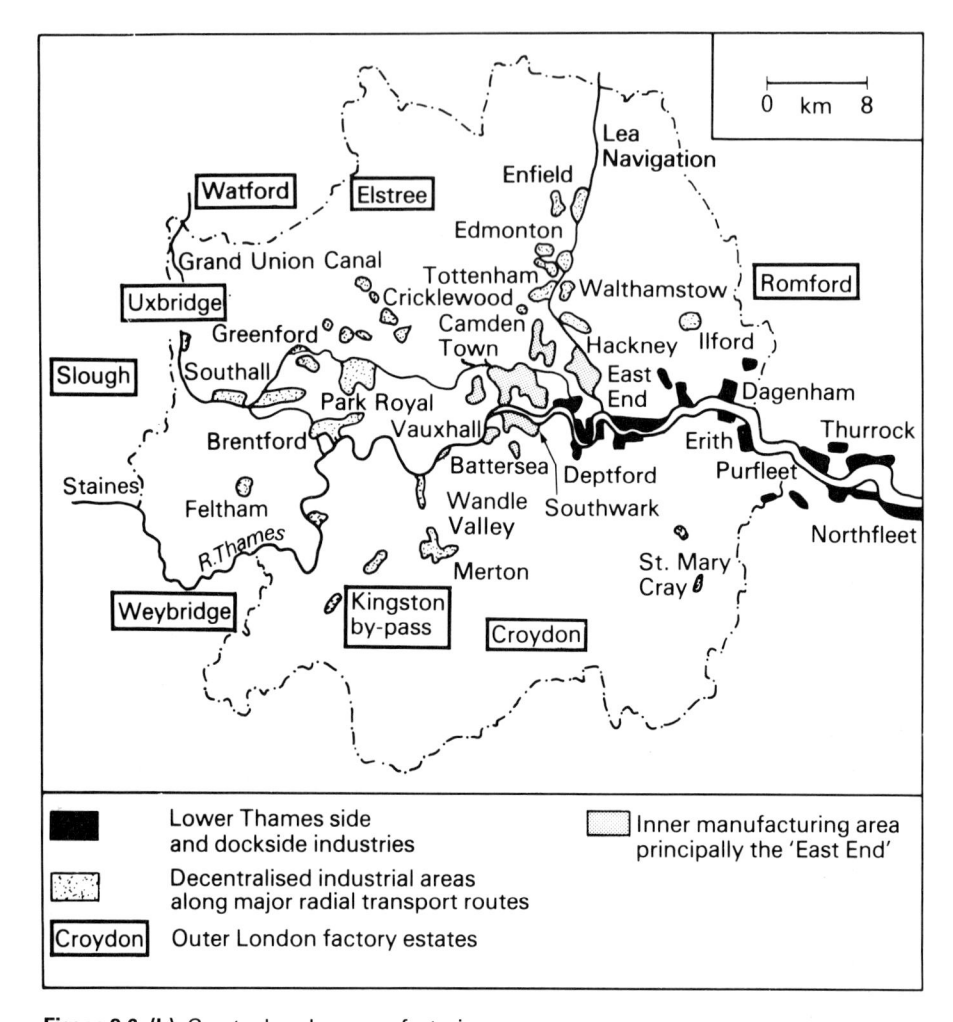

Figure 3.6 (b) Greater London manufacturing areas

1. **Central London.** The central areas of metropolitan cities contain activities which are a reflection of their position in national life. The manufacture of fashionable clothes, jewellery and other luxury goods, is often carried out very close to the marketing centre, as in the West End of London adjacent to Oxford Street and Bond Street. The publishing and printing of periodicals and newspapers is another specialised industry which is carried out in the City of London, although this has been subject to considerable migration out to modern sites in the redeveloped docklands (Wapping).
2. **Inner London, 'The East End'.** This was another nucleus of London's industry in the nineteenth century, with a crescent of small workshops from Whitechapel through Bethnal Green and Hackney to Finsbury, Clerkenwell and Camden Town. Sixty years ago the bulk of London's industry lay within this area, but considerable migration has taken place

out of the cramped quarters and decayed inner suburbs to outer London and the New Towns. The East End is, however, still a major element in London's industrial geography. The main industries are clothing and furniture, food, drink and tobacco, printing and specialised instrumental engineering. The most interesting feature is the distinctive pattern of 'quarters', as in Whitechapel, which has small-scale workshops in adapted and congested premises devoted to clothing manufacture.

3. **Thames-side.** This is a most distinctive zone with a role as the transhipment point for the primary processing industries. The docks have migrated down-river from the original port in the Pool of London above Tower Bridge past Woolwich to Dagenham and to Lower Thames-side at Purfleet, Thurrock and Northfleet. The evolution of this zone typifies that of all the European estuary-based port areas: the decline of the inner areas and the movement outwards to deep water terminals near the open sea. London docks are now at Tilbury, and the former docks from Wapping to the Royal group of docks, are the focus of a very dramatic redevelopment programme, the LDDC (London Dockland Development Corporation). The heavier and more noxious industries are downstream: vehicles at Dagenham, and cement, explosives, paper-making, oil refining, and petrochemicals out through Northfleet and as far as the Isle of Grain.

4. **Radial transport lines.** Decentralisation along main roads and railway lines followed the growth of transportation facilities. Significant industrialisation occurred between 1900 and 1914 in the Lea Valley, from Tottenham and Edmonton towards Enfield with the introduction of workmen's train-fares and the growth there of working-class housing. Much of the former East End industry migrated there, in a linkage process along the nearest line of communication. The move from cramped inner sites was also a powerful factor in the growth of these manufacturing sites along radial railway lines. Much of the industry in the Lea Valley is a natural continuation of that of the East End: furniture, clothing and shoes, supplemented by electrical engineering. Most of the other areas with factory concentrations, particularly those in West London, depend upon road transport: Colindale and Cricklewood on the Edgware Road; Park Royal, Perivale, Wembley and Greenford along the A40 and Western Avenue; and Brentford, Feltham, Yiewsley and West Drayton along the A4/M4 section to the west of the city. The Great West Road illustrates the great variety of planned new factories. Here, engineering products, particularly electrical, are dominant, but there are also consumer goods needing a large market, labour supply and efficient communications (e.g. branded foods, pharmaceuticals, cosmetics, plastic kitchenware, radio and scientific instruments, refrigerators).

5. **Factory estates in Outer London.** A more recent growth, these show many similarities to type (3), being usually along radial routes, but they have often been planned to take account of the movement of population out of London, and the associated labour supply. Croydon, Staines, Uxbridge and Elstree are examples, with mainly light and consumer goods factories. The movement of industry out into London's new towns is also part of this

62   THE NEW EUROPE

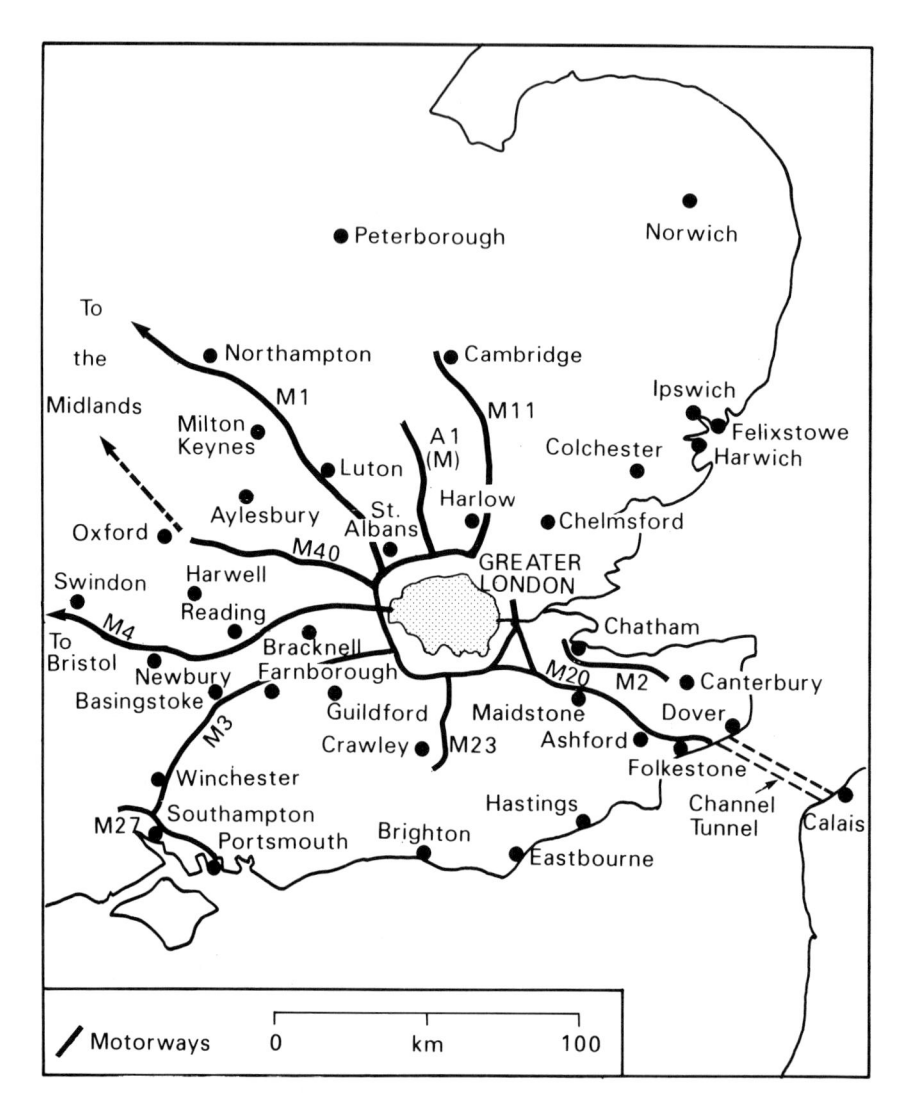

Figure 3.6 (c) Growth centres in South-east England

process. Perhaps the best example of an outer industrial estate is, however, Slough, one of the first trading estates established in the 1930s.

6. **New industry in the city region.** A tendency of an outward dispersal of industry into the smaller towns and rural areas both surrounding and accessible to the city is a feature of the contemporary industrial scene. Higher operating costs, shortage of space and expensive labour costs in the city have led to relocation of industry along the major motorways and to industrial growth in places such as Reading, Swindon, Winchester, Basingstoke, Newbury. Motorways such as the M4 are referred to as the 'sunrise strip' because they provide access to the facilities of London, proximity to

Heathrow airport, a high quality of life, and have a large catchment area for skilled workers. Computers, precision engineering, biotechnology and electronics are amongst the wide range of hi-tech industry which has been attracted to the area. The 'Cambridge effect', the grouping of hi-tech companies in 'Science Parks' around the research facilities of the university, has led to Cambridge becoming one of the fastest growing industrial towns in England (fig. 3.6(c)). The crescent of growth stretching from Portsmouth to Oxford, Cambridge and Felixstowe, will be supplemented by the effects of the Channel Tunnel in Kent, Sussex and Surrey.

The organisation, control and financing of industrial production

Basic structural changes in industry have often been accompanied by locational changes as the small workshops of the nineteenth century have become the large factories and the vast corporations of today. Many of the problems of industrial areas such as Lancashire or the Franco-Belgian coalfield arise from the decline of staple industries like cotton and their replacement by sophisticated industries such as radio and electronics, often not tied to the same location. The key to structural changes lies in the economies of scale which can be achieved by mass-production, integration of the means of production, and large-scale capital investment, research and technology in industry. The large units are exemplified by giant corporations such as British Petroleum, Imperial Chemical Industries or Volkswagenwerk (fig. 3.7). These giant corporations have an important geographical effect in that they tend to intensify core-periphery contrasts. Although they establish branch factories in peripheral regions, their principal investment tends to be directed towards 'leader' regions as they maintain their headquarters and market effort in core city regions such as London, Paris, Amsterdam, Milan and Cologne.

Mass production and the specialisation of labour

The many stages required by the assembly of components, the increasing substitution of labour by machinery, and the concentration of processes in large factories have led to high capital costs and a need for continuing large-scale investment. Integration of related processes occurs as a means of cutting costs of production and increasing efficiency. Horizontal integration is the control by one group of most or all of the productive capacity in one stage of manufacturing. Added to this is vertical integration where in many cases economies can be made by bringing together successive stages of manufacture. Substantial savings in fuel, transport and other overheads give increased efficiency, lower costs of production and economies of scale. Ford Motors (United Kingdom) have a vertically integrated plant at their Dagenham factory capable of producing cars through all stages of manufacture and assembly. The synthetic fibres group Courtaulds illustrates both aspects of integration very well. They have a 'Northern Textiles Division' to coordinate their control of over one-third of Lancashire cotton spindles. In addition, they also control production stages from the raw materials (wood pulp and chemicals) to the initial production stages (spinning, weaving, knitting and bonded synthetic fabrics) and finishing (dyeing, printing, hosiery, garments and marketing) as well as textile engineering.

The assembly line, which dominates the car industry, the aircraft industry and much of the electronics industry, requires thousands of components which are produced in separate specialist factories and then transferred to the assembly factory. Industrial linkage of subsidiary component factories by transport is therefore essential, with this type of industrial association being known as 'regional swarming'. One of the best examples is the West Midlands conurbation with its associated towns having a high proportion of their industry geared to producing components for the vehicle industries of Birmingham and Coventry. Older industries such as cotton textiles, iron and steel, and shipbuilding, have for decades had a tendency to cluster in their traditional areas, where there are similar linkages.

Giant industrial corporations: multi-nationals

Multi-national corporations illustrate the technological advantages of size: Imperial Chemical Industries of the United Kingdom has become a significant multi-national corporation which has almost a conglomerate structure. It controls many types of both basic and sophisticated chemical products (fig. 3.8). With over sixty manufacturing plants in the UK and over thirty wholly owned or subsidiary companies in the rest of the EC, it can claim to be European in outlook (fig. 3.9). Its world stature is illustrated by over 200 subsidiary and associated companies in the rest of the world.

Imperial Chemical Industries in the United Kingdom operates in both chemicals and synthetic fibres in a virtual monopoly situation. But the monopoly is within the United Kingdom and provided there is government regulation and protection of the consumer, no great problems arise. The severe competition which ICI faces is from its counterparts in the United States of America, Du Pont and Union Carbide, and in Europe from Montedison and Hoechst. An indication of the role and power of large companies in terms of total national capacity is in the United Kingdom where the fifty largest companies produce nearly half the national turnover. An even more extreme example is the Netherlands, where thirty-five per cent of production is from three companies: Philips Electrical (Eindhoven), Royal Dutch Shell and Unilever NV.

Dramatic changes in structure and size have occurred with increased global competition, the dismantling of tariffs and increased production and movement throughout the EC, creating direct competition and the need for enlargement and rationalisation. In France, particularly, there has been considerable activity as its industrial structure has developed. Two steel giants are between them responsible for the bulk of production: Sacilor and Usinor. In textiles, Lainiére de Roubaix and Filatures et Fréres of Tourcoing are now

Rank	Company	Headquarters	Main Activity
1	Royal Dutch/Shell Group of Cos	UK/Netherlands	Petroleum
2	British Petroleum Co Plc	UK	Petroleum
3	IRI – Istituto per la Ricostruzione Industriale	Italy	Petroleum, engineering, chemicals
4	Daimler Benz AG	West Germany	Motor vehicles
5	Siemens AG	West Germany	Electrical engineering
6	Volkswagen AG	West Germany	Motor vehicles
7	Fiat SpA	Italy	Motor vehicles
8	Unilever Group	UK/Netherlands	Food, chemicals
9	Deutsche Bundespost	West Germany	Postal services, telecommunications
10	Philips' Lamps Holding NV	Netherlands	Electronics and electrical engineering
11	Veba AG	West Germany	Electricity, chemicals
12	BASF AG	West Germany	Chemicals
13	ENI-Ente Nazionale Idrocarburi	Italy	Petroleum, engineering, chemicals
14	Renault (Régie Nationale des Usines)	France	Motor vehicles
15	Hoechst AG	West Germany	Chemicals
16	Bayer AG	West Germany	Chemicals
17	Electricité de France	France	Electricity production and distribution
18	CGE-Cie Générale d' Electricité	France	Electrical engineering
19	Elf Aquitaine (Sté Nationale)	France	Petroleum
20	Imperial Chemical Industries Plc	UK	Chemicals

Figure 3.7 EC top twenty industrial groups 1989 (Source – *Times 1000 Review*)

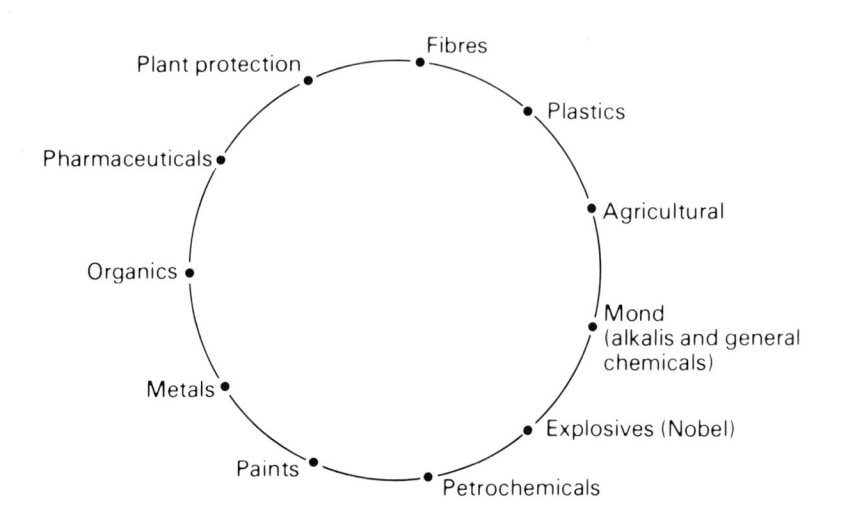

Figure 3.8 ICI Divisions (Imperial Chemical Industries)

66 THE NEW EUROPE

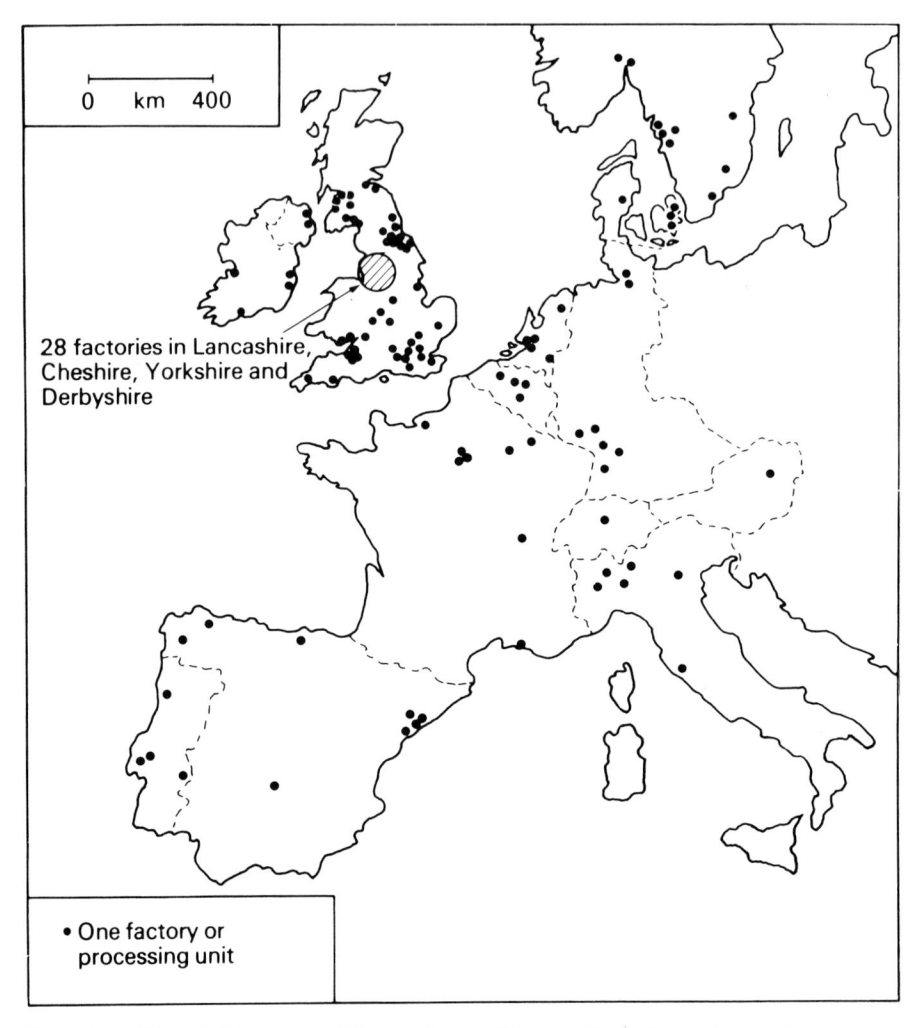

28 factories in Lancashire, Cheshire, Yorkshire and Derbyshire

• One factory or processing unit

Figure 3.9 ICI establishments in Western Europe (Imperial Chemical Industries)

merged into one of the largest companies in Europe. The table of the largest EC companies and comparable American companies (figs. 3.7 and 3.10) serve to illustrate several significant facts:

(a) American companies usually head the 'league table' by large margins, followed by Japanese companies in sectors such as cars and electrical goods.
(b) West Germany, the United Kingdom, the Netherlands, France and Italy have the largest companies in the EC, reflecting the mature industrial structure of the European core.
(c) Petroleum, chemicals, motor vehicles and electrical goods are the sectors with most large companies.
(d) Ireland, Greece and Portugal have generally small-scale industry reflect-

CHEMICALS		CARS	
Rank	Name	Rank	Name
1	Du Pont (USA)	1	General Motors (USA)
2	BASF (West Germany)	2	Ford (USA)
3	Hoechst (West Germany)	3	Toyota (Japan)
4	Bayer (West Germany)	4	Daimler-Benz (West Germany)
5	Imperial Chemical Industries (UK)	5	Volkswagen (West Germany)
6	Union Carbide (USA)	6	Nissan (Japan)
7	Dow Chemicals (USA)	7	Fiat (Italy)
8	Montedison (Italy)	8	Renault (France)
9	Rhone-Poulenc (France)		

ELECTRICAL AND COMPUTERS		AIRCRAFT/AEROSPACE	
Rank	Name	Rank	Name
1	IBM (USA)	1	Boeing (USA)
2	General Electric (USA)	2	McDonnel Douglas (USA)
3	ITT (USA)	3	Lockheed (USA)
4	Siemens (West Germany)	4	British Aerospace (UK)
5	Philips (Netherlands)	5	Aerospatiale (France)
6	Western Electric (USA)		
7	Hitachi (Japan)		
8	Toshiba (Japan)		
9	Westinghouse (USA)		
10	RCA (USA)		
11	GEC (UK)		
12	Bosch (West Germany)		

Figure 3.10 Table of industrial comparisons 1989 (Source – *Times 1000 Review*)

ing their peripheral situation and under industrialised structure.
(e) Belgian industry is on the whole small-scale with the largest company, Petrofina, ranking twenty-first.
(f) Spain, too, has a relatively under developed industrial structure, with state holding companies controlling large sectors of the economy. The largest company ranks twenty-sixth.
(g) Some giant companies dominate the industrial structure of Italy: Fiat (cars), Montedison (chemicals), Snia-Viscosa (synthetic textiles), Finsider (steel), ENI (chemicals and engineering) and IRI (state holding company for industrial reconstruction).

The survival and revival of small-scale industry

Industry in the EC is not all controlled by large companies. In fact, much of the industrial structure is small and fragmented, at least by American standards. In the newer member countries of the periphery, there is still a predominance of small production units, many family firms, a lack of rationalisation and the survival of small workshops. This reflects the fact that the industrial revolution in these countries has been slow, late and incomplete. However, it must be stressed that the survival of small-scale industry has been looked upon much more favourably by governments in recent years. Although the great corporations have advantages of scale and technology, nevertheless they have problems such as labour and management relations. Small companies have flexibility, fewer labour problems, and have a major part to play in the specialised production of components, in light industry, and in providing employment in rural areas. The British Government is actively encouraging small factories and workshops in rural areas, through rural development agencies. Many of the new 'sunrise industries' of electronics, computers, data-processing and information technology are in relatively small factories with a limited labour force. They often deliberately locate in small towns and rural areas with a higher quality of life. The agglomeration factor in industry may now be in reverse in many sectors of the economy.

Investment and technology: research and development

Industry has moved into a sophisticated stage of development during the mid-twentieth century. The replacement of many traditional raw materials with synthetics, particularly in plastics and textiles, is one example of this. More important is the influence of high technology with certain sectors of the economy assuming key roles: space satellite technology; aero-engines and aircraft; radio and electronics; nuclear technology; automation systems; information technology; biotechnology and computers. These require vast expenditure on research, which has led to the existence of a 'technological gap' between Europe on the one hand and Japan and the United States on the other and to the invasion of Japanese and American industry into Europe in these key sectors in an attempt to control the fastest growing industries. Comparisons can be made between the smaller national European companies and the larger Japanese and American ones (fig. 3.10). In the advanced sectors of the economy, four factors are crucial; the relatively small home market of the EC country and the financial difficulty of sustaining projects to the stage of commercial viability; the reluctance of governments to pay indefinitely for projects costing millions of pounds; the smaller European market in military weapons; the difficulty of countering the sales capacity of the Americans and Japanese.

European aircraft manufacturers have suffered in particular because of the great market benefit reaped by the Americans from their large military commitments and widespread use of civil aviation. The United States airforce has twice as many aircraft as all the countries of Western Europe. The United States produces 60 per cent of the world's aircraft. Nevertheless, the British, Spanish, West German and French governments co-operate in the 'Airbus' project. This is now successful enough to be able to compete with the American aircraft industry. Also the 'Ariane' project is a European co-operative venture in the field of space satellites.

In nuclear technology the position is better. In Europe generally there has been a realisation of the immense benefit that this energy source can give, and co-operation in the form of Euratom has been, though only partially successful, a valuable pointer to the future and to the possibilities of energy from nuclear fusion. The recent developments in nuclear power in Western Europe (chapter 2) are an example of a considerable pre-eminence in the production of what may well ultimately be the cheapest form of electricity. The UK, France and West Germany have now embarked on a large nuclear power programme (fig. 2.9).

The electronics and electrical engineering industry is one of the fastest-growing sectors of the economy. American companies dominate the field, and command about 80 per cent of the world sales in electronics, computers and telecommunications. The large European companies are Philips of the Netherlands, Siemens of West Germany, and GEC of the United Kingdom, but even they are small compared to some of the United States giants. Europe's computer companies need to co-operate to survive. International Computers Ltd., the United Kingdom company, has three per cent of world sales, whereas IBM, the American company, has 75 per cent. It is in this sector, with its telecommunications and automation systems, essential to the sophisticated economy, that Europe lags considerably behind the USA and Japan.

Related directly to this is the invasion by American industry, by the injection of capital and extension of ownership and management. This investment is generally advantageous with the stimulation it brings in new factory growth and employment, higher wages and rationalisation. The real problem is when this leads to excessive outside control of key industries with its implications for dependence upon foreign technology. In the United Kingdom, American companies own ten per cent overall of British production, but six key industries - electronics, cars, drugs and medicines, petroleum, computers and tractors - have absorbed seventy per cent of total American investment. Throughout the EC the position is similar. Cross-frontier EC mergers to give European industrial units of international size and stature serving a home market of 320 million people are necessary. Of course, investment is a multi-layered process, and EC countries, particularly the UK, are now investing in and owning large sections of American industry.

There are significant economic reasons and geographical implications for cross-frontier operations of multi-national companies. Western Europe is an affluent, mature market, with abundant skilled labour. Outside corporations need to overcome EC tariff barriers by manufacturing within its market area. Much American investment lies within the UK and the 'Golden Triangle' or leader regions of the EC. The UK gains because of its 'cultural' proximity to the USA, and the absence of language barriers. Belgium is seen as a 'core' country and the Belgian government has enthusiastically supported American investment. By way of contrast, since the recession of the mid 1970s, many multi-national subsidiary plants previously established in peripheral areas of the EC such as Spain and southern Italy, based upon cheaper labour costs, have been closed. Production has been transferred to cheaper labour areas of the Third World. Spatially, the operation of multi-national companies illustrates the different economic potential of the core and periphery.

Government intervention and supra-national factors

Government intervention has occurred principally because of the need for adjustment in the complex economies of the EC countries, and has three basic aims: to control areas of production which are basic to the economy: to support prestige and high technology industries; and to support declining industrial areas. State intervention reached its maximum extent in the 1970s in the United Kingdom, France, Italy and Spain, where extensive government control was seen as the best way to guide the economy.

Nationalised (state-controlled) industries

In the UK nationalised industries include some of the largest employers in the country. In certain areas basic to the national economy, there are advantages in state control. In theory at least, an overall planning view can be secured, national resources are available for investment, and large-scale operations can be funded nationally. An alternative view is that state companies are wasteful of public money and inefficient. Thus there has been a major reduction of the state controlled sector during the 1980s, with many production industries and service companies returned to the private sector. These include:- Jaguar Cars, British Airways, British Telecom, British Steel, Rolls Royce Aero-engines, British Shipbuilders, British Aerospace, British Gas. The electricity industry is in process of privatisation (1989). The major areas where state control is still operative are in the coal-mining industry (British Coal) and the railways (British Rail), but even these are seen as possible private sector companies in the future.

In France the state owns and operates large sections of the economy: the Charbonnages de France (coal-mining, power stations and chemicals); the steel industry; Electricité de France; Gaz de France, SNCF (French Railways); SNECMA (aero-engines). A variation on direct control is the Renault Car Company where government control is vested in a director-general and day-to-day running is left very much in the hands of the company management. In Italy IRI (Instituto Per La Reconstruzione Industriale) originated as a financial rescue operation in 1933 and has since grown into a state corporation which controls the telephone system, Alitalia Airways, Alfa Romeo cars, the Autostrada del Sole, Finmare (the merchant marine), Finsider (steel), and other engineering, textile and chemical companies. A more recent development has been ENI (Ente Nazionale Idrocarburi), set up with the object of capitalising on the oil and gas reserves in the Po valley and having a monopoly of prospecting in Italy. With its headquarters near Milan, it now controls oil refineries, petrochemical plants, tanker fleets and oil pipelines from Pegli and Trieste to Ingolstadt in Southern Germany. In Spain INI is a state holding company set up during the rapid industrial development of the 1960s and '70s.

Active intervention and guidance in industry

Active intervention is even more widespread than direct government control. Where larger units have been necessary and the national interest has been at stake, persuasion and guidance has been used. In the United Kingdom aircraft industry, famous names such as Fairey, Avro, Bristol, Supermarine, Vickers and De Havilland were merged into two giants, the British Aircraft Corporation and Hawker Siddeley, subsequently nationalised and now privatised as British Aerospace. Rolls-Royce, the largest British aero-engine manufacturer could not be allowed to disappear and so the company's bankruptcy in 1971 was annulled by the expedient of a financial takeover by the government. This has now been returned as a profitable company to the private sector. The United Kingdom government was a controlling shareholder in British Petroleum, and acquired effective control of British Leyland during the 1970s. The privatisation programme of the 1980s has seen British Petroleum become entirely private, and the Rover Car Group (formerly British Leyland), after a major slimming and rationalisation programme, was sold to British Aerospace in 1988. In industries where there has been persistent intense foreign competition, and an industry has remained at an inefficient level of production with too much labour, or organised in small units more akin to the nineteenth century, then regrouping and controlled reduction of output is necessary. Two severe cases have been the United Kingdom textile and shipbuilding industries. The Charbonnages de France has established a gradual reduction target in the French coal industry. The French Government took control of, and regrouped the steel industry to save it from bankruptcy. The establishment of British hi-tech companies such as INMOS (computer technology) has been another aspect of government operation as was the intervention to assist the Westland helicopter group in late 1987.

Support for declining industrial regions and relocation policies

Specific areas of severe decline have been the major preoccupation of European governments during the past four decades. The efforts of the Italian government to promote industrialisation and to raise living standards in the south have been made through IRI development projects such as integrated steelworks, paperworks, cement works and various land-improvement and marketing schemes. The most important move, however, was the setting up of the Cassa per il Mezzogiorno in 1950 (chapters 10 and 17). In France the Sociétés de Développement Regional (SDR) are aimed at stimulating investment and attempting to persuade industry to move to those areas with contracting staple industries and those which are shedding excess agricultural workers. The Nord/Pas de Calais fits the first category, and the south and west of France the second. In the United Kingdom the designation of development areas of three grades of severity has been accompanied by fiscal measures to improve their attractiveness and stimulate industrial growth. Trading estates, and enterprise zones, are all part of the complex operation required to arrest industrial decline. The concept of the regional problem is, however, much wider and is dealt with more fully in subsequent chapters.

European Community Industrial and Competition Policy

This area of operations dates from the early days of the Treaty of Rome, but has become much more interventionist and powerful since industrial problems arose in the 1970s and 1980s. Competition policy-making now has a high profile as the integrated single market of 1992 approaches.

The principal aspects of this as it affects industrial policy are:

1. free movement of all industrial goods and services throughout the market;
2. the removal of technical barriers to trade resulting from different national provisions covering quality, measurement, composition, packaging or control of goods;
3. removal of legal and fiscal barriers, and tax discrimination throughout the market;
4. harmonisation of taxes, excise duties and indirect taxation;
5. convergence of economic and monetary policies within the EMS (European Monetary System);
6. implementation of research and technical development programmes to strengthen the scientific basis of industry;
7. the establishment of a European company statute, integration of company law and a European trade mark and patent system;
8. freedom of movement of capital, freedom of establishment, and the right of persons to engage in business or professions throughout the Community;
9. the opening-up of public markets so that public and local authority contracts will be open to the Community and not just to national suppliers.

In spatial terms these will assist:

(a) cross-frontier mergers to form companies which can trade in all parts of the EC as a single unit, taking advantages of economies of scale. There has so far been little progress in this direction. The Leyland-Innocenti Vehicle agreement ended in failure. Arbed, the Luxembourg steel group, is one of the more successful examples, having additional control of Belgian and Saarland steelworks. Specific co-operation projects are an alternative as in the successful Airbus consortium of French, British, Spanish and German Aerospace companies.

(b) the restoration of competitiveness in the old declining industries of textiles, steel, footwear, and shipbuilding at a reduced level of production by rationalising plant, and the resolution of social problems by the creation of new employment in the affected areas. There are limitation agreements on imports of textiles, production quotas for

steel and retraining of workers whose jobs have been lost.
(c) reduction in disparities between regions by means of the structural funds such as EAGGF, the ECSC Fund, The European Investment Bank, the ESF and the European Regional Development Fund, particularly in areas with severe unemployment.
(d) investment in high technology industries of aerospace, electronics, data-processing, telecommunications and information technology, sophisticated capital equipment such as machine tools, and new energy sources such as the nuclear research programme and the JET project for thermo-nuclear energy established at Culham in Oxfordshire. New zones of high technology industry are fast developing, often in city regions with good motorway access such as the London-Bristol M4 corridor, Greater Paris, and the Rhineland cities of Frankfurt and Stuttgart. These 'sunrise industries' have growth potential for the 1990s.

The Tertiary or Service Sector

From the mid-1950s to 1973, industrial output in the European Community rose two and a half times. Industry provides employment for about a third of the working population, and the Community is an industrial giant alongside the United States and Japan. Nevertheless since the mid-1970s a crisis has developed in large parts of the Community's industrial structure. Large-scale unemployment is widespread and many sectors of industry are under severe pressure. The rise in energy costs since the oil-price rise of 1973, ensuing inflation, rapidly rising labour costs, falling productivity and exchange rate fluctuations mean that the market share has fallen. At the same time competition from recently industrialised countries, often in the Third World, with low wage levels and modern equipment, has risen dramatically. This is the new international division of labour which is now taking place. Certain old-established sectors have suffered most, including the steel industry, textiles, clothing, footwear and furniture, paper and pulp manufacture and shipbuilding. Even the automobile industry has suffered from international competition and depressed demand.

The industrial workforce has declined dramatically, particularly in the older industrial centres of the Community such as the UK and West Germany. From 1971 to 1984 West Germany lost 2.9 million industrial jobs, and the UK lost 3.3 million during the same period. However, recently industrialising countries in the southern periphery, with lower labour costs, such as Greece and Portugal, were still increasing their labour force in industry.

Although there are major variations across the European Community, the tertiary or service sector is now the majority employer in all cases (fig 3.11). Levels in excess of 65 per cent in Belgium, Netherland, Denmark and the UK compared with the relatively low levels of Greece and Portugal, indicate major contrasts in levels of development.

Theorists such as Kondratieff have explained this in terms of a series of fifty year waves of economic activity, where the innovation and expansion of the post-war period lost strength and was followed by a down wave of activity which culminated in the recession and inflation of the 1970s and 1980s. One solution to this is the move into innovative hi-tech industrial functions including electronics, telecommunications, office automation, and aerospace sited in science parks in pleasant, amenity-rich world-city regions with skilled workforces and high purchasing power. The fringes of South West Paris, and the Southampton–Oxford–Cambridge crescent around London are examples of this.

The other explanation is that in the late capitalist phase in developed countries there is a shift of employment into the service sector and a blurring of the distinction between secondary, tertiary and quaternary employment. From 1971–1984 the service sector in the European Community saw a net increase in employment of 18 million people. The centres of tertiary employment are in the national core areas and major cities. These cities are prestige centres, often a capital city such as London, Rome or Paris or the regional cities of Frankfurt, Munich, Barcelona or Milan. They have a large pool of skilled labour, high proportions of well-qualified graduates and highly qualified technical, scientific and professionally trained people. They are the centre of the transport network. Centrality exerts a major influence upon the location of administrative offices, linkages between businesses, ease of attracting staff and the availability of retail and social facilities are other aspects of the positive advantages of cumulative causality. Young people are attracted to the city and look for 'white collar jobs'. The city acts as the administrative, commerical and cultural heart of the country or region.

	Agriculture	*Industry*	*Services*
Eur 12	8.0	32.9	59.1
Belgium	2.8	26.5	68.5
Denmark	6.5	26.5	67.1
West Germany	5.2	40.5	54.2
Greece	27.0	28.0	45.0
Spain	15.1	32.3	52.5
France	7.1	30.9	62.1
Ireland	15.4	28.1	56.5
Italy	10.5	32.6	56.8
Luxembourg	3.7	32.6	63.8
Netherlands	4.7	27.1	68.2
Portugal	22.2	34.9	42.9
UK	2.4	30.2	67.4

Figure 3.11 Employment by sector 1987 (percentage)

The range of tertiary employment may be classified as follows: insurance, banking and commerical services, professional, medical and law services; trade storage distribution; retail services; transport and shipping services; public utilities, publicity and sales activity; scientific, education and research establishments; national and local government; cultural, travel and tourist services; offices of private industry and head offices of international corporations.

London's Central Area

Central London is defined by Victoria Station and the Houses of Parliament on the south, Hyde Park and Paddington Station on the west, Regents Park and Euston Station on the north, and the Tower of London on the east (fig. 3.12). This large area, one of the greatest foci of commercie in the world, attracts companies because of the prestige of the address, the close contacts with other forms of business, and the facilities for national and international trade provided by road, rail and air transport. The most important area is the square mile of the City of London around the Bank of England which is devoted to banking, commerce, insurance, shipping and the commodity markets. There is also Fleet Street, the Inns of Court and the Temple, and the theatre area of Piccadilly and Leicester Square. By the river lies the distinctive administrative zone of Whitehall and St. James with its public buildings, parks and palaces. Mayfair with its embassies and squares, the 'West End' retail zone of Oxford Street, Bond Street and Regent Street, and the University of London in Bloomsbury add to the picture. London is a world centre for commerical, administrative, legal, professional, cultural and diplomatic purposes. It is also a world famous tourist city, with facilities

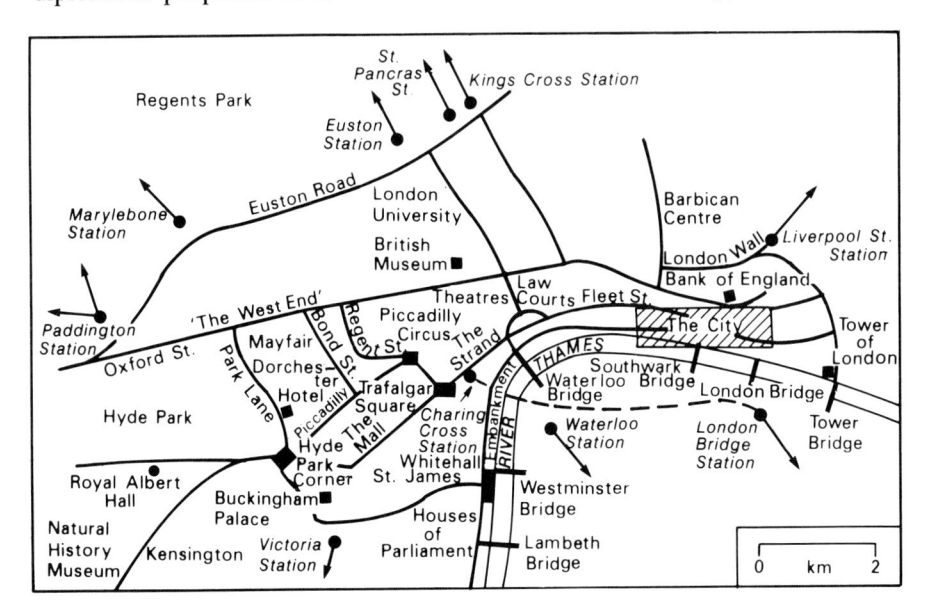

Figure 3.12 Central London

including the museums of South Kensington, the Tower of London and the hotels of the West End. London's employment pattern is dominated by its tertiary occupations.

Tourism

Travel has been a long-term element in human activity for leisure as well as business. However, tourism, as such, is of relatively recent origin. Tours by English Victorian ladies began in the 1860s, but the increasingly mass character of tourism dates from 1945 and has moved rapidly from a national to an international scale. The most rapid growth in numbers of people was during the 1970s at a rate of over twelve per cent per annum, and by 1975 over 80 million people were visiting the Mediterranean coasts, where by far the greatest volume of tourism takes place.

The reasons for this dramatic increase fall into several categories. Rising living standards in the European Community including both increasing amounts of surplus income and increased leisure time are important. They are linked to important changes in psychology and social habits fostered by the effects of mass media. There is the need to escape from day-to-day pressures to a range of holidays which include the family holiday at the seaside, the cultural and historic tour, the cruise, and the physical pastimes of mountaineering, trekking and winter sports. The seasonal rhythm has been extended massively from the summer into the winter period. Infrastructural improvements are also important, ranging from hotel accommodation to self-catering and camping facilities. The single most important factor by far, however, is increased access. The railways, once crucial to the Victorian tours, have been overtaken by car ownership, the cross-channel ferry services, and the European highway network, so that the French Riviera is easily accessible from the UK within 24 hours. Even more so, however, the development of mediterranean tourism, particularly to Spain, Greece and Portugal, is related to the expansion of air transport and particularly the charter flight and package tour. The 'fly-drive holiday' is an interesting variation on these themes.

The location of tourist centres in Western Europe may be summarised by (fig. 3.13) and classified as well-established historical and cultural cities such as Florence and Seville; capital cities such as London, Rome or Paris; wintersports in the Alpine region; scenic walking areas such as the Auvergne, Black Forest and Ardennes; the south-facing coastlands of the Mediterranean which stretch from Malaga to the Greek Islands. The great diversity is based upon the dramatic physical and climatic ranges within the environment, and the variations in holiday demand.

Tourism is significant in terms of the core-periphery division in the European Community. It is responsible for large-scale capital transfer from the tourist demand areas of the north-western core to the peripheral mediterranean coastlands. Employment generated by tourism in Spain is over 9 per cent, Portugal 8.6 per cent, Greece 6.9 per cent and Italy 6.7 per cent.

The City of London: European and global financial and commercial centre

Development has thus been encouraged in areas which have no other economic value beyond their specific environment of sun and beach or snow and mountain. The development that has taken place involves the multiplier effect with the reconstruction of the natural environment with hotels, water-supply, sewage and drainage facilities, transport infrastructure and a range of other buildings and services. Governments have used these to positive advantage in the stimulation of regional development. In spite of the advantages of tourism in regions of little alternative economic value, there are disadvantages to the local region from the tourist economy. Employment is seasonal, old social amenities have been broken up, and there is congested high-rise ribbon development along the coast in many parts of Spain. The international system of multi-national hotels, air charter companies and travel agencies are responsible for the influx of tourist cash, but often the profits of the enterprise are returned to the core and do not remain in the tourist region.

Tourism in Mediterranean France (fig. 3.14)

Provence-Côte D'Azur has been an important tourist area since the mid-nineteenth century when the French Riviera became an attraction for wealthy English tourists. The opening of the railway from Paris in 1865 was a major stimulus to the Riviera and the development of the now old established resorts of Nice, Monte Carlo, Cannes and St. Tropez. In addition in Provence are the Roman cities of Arles, Orange and Nîmes and the medieval cities of Avignon and Aix-en-Provence with their cultural and historic attractions, and Roman sites like the Pont du Gard aqueduct. The Camargue, the marshy

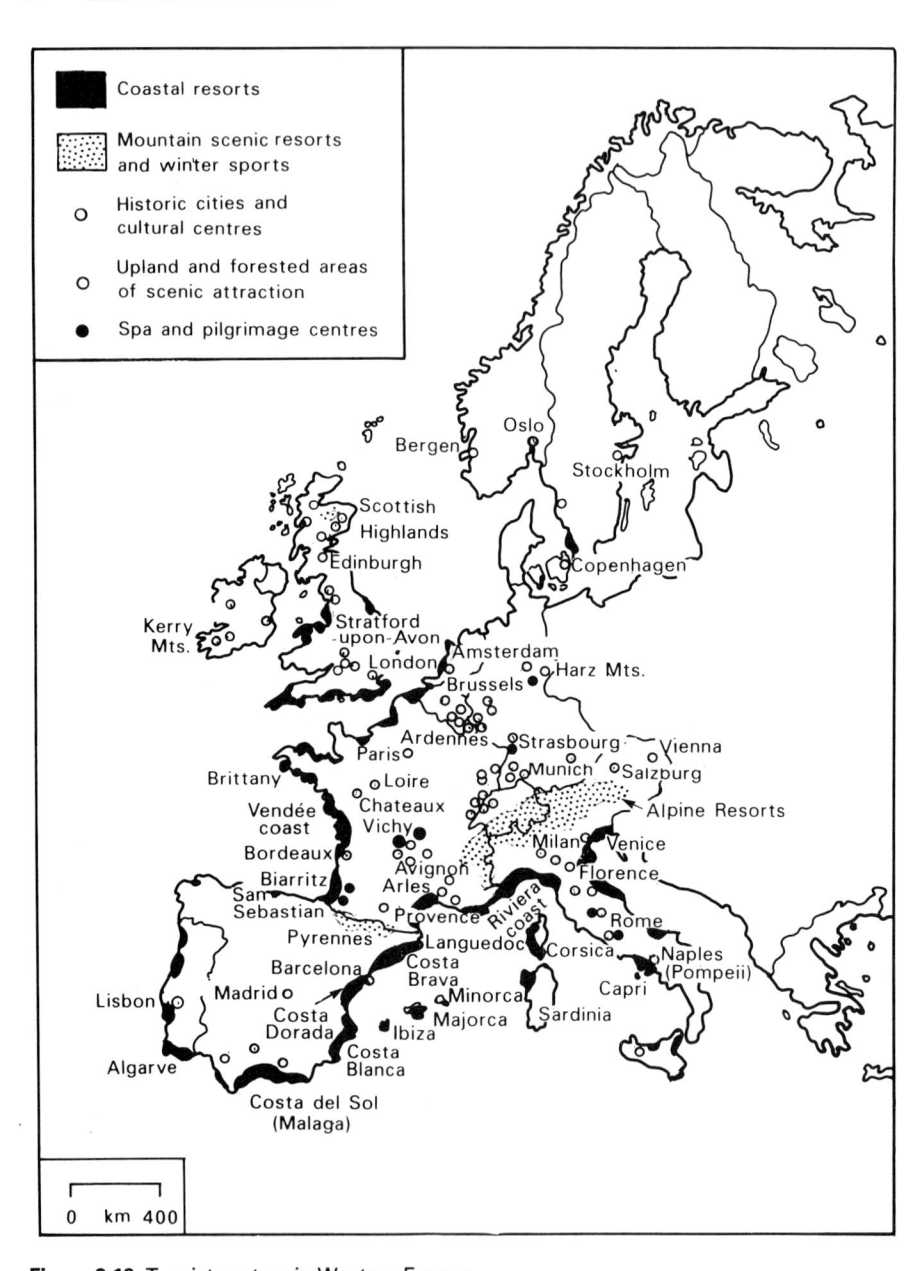

Figure 3.13 Tourist centres in Western Europe

section of the Rhône delta, has been recently established as a regional park. Tourism has developed along the whole of the Riviera coastline and extends eastwards into Liguria in Italy. Provence-Côte D'Azur has been one of the major growth areas of France in the last forty years, with some forty per cent of the 'second homes' in the country, and it has been able to exploit its greatest advantages. These are a scenic, mountainous coastline with sheltered bays, hot sunshine, a southerly aspect, and proximity to the wealthy areas of northern Europe with good, well-developed communications. There are over fourteen car sleepers weekly from Paris, and excellent motorway links via the Rhone valley. The traditional hotel accommodation has been supplemented by caravans and camp sites and the Riviera coast has become badly congested. The Avignon-Nice autoroute, 'La Provençale', now helps to alleviate traffic congestion.

Congestion on the Riviera coast and in Provence first turned attention towards the western side of the Rhône valley, the Mediterranean coast of Languedoc, an under-developed area with over 200 km of empty sandy coastline. There were few roads and few towns of any size except Montpellier. The whole region was underdeveloped, depending upon monoculture production of large quantities of poor quality wine. From 1963 onwards, Languedoc was the focus of a French Government planning project for coordinated development. The development company for Lower Rhône-Languedoc (CNABRL) has controlled the development of Languedoc. The modernisation and diversification of agriculture has involved irrigation schemes using the Rhône near Arles, and the Herault and Orb rivers. There are major irrigation projects around Beziers, Montpellier and Nîmes. There is a great increase in cultivation of intensive crops such as apples and market garden crops, and a reduction in vines. A most important development has been the construction of the motorway, the 'Languedoccienne' which has ended the relative isolation of the coast. There have been pest eradication programmes, afforestation projects, and an improved water supply system. Six tourist units have been developed with accommodation for 450 000 tourists. Local development corporations construct local roads, recreation

Cannes on the Côte D'Azur

80 THE NEW EUROPE

and urban services, parking facilities, marinas and hotels, camping villages, and apartment accommodation. Two large projects are most important. Leucate near Perpignan and La Grande Motte-Port Camargue (fig. 3.14). The latter is next to the regional park of the Camargue, and the airports at Nîmes and Monpellier. It has accommodation for 42 000 people and a marina for 1000 boats. The transformation of this region into a tourist attraction has been described as the creation of the 'French Florida'.

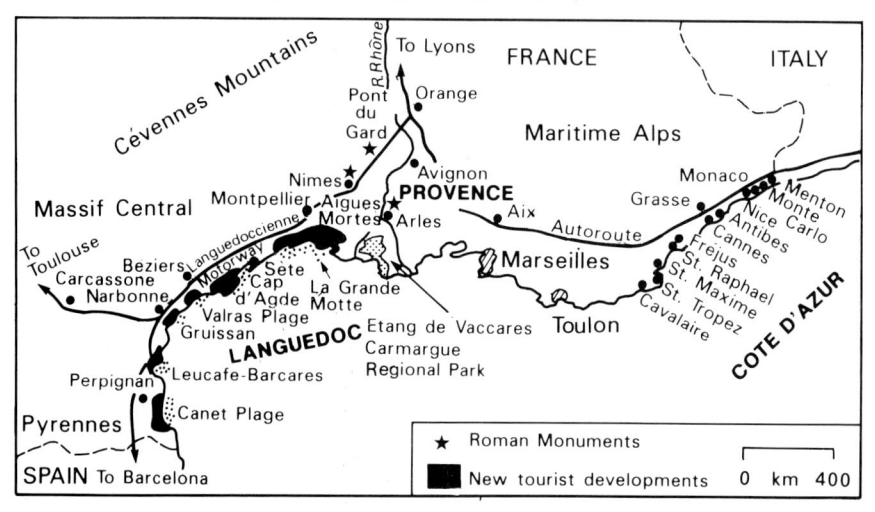

Figure 3.14 Tourism in Mediterranean France

The Roman Arena at Arles in Provence

4

The iron and steel industry: integration and rationalisation

This chapter and the next will give some detailed consideration to particular industries and their changing locations and structure. The steel industry, automobiles, textiles and chemicals have been selected because of their very significant contribution to the Western European economy, and because they exemplify many of the processes of change which have been discussed in chapter 3.

The steel industry is an important index of the relative wealth and stage of development of any particular country. It is traditionally a major employer of labour and requires high capital investment, and therefore has a tendency to locational inertia. Steel is a basic material which is used in almost every part of industry, transport of all types, construction and engineering, both heavy and sophisticated. It is therefore greatly involved in the complex patterns of modern industry.

The EC is one of the world's major steel producers (fig. 4.1) and this chapter analyses its development in terms of four themes: the growth of national industries up to 1945; government intervention and control; the supra-national factor; contemporary locational and structural changes.

The development of the United Kingdom steel industry

The early nineteenth century

During the industrial revolution coal was the critical determinant of location. Those coalfields developed which had coking coal, clayband iron-ores and limestone available. These raw materials were heavy and of low value in proportion to bulk and were costly to transport very far. In addition, early iron and steel technology was relatively inefficient, requiring large quantities of coal in proportion to iron-ore, and the means of transport available could not cope with such a situation. For these reasons the coalfields were unrivalled sites for steel-making, amongst them South Wales, Durham, and the 'Black Country' of the West Midlands.

The 1870s to 1930s

Improvements in iron-smelting techniques, allowing a more efficient use of smaller quantities of coal and, therefore, lower fuel costs, made it less

	1964	1971	1973	1974	1976	1977	1981	1987
West Germany	37.3	40.3	49.5	53.2	42.4	39.0	41.6	36.2
France	19.8	22.8	25.3	27.0	23.2	22.1	21.2	17.4
Italy	9.8	17.5	20.9	23.8	23.4	23.3	24.8	22.9
Belgium	8.7	12.4	15.5	16.2	12.1	11.3	12.2	9.8
Luxembourg	4.6	5.2	5.9	6.4	4.6	4.3	3.8	3.3
Netherlands	2.6	5.0	5.6	5.8	5.1	4.9	5.5	5.1
EC (Six)	82.8	103.2	122.7	–	–	–	–	–
UK	24.7	24.2	26.6	22.4	22.4	20.5	15.3	17.1
Denmark	neg.	0.5	0.5	0.5	0.7	0.7	0.6	0.6
Ireland	neg.	0.1	0.1	0.1	0.1	0.1	0.3	
Greece							0.9	0.9
Spain								11.7
Portugal								0.7
EC (total)	–	–	149.9	155.4	134.0	126.2	126.2	125.7

Figure 4.1 EC steel production (million tonnes)

essential for steel-making to be tied to the coalfields after the mid-nineteenth century. Two new steel conversion processes also contributed to this trend. The Bessemer converter in the 1850s was modified by Gilchrist-Thomas so that high-phosphorus iron-ores could be utilised thus allowing increasing use of the abundant, though lean, ores of the Jurassic scarplands of Cleveland, Lincolnshire and Northampton. In 1873 the open hearth furnace enlarged the scale of steel-making and was to dominate the scene until the 1950s. The iron-ore fields became location sites for new steel-works, established on 'greenfield sites' (entirely new locations with no previous industrial characteristics). Corby, opened in 1935, was a good example of this, and together with Scunthorpe, became a major iron and steel manufacturing zone in the first half of the twentieth century.

The increasing complexity of location was compounded by changes in the supply of raw materials which began during the early twentieth century. Imports of foreign ores from Sweden, Canada, Spain and North Africa increased rapidly, partly because of larger ore-carriers, and partly because of the higher ore-content of foreign ores, two or three times as high as the lean Jurassic ores of Northampton. There was a growing tendency for steelworks to be located at ports and estuary sites around the coast. The natural growth points became the South Wales coast between Cardiff and Swansea; Teesmouth; the Dee estuary and the Manchester Ship Canal. These coastal zones adjacent to already existing coalfields and iron and steel producing regions were in an excellent position for the low-cost assembly of all necessary materials. Teesmouth was a good example, having an abundant supply of ore in the nearby Cleveland hills, coking coal from south-west Durham, 40 km away, limestone from Weardale, flat estuarine land for the construction of large factory areas, and, most important, the estuary, which allowed the import of rich Spanish and Swedish ore.

The decline of formerly important areas was beginning, but was complicated by the effects of the great depression of the 1930s. Inertia kept the original locations on the coalfields in existence but their long-term operating efficiency was low. They were often in small congested works, with exhausted iron-ore and uneconomic coal, and requiring imports with high transport costs. The Ebbw Vale Steelworks was one such case, closed during the 1930s slump and reopened in 1938, largely as a social measure to help combat the serious long-term unemployment problem in the mining areas of South Wales. This structural problem of reconciling the development of new growth areas, whilst maintaining employment in run-down regions as a social service, has continued to affect the steel industry to the present day.

Post-1945

By 1945 the UK steel industry had completed the second stage of its evolution. It was a large-scale industry with a network of factories in seven traditional regions (fig. 4.2). There was, however, an unintegrated structure and much duplication of products. Control was in the hands of many private companies, often in small factories, undercapitalised, and with obsolescent

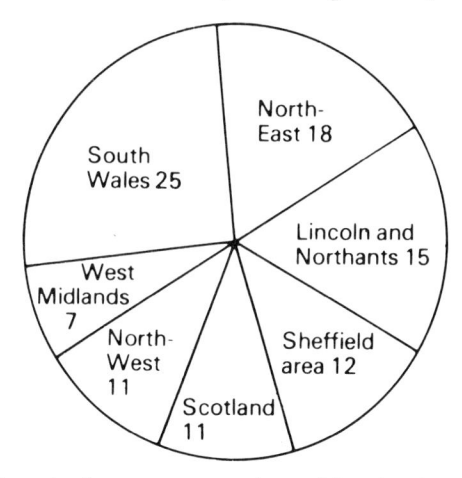

Figure 4.2 UK steel production - percentages by traditional regions 1970

equipment. The processes of change begun in the 1930s were continued and intensified; the movement to the coast; technological change and increasing advantages of size gained by amalgamation and concentration in large factories; and the residual areas of steelworking maintained by inertia and increasing state intervention. Port Talbot and Llanwern, in particular, are examples of favourable sites for large integrated steelworks built on extensive areas of flat land and using low cost imports at deep-water access points.

Inland sites have appeared increasingly uneconomic in recent years, a conflict of interest emerging over many old steel-working areas where government intervention has often delayed the closure of unprofitable works for social reasons. Ebbw Vale, Consett, Corby, Bilston and Workington are all examples of these. On the other hand, some areas have adapted by making special, high quality steel. Sheffield is such an example.

New processes

Technological progress has continued rapidly with two new factors: the basic oxygen process and electric arc furnace. The separate processes of iron-making, steel-making and finishing are increasingly being integrated in large works. Internal costs are lower and economies of scale and greater efficiency are achieved. The open hearth furnace has now largely been replaced by the basic oxygen converter which combines three great virtues: the rapidity of the Bessemer converter; the scrap-consuming ability of the open hearth furnace; and a facility for continuous casting and large-scale production. Two such converters can produce four million tonnes of steel per year. The entire UK capacity could thus be obtained from ten converters in five steelworks. The likely pattern for the next decade is therefore concentration in larger units. In 1969 open hearth furnaces still accounted for 51 per cent of capacity, and the basic oxygen converters only 34 per cent. By 1986 75 per cent of steel was produced in oxygen converters, with electric arc furnaces, used for very high quality and alloy steels, providing the remaining 25 per cent.

Nationalisation

The final element is Government intervention. There were thirteen major steel companies, some very large, such as the Steel Company of Wales, but there was too much product duplication, wasteful internal competition and a lack of scale required for international competition. In 1967 the steel industry was nationalised and became the British Steel Corporation, with a few private sector companies concentrating on alloy and special steel products. A major reorganisation plan was carried out by the British Steel Corporation, including the following:

1. Reorganisation on a product basis, with area specialisation creating greater efficiency. Wales has traditionally specialised in strip and sheet metal steel. South Teesside is headquarters of general steels, and Sheffield of alloy and special steels.
2. A major investment programme has concentrated upon modern plants such as the Teesside complex and coastal terminals like Hunterston.
3. A large new fleet of ore-carriers has been developed to gain from the economies of large-scale sea transport; Port Talbot (150 000 tonnes capacity), Immingham (100 000 tonnes), Redcar (200 000 tonnes) and Hunterston (350 000 tonnes) have been developed as ports for this purpose.

4. Rationalisation and modernisation means the retention of five main heritage (traditional) bulk-steel-making areas. There are five large integrated plants at Teesside, Scunthorpe, Port Talbot, Llanwern and Ravenscraig. At Sheffield and Rotherham the capacity for stainless steel, alloy and special steels is maintained. Although closure of bulk-steel making has occurred at Shotton, Ebbw Vale, Corby and Hartlepool, nevertheless important finishing processes are maintained. Corby has a major tube-making works and Shotton cold rolling mitts and a steel coating plant.
5. A reduction in the workforce is a necessary corollary of modernisation. This has involved a substantial reduction from 257 000 (1967) to 196 000 (1973), 88 000 (1981) and 53 000 in 1988.
6. The modernisation of steelmaking plant has affected some areas adversely, a continuation of the problem first seen in the 1930s. East Moors (Cardiff), Consett, and Bilston were old obsolescent plants which have been closed. In terms of traditional regions (fig. 4.2) the West Midlands and the North-West have been phased out from steel production.
7. In the considerably changed world economic conditions since the mid 1970s, the original British Steel Corporation plans for expansion have been significantly modified. For much of the period since 1975 the industry was operating well below capacity and by 1988 the industry had undergone a major rationalisation and privatisation programme, combined with investment in new plant and technology, but capacity has been reduced in line with current demand. Production in 1988 was 15.4 million tonnes.

The British Steel Corporations Anchor Steelworks at Scunthorpe

Steel on the Continent: the dominance of the 'Heavy Industrial Triangle'

Germany: the Ruhr

Prior to the establishment of the Community, the steel industry on the Continent was related almost entirely to the central coal belt and its associated supplies of iron-ore. The original definition of the 'Heavy Industrial Triangle' (fig. 4.3) is based upon the Ruhr, the Nord/Pas de Calais and Sambre-Meuse coalfields and the southern apex of Lorraine. The Ruhr rose to importance during the middle and late nineteenth century with large quantities of high-grade, easily extracted coking coal, and clayband iron-ore. The Siegerland iron-ore and limestone was close by, and after the Franco-Prussian war of 1870-71, when Lorraine was under German occupation, the Ruhr used large quantities of Lorraine iron-ore. The river Rhine and its extensive canal system including the Dortmund-Ems, was an excellent vehicle for cheap imports of iron-ore. Finally, the growth of the Ruhr as a major urban and industrial area helped generate its own market of numerous steel-using industries such as manufacturing locomotives, armaments, rolling stock, heavy engineering products and light engineering products such as cutlery. By 1938 Germany produced 22 million tonnes of steel, nearly 70 per cent of which came from the Ruhr. By 1974, in spite of a setback immediately after World War II, the Ruhr was producing 34 million tonnes of steel. Even with the reduced production levels since 1974, it still produces over 60 per cent of West Germany's total. Some idea of the concentrated nature of the German steel industry in the Ruhr is given by the fact that this great volume is produced in a conurbation of some 60km by 16km in extent, mainly in large integrated plants at Oberhausen, Rheinhausen, Duisburg and Dortmund.

France: Nord/Pas de Calais

The French steel industry is much less concentrated, but one major area is the Nord/Pas de Calais coalfield which is responsible for about one-third of French production. There are three traditional districts: Denain-Valenciennes, the centre of the heavy industry; Boulogne and Isbergues on the Aire canal; the minor production centres of Maubeuge and Hautmont in the Sambre valley. These small-scale steel-works are in the congested older areas of the coalfield and have become relatively insignificant in recent years when compared with the large integrated works at Dunkerque, which now has 80 per cent of the region's steel-making capacity.

Belgium

Across the Belgian frontier lies the Sambre-Meuse coalfield. Both the coal and haematite iron-ores outcropping along the valley were the initial location factors for a very considerable heavy industrial zone. La Louvière-Charleroi in the centre and Liège-Seraing at the eastern end of the coalfield retain a large proportion of Belgium's steel industry, now in a few large integrated works.

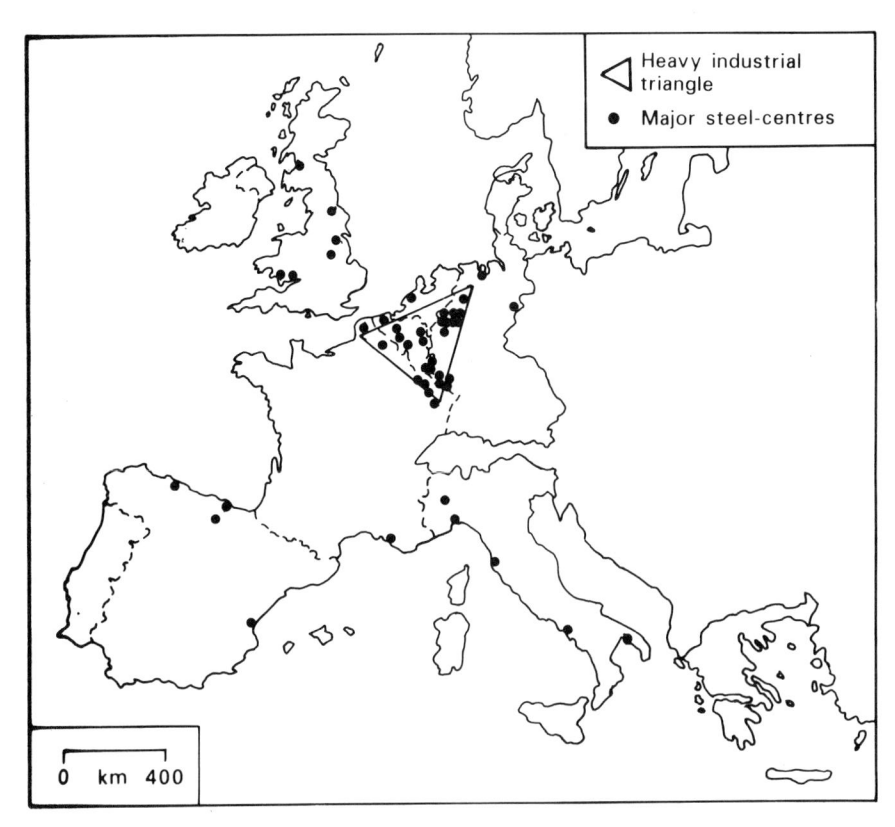

Figure 4.3 Major steel-producing centres in the EC 1986

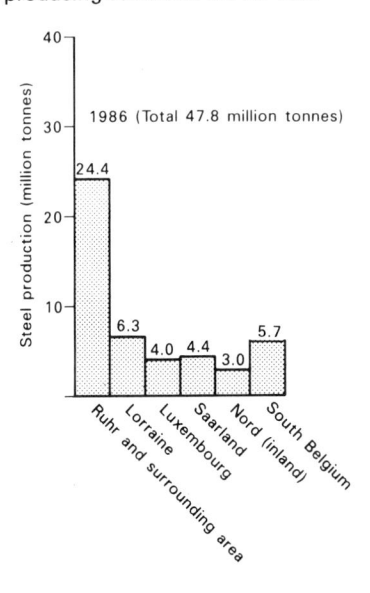

Figure 4.4 Steel-making areas in the 'Heavy Industrial Triangle' 1986 (total 47.8 million tonnes)

Lorraine, Luxembourg and the Saarland

The major home iron-ore producing region in the Community is Lorraine (fig. 4.5). This developed after 1870 based upon the exploitation of the Jurassic iron-ores. These lean ores (up to thirty per cent iron), with their high phosphorus content, became economic only with the invention of the Gilchrist-Thomas process. This stimulated the movement of steel manufacturing to the iron-ore field by the early twentieth century, similarly to that of the United Kingdom. Lorraine is a major French steel-making area with one

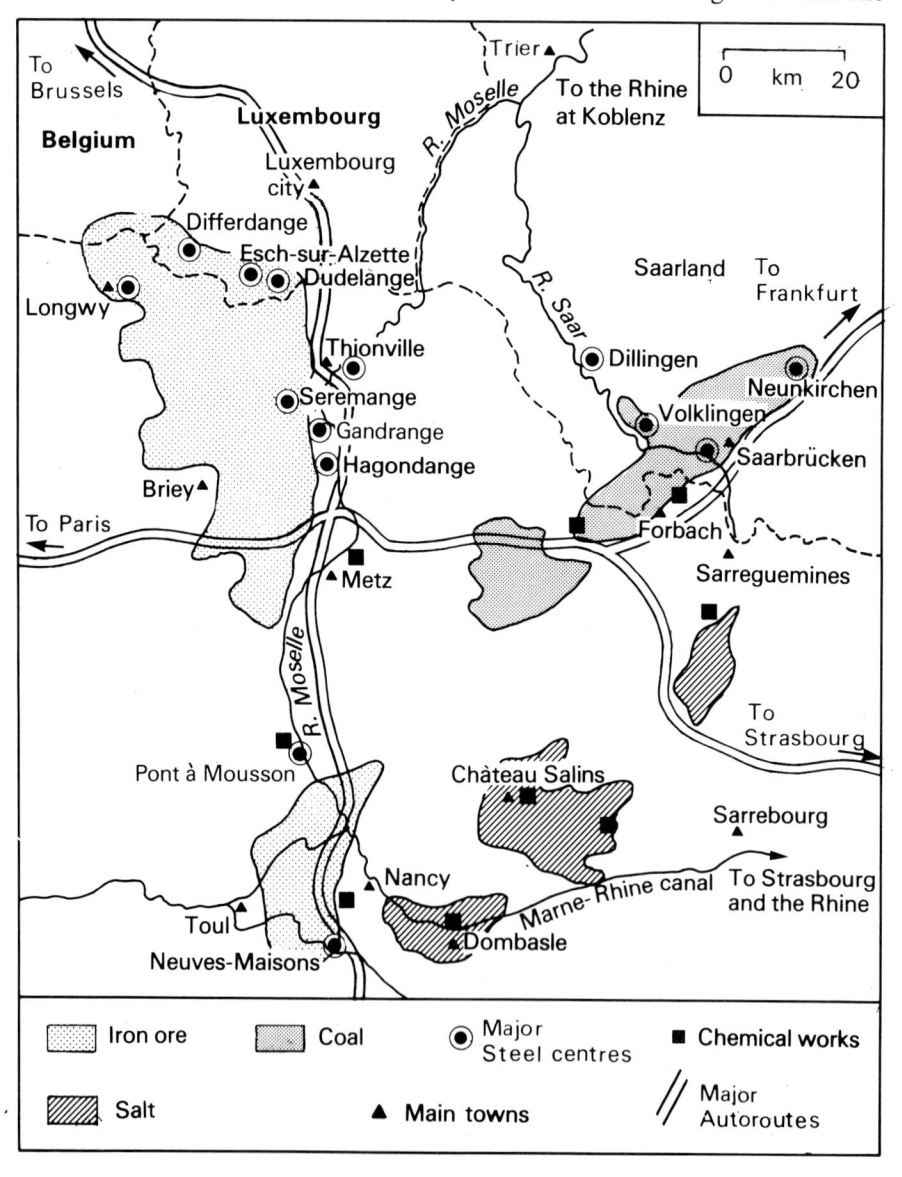

Figure 4.5 Lorraine, Luxembourg and the Saarland

third of total production. Considerable rationalisation and modernisation of small factories has occurred. Iron-ore output has fallen to fifteen million tonnes per year (fig. 4.6), and now provides only a partial supply of the raw material for steel-making. There is also the small but productive Lorraine coalfield. Even so, Lorraine imports coal from the Ruhr. The export of iron-ore to the rest of the community is increasingly difficult because of competition from imported high-grade ores from Africa and Sweden. The Ruhr, Belgium and the Nord no longer use Lorraine ore and its use is confined to its own region, the Saar and Luxembourg. The Lorraine steel-making area stretches from Nancy in the south through Hagondange and Thionville to Longwy. In Luxembourg, a large steel industry for a small country is based upon the northern extension of the Lorraine ore-field. There are three steelworks near the southern border with France (fig. 4.5), but large imports of coal are needed to sustain the industry. The Saarland coal-basin is the third major steel-producer in the region. It produces ten per cent of West Germany's steel at Saarbrücken, Volklingen and Neunkirchen.

Other steel-making centres away from the Rhinelands

The areas so far discussed in the heavy industrial triangle are the traditional national steel-making regions of West Germany, France, Belgium and Luxembourg, which although adjacent to one another, developed entirely separately. They are all based upon highly concentrated reserves of coal and iron-ore, but historical rivalries over the national frontiers which dissected this natural resource zone created fears during the nineteenth century that each steel industry was too close to the frontier for strategic safety. The best example of this geographical unity but historical separation is the essential complementarity of the Ruhr and Lorraine, whose logical exchange of raw materials was for long prevented not only by transport problems but by national rivalry and high tariff barriers. One result of this has been the development of steel-making capacity, albeit on a small-scale, in regions away from the frontiers which were felt to be strategically safer, particularly in France and Germany. In France the Massif Central has scattered deposits of coal and iron-ore and had the added strategic advantage of being within the heart of the country, safe from attack. The St. Etienne region around the Loire coalfield was the cradle of the industrial revolution in France and, until the end of the nineteenth century, was the leading area of steel production. To the north lies the small Blanzy coal basin with Le Creusot as another nineteenth-century steel and engineering area. Other small coal basins with industrial pockets lying around the Massif Central are Commentry and Decazeville. An interesting example of these small, dispersed and often very specialised industrial pockets is the cutlery centre of Thiers, between St. Etienne and Clermont Ferrand. This pattern of inland dispersal was accentuated by the growth of centres such as Caen, where a steelworks was based upon local iron-ore, and Grand Quevilly below Rouen on the Seine, using imported iron-ore. In the Upper Isère and Arc valleys between Grenoble and Albertville, a specialised steel industry using hydro-electric power has developed. In West Germany the steelworks at Peine and Salzgitter in Saxony had been originally developed in 1938, partially upon the iron-ore and coal of the Harz foreland of Lower Saxony. These areas maintained their position during the inter-war years because of favourable government policies, but have become increasingly insignificant in terms of contemporary steel making capacity.

By the 1950s two new factors similar to those already referred to in the United Kingdom, were to play an increasingly important part in further development. These are the ECSC, and technological changes which together have substantially changed patterns of location and structure.

The European Coal and Steel Community (ECSC)

The European Coal and Steel Community (ECSC) came into effect in July 1952 primarily as a common market in coal, steel, iron-ore, scrap, pig-iron and coke. Its effect upon the coal industry was referred to in the previous chapter, but in the steel industry it has helped to give a European perspective and a long-term view of supply and demand. It has assisted in lowering freight rates; ensured conditions of equal competition and a regular supply of raw materials; created a single market and price levels; led to modernisation, rationalisation and expansion of production; raised the living standards of workers; assisted new steelworks development projects with investment grants; and encouraged intra-community mergers.

Sidmar Zelzate steelworks, Ghent, showing a large converter in operation

The abolition of frontier tariffs, national subsidies, and transport-rate discrimination has greatly increased interdependence and intercommunity trade. Lorraine, for instance, had a natural geographical advantage for marketing its products in the adjacent parts of South Germany, and from 1956 to 1961 it increased its sale of rolled steel products there fourfold. The transport costs of Ruhr coke fell by thirty per cent in the ten years up to 1962. In 1952, just before the ECSC began, scrap cost 22 dollars per tonne in the Netherlands, but in Italy up to 55 dollars per tonne. The ECSC has been responsible for large reductions in transport charges across frontiers and for smoothing out cost variations between member states.

The stability and large home market within the ECSC encouraged long-term planning and an increased growth potential, and there was a huge expansion of production. Between 1952 and 1964 the Community steel output more than doubled to 83 million tonnes and, by 1973, the enlarged EC of nine produced 150 million tonnes, substantially more than USA. The investment loans, research activities and regional social policies of the ECSC were together partially responsible for guiding the location of at least two new steel plants (fig. 4.6). The Sidmar steelworks on the Ghent-Terneuzen canal at Zelzate in East Flanders was built in 1962 in an area of economic depression. The Taranto steelworks in Southern Italy was constructed with a view to it becoming an initial growth pole for metal fabricating industries and for industrial employment in this underdeveloped area.

The large EC market of 320 million people has replaced raw materials as the dominant location factor.

The contemporary pattern of steel-making in the EC

The locational attraction of the coalfields has diminished with better iron-ore sintering and preparation, and with improved fuel economy there has been a fall in the amount of coke needed per tonne of pig-iron. From the early 1960s it became progressively less economic to use home iron-ore, which has a low iron-content, and there was a great increase in imports of cheap, high-grade foreign ores of some 60 per cent iron purity. Siegerland production ceased in 1966 and Salzgitter ore is now used only locally. Lorraine, and the Basque coastlands in Spain are the only substantial home ore producers left (fig. 4.6). West Germany finds it cheaper to import most of her iron-ore from Sweden, and supplies of cheap American coal via the Rhine waterway. The Netherlands and Italy also depend upon imported coal. The coastal steelworks at Dunkerque consistently uses cheap imported coal from Poland and USA rather than local coal from the Nord coalfield. For the EC as a whole imports of iron-ore from external countries now dominate supplies. The use of bulk carriers enables these raw materials to be carried cheaply by sea. The increasing use of scrap, particularly in those heavy industrial districts which produce a surplus for re-using, is shown by the fact that it is used for a third of West Germany's, and nearly 60 per cent of Italy's, steel.

Coastal locations

The continuous wide-strip steel mill and the basic oxygen process were the two other very significant developments which assisted the cumulative movement to the large integrated coastal steelworks (fig. 4.6). These new locations have developed as a response to the increasing need for imports and deep-water ore-terminals, together with the availability of large bulk-carriers, and the need to eliminate transport costs to the interior. There are, including the five large plants in the UK, fifteen integrated coastal steelworks in the EC, producing over 43 million tonnes (fig. 4.8). They are all large and many have a potential capacity of up to 8 million tonnes with berthing facilities for large ore-carriers. The first concentration of these is along the North Sea coast from Dunkerque to Bremen, and the second along the Mediterranean coast from the Rhône delta to Italy. Most of the Dutch steel production comes from the Ijmuiden works, built at the end of the North Sea canal, with a capacity for 6 million tonnes of ingot steel, rolling mills and a tinplate factory. It has a dock capable of handling 80 000-tonne ore-carriers. The Sidmar plant at Ghent (Zelzate) currently produces four million tonnes of steel and is capable of expansion. The only limitation here is that the Ghent-Terneuzen canal, by which it is linked to the West Scheldt, can take only 60 000-tonne ore-carriers. The Bremen works uses the Weserport

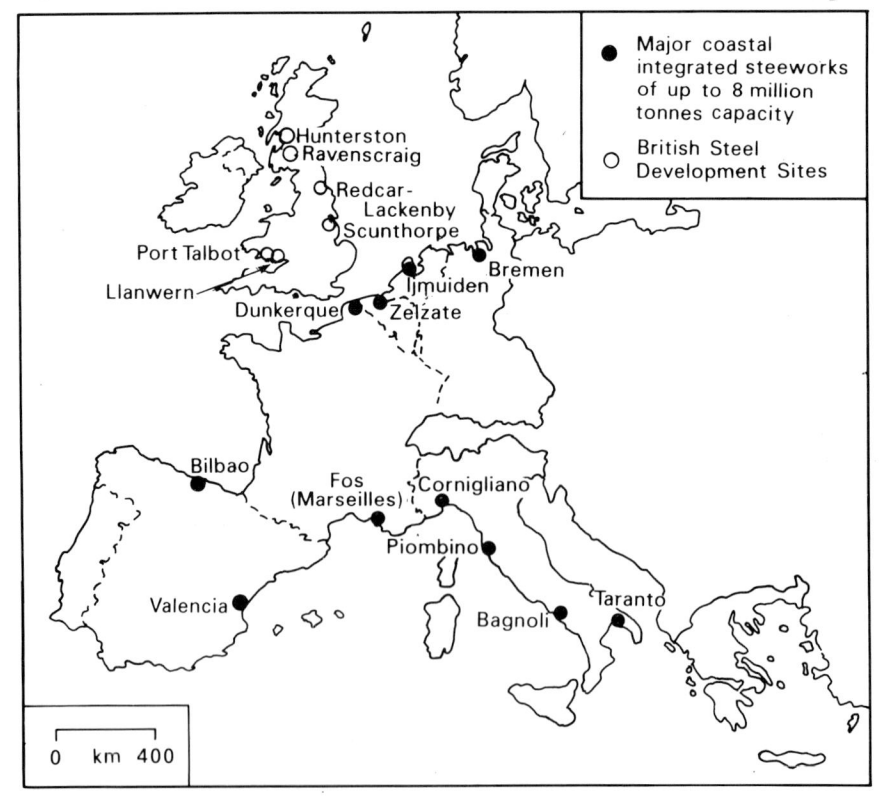

Figure 4.6 EC Coastal steelworks

	Home production			Iron content of home ores (%)
	1976	1981	1986	
West Germany	3.0	1.6	1.0	27
France	45.5	21.8	14.7	30
Italy	0.6	–	–	–
Spain	–	–	6.4	–
Luxembourg	2.1	0.4	–	27
United Kingdom	4.6	–	–	26
Total EC	55.8	23.8	22.1	

Figure 4.7 Iron-ore production in the EC (million tonnes)

facilities 55 km downstream for bulk ore-carriers. The Dunkerque works (6 million tonnes of steel) has facilities for 100 000-tonne carriers and, with Ijmuiden, these two have the greatest potential in a situation where import handling efficiency is critical.

Italy

The Italian steel industry vividly illustrates the dramatic growth and locational changes of the past 30 years. Prior to 1945, with negligible raw materials, steel-making was confined largely to northern Italy near the sources of hydro-electric power and the steel-using industries for scrap. This area has maintained its importance through heavy investment and the large markets of the Italian industrial triangle based upon Turin and Milan. Sesto San Giovanni, Turin, Bergamo and Brescia have a number of specialised steelworks producing quality steel in electric furnaces. There was a small works at Piombino, adjacent to Italy's only supplies of iron-ore, on the island of Elba.

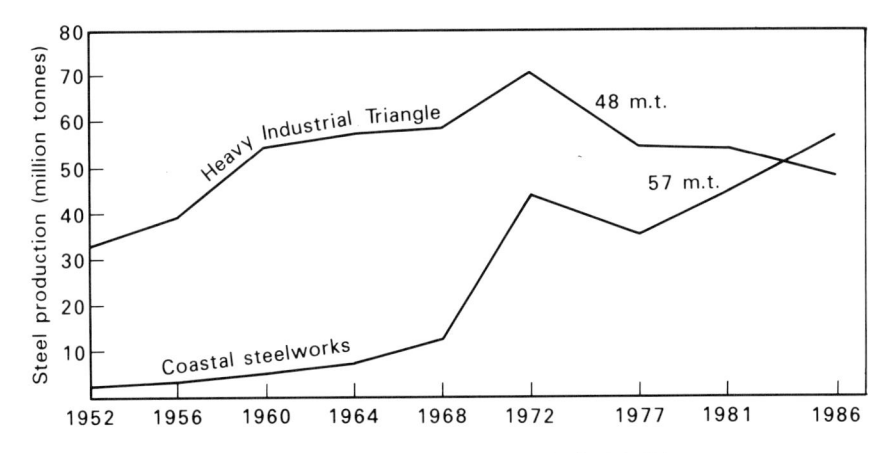

Figure 4.8 The growth in importance of coastal steelworks (1952–86)

94 THE NEW EUROPE

Total Italian production in 1953 was only 3.5 million tonnes. Since the early 1960s, however, production has increased at a phenomenal rate: in 1964 production had reached 9.8 million tonnes per year and by 1968 17 million tonnes. Production figures for 1986 rate Italian steel production substantially higher than that of France and the United Kingdom (fig. 4.1). Post-war reconstruction patterns have involved heavy state participation and control. The state has taken a controlling interest through the holding company of Finsider, which now controls some 60 per cent of total production. The unemployment problem of Southern Italy was a major factor in the opening of the Taranto steelworks in 1964, with handling facilities for 100 000 tonne bulk ore-carriers, and it was planned that this should be the nucleus for the metal fabricating industries in Bari and Taranto. The programme of development has been based upon coastal steelworks which import iron-ore, scrap and coal very cheaply, and now four large integrated plants, Cornigliano, Piombino, Bagnoli and Taranto, produce the bulk of Italy's steel (fig. 4.7).

Spain, Portugal and Greece

The Spanish steel industry was largely developed in the nineteenth century upon the coal and iron-ore reserves of the Basque coastlands, but increasingly now relies upon imports. Development was small because of the peripheral nature of the Spanish economy in relation to Western Europe, but there has been intensive development of the steel industry since the Second World War. Major investment was made during the 1960s and 1970s by the regime of General Franco, assisted by West German and American investment with the aim of developing a modern steel industry. There are integrated steel plants at Bilbao, Oviedo, Gijon and Aviles. Bilbao is able to import scrap iron ore

Piombino coastal steelworks, Livorno, Italy

and coal cheaply by sea. There is also a modern integrated plant at Valencia on the east coast. Spain, now producing 11.9 million tonnes of steel per year, is a good example of heavy state involvement in the industry. The Portuguese and Greek governments have also created small steelworks at Seixal near Lisbon and Piraeus near Athens respectively, as part of the national development programmes and for import substitution purposes (fig. 4.1).

The Heavy Industrial Triangle

Although the movement to the coast has been very substantial, the Heavy Industrial Triangle still accounts for a major proportion of steel-making on the continent, and of West German, Belgian, Luxembourg and French national output (fig. 4.4, 4.8). The tremendous concentration of steel-making capacity in the Ruhr has substantial advantages with its large coal reserves, adjacent consumer markets, dense communications networks and local sources of scrap. Lorraine remains a major factor in French production plans. A new steelworks has recently been built at Dillingen in the Saar. Enormous capital investment is required for the development of entirely new 'greenfield sites', and this is another reason for the maintenance of traditional steel-making regions. They have, in addition, a positive advantage which accrues directly from the establishment of the ECSC and Common Market. The triangle now has a new role. It is the most centrally placed industrial area within the community with good, short, internal lines of communication, the principal source of raw materials and a huge urban market and supply of labour. The canalisation of the Moselle has benefited Lorraine (fig. 4.5), particularly with its dependence upon Ruhr coke via a waterway from which frontier tariffs have now disappeared. The effective disappearance of the frontiers has given this region a new advantage: the natural unit which has always been prescribed by geography but denied by history. Its complementary resources can now be used in a coherent manner, focused upon the major artery of the river Rhone and its tributaries. The advantages of this Rhineland core region, with increased integration, have encouraged its maintenance as the single most important steel-making region of the EC (fig. 4.3).

Changing Geography

However, the geography of the steel industry has changed considerably. The old core of steel making in the Heavy Industrial Triangle has been supplemented and extended northwards towards the coastal regions around the North Sea, and southwards to the Mediterranean coasts.

State involvement

Considerable importance in decision-making still rests with the national governments and there is an increasing state involvement in both control and planning. Modernisation and rationalisation has involved relocation and amalgamation of companies, larger steelworks and an increasing domination by large corporations (fig. 4.9).

96   THE NEW EUROPE

France

Three specific French projects have been significant:
(a) substantial investment in the Lorraine steel industry for manpower reduction, rationalisation and modernisation, including raising the capacity of Gandrange to four million tonnes.
(b) the doubling of the capacity of Dunkerque to eight million tonnes.
(c) the construction of the Solmer plant at Fos-sur-Mer near Marseilles, with eight million tonnes capacity, as part of the development plan for the Lower Rhône region.

This has completely changed the locational balance of the French steel industry, which is now dominated by two large coastal steelworks. Its structure is also changing rapidly with rationalisation: it is now dominated by two large groups, Usinor and Sacilor, which account for over 90 per cent of bulk steel output. There are still a number of small companies as a legacy from the days when the isolated metallurgical centres of the Massif Central and Alpine region flourished. Although many closures have been inevitable, survival has been based upon specialisation in special steels. Pechiney-Ugine and Creusot-Loire produce much of French special and alloy steels. The French steel industry ran into considerable difficulties during 1978 and the French Government was forced to nationalise it in an effort to rationalise its

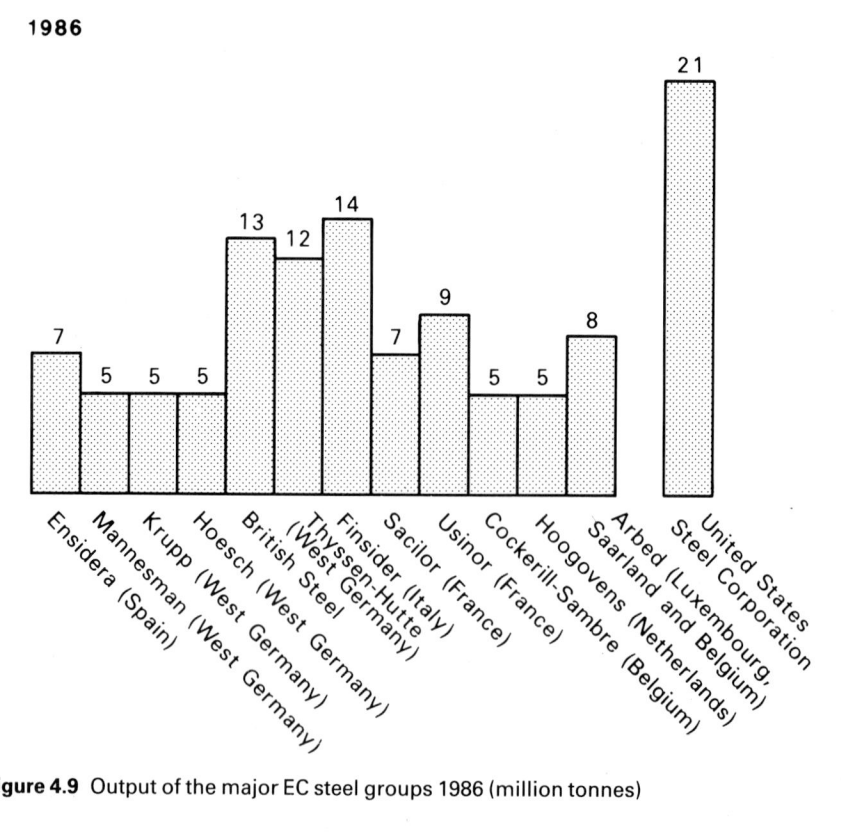

Figure 4.9 Output of the major EC steel groups 1986 (million tonnes)

structure and increase productivity standards. This involved closure of inefficient plants and a large reduction in the labour force. Lorraine, heavily dependent upon the steel industry, has been badly affected, particularly in towns such as Longwy and Thionville, which have become depressed areas.

Mergers and large steel groups

The largest steel groups in the Community are Thyssen-Hutte of West Germany, British Steel, and Finsider of Italy. Other member countries have experienced reorganisational change in varying degrees. The Luxembourg-based company Arbed also controls integrated steel plants in the Saarland and Belgium and has become a multi-national steel and engineering company which produced over eight million tonnes of steel in 1986. West Germany, the largest producer of steel in the community, has a number of other major companies: Mannesman, Krupp, Hoesch and Klocknerwerke. In Belgium Cockerill-Sambre emerged as the largest group with fifty per cent of production.

The steel crisis from 1975 onwards

The rapid expansion of the steel industry in the 1960s was continued into the first half of the 1970s. This was accompanied by basic locational changes, and an increasingly integrated structure and larger size of steelworks. From no less than 113 companies in 1958, sixty per cent of production is now controlled by ten large companies. Modernisation continued with new investment and the closure of almost all Open Hearth and Bessemer furnaces. In 1986 75 per cent of EC steel was produced from basic oxygen converters, and 25 per cent from electric arc furnaces.

The steel industry however, was badly affected by the recession which began with the oil crisis of 1973. Steel output fell from a high-point of 156 million tonnes in 1974 to 126 million tonnes in 1986 (fig. 4.1). The projected EC capacity for the 1980s was originally over 200 million tonnes, but this was completely unrealistic. In 1981 EC steel production was running at only 63 per cent of capacity, and each country has experienced severe problems with closures of steelworks, and labour redundancies. Even the Ruhr has reduced production by 25 per cent. British Steel has slimmed its capacity most of all and closures of outdated plant have proceeded, with gross capacity of eighteen million tonnes being concentrated at five integrated steel plants. It has now returned to profitability and to private ownership (1988). The Belgian/Luxembourg steel industry has been badly affected. These are two small countries with a high per capita steel production which is very important to their economy. There are too many small producers and much of the steel industry is in inland locations which tend to be more costly. The industry is to be restructured around two major companies: Cockerill-Sambre, which is located in the old heavy industrial zones of Liège and Charleroi, and Arbed, which is based in Luxembourg.

As an example of the process of integration into the declining EC steel industry, Spain in particular has encountered great problems. The state steel company ENSIDERA has been granted EC aid to restructure and modernise, reducing capacity by four million tonnes. This has involved job losses of 25 000 since 1983 and more are to follow, particularly in the northern coastal provinces and Bilbao.

The causes of decline in steel-making are many and complex. There had been over-investment in the industry during the boom years of the 1960s. There was massive surplus capacity in the economic recession from 1973 onwards as demand fell sharply. European steel is high-cost compared to many of the newer low-cost Third World producers of steel. The dominant position of steel in the economy of developed countries has been severely eroded because of new raw materials and substitutes which can now replace steel. The Japanese steel industry, with investment in robotics, is a major competitor. High technology industries require less steel than the older industries. Steel has followed industries such as textiles and shipbuilding into the category of a 'declining industry'.

The European Commission has taken steps to reorganise the industry. It favours the amalgamation of all bulk steel-making into a new pattern, which would involve extensive cross-frontier mergers and result in a few major steel groups each producing up to twelve million tonnes per year. Their probable locations should suggest themselves. The Davignon Plan in 1978 has been the basis upon which action is based. The ECSC now has a new role: guidance of restructuring; modernisation and amalgamation to improve competitiveness; reduction of capacity and controls and mandatory production quotas for each member state to cushion the decline in production; a planned reduction of the labour force; short term subsidies and social measures, such as retraining, and the creation of new jobs to assist redundancy; levies on low-price steel imported from outside the EC and negotiation with external countries on limitation agreements for their exports of steel.

The number of workers in the steel industry has dropped from 778 000 in 1973 to 397 000 in 1986. By the end of 1986 production capacity still exceeded requirements by nearly 30 million tonnes. Production quotas were abolished in July 1988, but a review of the situation takes place annually. It is envisaged that another 80 000 jobs will go by 1990 as over-capacity is reduced to an acceptable level. The European Commission has stated that a top priority is the successful modernisation of the steel industry.

5

Other industries: automobiles, textiles and chemicals

The Automobile Industry

The automobile industry has a central role in the complex industrial economies of Western Europe. It is a major user of raw materials and a large-scale employer. In France it absorbs 50 per cent of the production of rubber, 50 per cent of the shaped aluminium, and 21 per cent of sheet steel and machine tool production. Nearly two million people gain their living from the industry. Its real growth has been since 1945 as a reflection of consumer demand, and as a symbol of prosperity it has been seen as a propellant or growth-pole industry in the city regions where it has a significant multiplier effect. It is an assembly line and component industry with horizontal integration resulting in a few large companies dominating production. The geographical location of the industry near large centres of population does not usually correspond to the older heavy-industry areas (fig. 5.1).

Development of the industry

The European motor industry originated in the 1890s with names such as Daimler, Lanchester and Panhard. In 1896 the Daimler Company began production in Coventry, but the industry had a slow undistinguished growth up to 1914, when it was primarily a small-scale producer of high-cost goods for a restricted market. Even until the slump of the 1930s it remained best known for a large number of specialist quality companies such as Lea Francis of the UK and Ferrari of Italy.

But, during the 1930s, the character of the industry was changing. The economic depression wiped out many firms and mergers had occurred on a large scale. In the UK the number of manufacturers declined from 88 in 1922, to 31 in 1931. By 1937 the UK had become the second world producer next to USA. Mass production methods were introduced during the 1930s and the growth of per capita income was another indication of the imminent development of the consumer boom which was to come after 1945. The very rapid growth since 1945 corresponds to the initial demand for, and development of the family car. Since then there has been replacement demand and now a third stage with the proliferation of the second family car. There is now one vehicle for every three people in the EC. The industry developed very rapidly between 1950 and 1980 and is consequently modern in characteristics. Factories are usually housed in one-storey modern buildings, and, with the large amount of space needed for storage and parking, usually cover a wide area. They are often on 'greenfield sites', (e.g. Wolfsburg, West Germany) and some distance from city centres where space is readily available. Communications by road are invariably good and siting on by-pass roads is common. The car assembly factories on the A45 at Coventry and on the Rocade highway at Rennes in Brittany are good examples of this.

Structure of the vehicle industry

The assembly line involves the concentration of a wide range of supplies which come from subsidiary manufacturers. These provide basic materials such as pressed steel, rubber, glass, etc. and a host of manufactured components and accessories including electrical equipment and brakes. These specialist sub-contractors have traditionally been independent of the vehicle companies which are their main outlets and, therefore, vertical integration in the industry is almost non-existent. There are exceptions to this. Renault and Fiat have their own steel and metallurgy supplies; and Krupp, essentially a steel-maker, also produces commercial vehicles. In general, however, the major component manufacturers are independent and exert a considerable monopoly. Solex supply two-thirds of the French, half the German and one-third of the United Kingdom market for carburettors. With a great number of independent suppliers there is therefore a real necessity for efficient distribution. This leads to a tendency for a number of vehicle industry zones to develop, made up of linked component manufacturers and car assembly plants concentrated in and around city regions. This regional swarming has the advantages of short distances for the transport and distribution of components, a pool of skilled and semi-skilled labour, and easier technical co-operation. It is seen to best advantage in regions such as the West Midlands, Saxony, the Paris region, the Rhinelands and Milan-Turin.

Location

Each nation's vehicle-producing areas are usually city regions in national core areas whose large populations provide the natural market (fig. 5.1). There is a distinct tendency for factories to be sited on the outer fringes of the city, away from the congested centres. In the Greater Paris region the three largest companies are all in the suburbs or satellite towns. Renault have their main works at Billancourt, but have other factories at Nanterre and Clichy to the north-west, at Rouen and Le Havre on the Lower Seine, and Orleans and Le Mans. There is also Peugeot-Citroën at Reims and Poissy. In Italy ninety per cent of vehicle production is around Turin and Milan and the United Kingdom has a traditional concentration in the West Midlands, particularly at Coventry and Birmingham. There are also anomalous situations: the Peugeot

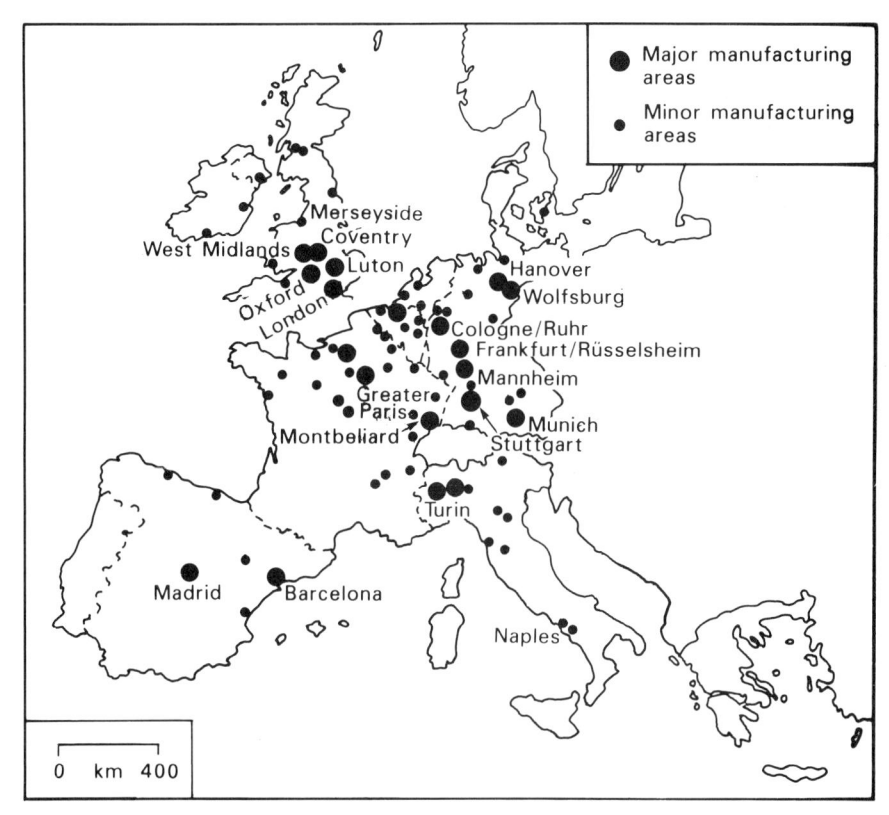

Figure 5.1 The motor vehicle industry - principal locations

factories are at Montbeliard in the Jura and Mulhouse in Alsace, and they continue to operate successfully from there, high managerial and technical skills overcoming their apparent disadvantage of location.

West Germany illustrates a modification of this pattern: Daimler-Benz at Unterturkheim and Sindelfingen near Stuttgart, and at Mannheim and Gaggenau in the middle Rhineland; General Motors (Opel) at Russelsheim (Frankfurt); BMW at Munich; Porsche at Stuttgart; and Ford at Cologne. All these indicate the importance of the cities of south Germany and the Rhinelands. The largest single vehicle enterprise, however, Volkswagen, does not conform to this particular model. Wolfsburg was an early example of re-location of industry in 1937 to the then centre of Germany, partly for strategic reasons and partly as a policy of dispersing Ruhr industry. It was located next to the Mittelland canal, the main waterway from the Ruhr to Berlin. Now, however, since the post-1945 political division of Germany, Wolfsburg is very near the East German frontier and on the periphery of the EC. The reputation and efficiency of the company have allowed it to expand its operations, the Volkswagen factory now employing over 50 000 people. It has expanded to nearby Hanover, Brunswick and Kassel, and Saxony has become the country's second vehicle region with forty per cent of output, equivalent to that of the Rhinelands.

To summarise, the EC vehicle industry is principally located in the following areas: the Midlands and south-east of England; the Paris region; Milan and Turin; Barcelona and Madrid, Saxony and the Rhinelands. Nevertheless, the industry has been encouraged to move into areas of high unemployment in certain cases. Governments have used the car industry as a target for decentralisation and to aid the restructuring of old industrial regions. The Italian Government has used the car industry in its industrialisation programme for the Mezzogiorno. In the UK this tendency is well established with factories on Merseyside, in South Wales, and at Bathgate in Scotland. In France and Germany there is also a movement to the older coalfield areas which need a new industrial infrastructure. Peugeot-Citroën have factories at Lille and Valenciennes in the Nord region and they have also set up two factories at Rennes, the development pole for Brittany. Opel have a factory at Bochum in the Ruhr, and Ford at Saarlouis. It is often politically useful but not geographically logical to move industries out of their traditional centres. Government intervention often runs in the face of the advantages of regional swarming.

The major vehicle companies

Horizontal integration has become almost complete. The assembly lines and mass production units are dominated by a few large companies, with resulting economies of scale. In the United Kingdom four groups produce practically all the motor vehicles, including commercial vehicles. British Leyland incorporated many famous car companies - Austin, Morris, Standard-Triumph and Rover, as well as the commercial vehicle company, Leyland Motors – but the decline of the group during the 1970s has left it with one major car group, Rover, which now has a much smaller percentage of the market than before. Two others are American subsidiaries Ford Motors (UK) and Vauxhall, controlled by General Motors, and the third at Coventry is part of the French group Peugeot. There are several small, but significant, luxury manufacturers, including Rolls-Royce and Jaguar cars. On the continent there is a similar picture. In West Germany Volkswagen dominates the market, with nearly half of total production and a very considerable export market, followed again by Opel (General Motors), Ford and Daimler-Benz. France has two major groups, Peugeot, incorporating Citroën, and the government-controlled Renault company. Italy, however, has the most marked concentration: one company, Fiat, produces 85 per cent of Italy's cars, and controls Lancia and Ferrari. Other producers are Alfa Romeo and Maserati.

The EC car industry in its global context

Motor vehicles are an important element in the EC's trade. West Germany and France are the most successful exporting countries with about half their total production exported. With the progressive abolition of tariffs during the 1960s and 1970s intra-Community trade increased spectacularly and it has now risen to over 30 per cent of total production. However, during the last

The Volkswagen factory at Wolfsburg, West Germany

decade there has been increasing import penetration from Japan, which now takes ten per cent of the EC market. American involvement in the European vehicle industry is very significant, but varies from forty per cent control in the United Kingdom and thirty-five per cent in West Germany, to a minimal level in France and Italy. This American involvement, the Japanese export assault upon the European market, and the advantages of scale mentioned earlier, are the principal reasons for the existence of a few large EC companies which can compete in the world scene. (fig. 5.3).

The vehicle industry has entered an uncertain period. The market is increasingly saturated and competition is severe. The instability underlying the oil industry in the 1980s has led to a search for more economical cars. Production of cars is now approximately the same as in 1973. In West Germany and France production has stagnated, whilst in Italy and the UK it has fallen substantially. The high cost of car manufacturing in Western Europe, particularly in the UK, with labour problems, low productivity, high wages, weak management and technical obsolescence during the 1970s, led to an inability to withstand foreign competition. By contrast, Japanese exports made major inroads into the EC markets. Their productivity was three times that of Western Europe and automation and robotisation were substituted for labour at an early stage. Spain alone in the EC has experienced a rapid development programme. Output has doubled since 1970 and Spain is now the Community's fourth largest producer. The principal reasons for this are an import tariff barrier of 37 per cent before joining the Community and an aggressive government sponsored expansion policy. The latter is aimed at import substitution with the aid of foreign investment particularly from the United States and West Germany.

Currently, there is an overcapacity of twenty per cent and the EC car

	Cars			Commercial vehicles	
	1967	1973	1985	1967	1986
West Germany	2296	3642	4165	187	266
France	1777	3202	2817	233	423
Spain	–	–	1230	–	132
Italy	1439	1823	1389	103	162
Netherlands	49	94	106	7	14
Belgium*	164	260	987	25	50
United Kingdom	1560	1747	1048	384	225
Total			11742	939	1272

* vehicle assembly by subsidiary companies

Figure 5.2 Motor vehicle production and assembly (thousands)

industry is still organised in a series of national markets rather than as a European entity, leading to loss of economies of scale. One successful example of EC integration is in Belgium, where there is a substantial assembly and components industry, with over a dozen companies operating production lines. Belgium's central position within the EC transport system, makes this advantageous for marketing purposes. The West German car industry, the most successful of all, produces a wide range of high quality models. Volkswagen have taken over SEAT, the Spanish company, in a successful example of intra-EC integration. Low cost labour in Spain will benefit Volkswagen. Spain is a useful base for exporting purposes, and SEAT needed major capital backing. With acquisitions in the United States, Brazil and Argentina, Volkswagen has been transformed from an essentially European company into an international corporation. The UK Government has eventually successfully rescued the car conglomerate, British Leyland. The commercial vehicle section has been merged with DAF/Volvo Trucks of

Fiat	14.9
Volkswagen	14.9
Peugeot/Citroen	12.9
Ford (USA)	11.3
General Motors (USA)	10.3
Renault	10.1
Rover	3.9
Others	21.7

Figure 5.3 EC car companies 1989 (percentage of EC market)

The Netherlands. Jaguar cars is now a private company, whilst the public sector company, Austin Rover, now renamed the Rover Group, has been modernised and sold to British Aerospace. New investment has been encouraged and Nissan, the Japanese group, have set up a new production plant at Sunderland in the North East. The Community produces 38 per cent of world car output (fig. 5.2), but heavy subsidies of the protected national car markets, particularly in France, Spain and Italy, have retarded the reorganisation of the industry on a European scale. As the integrated market in 1992 approaches, there are signs of recovery in confidence. The UK car industry has begun to recover from the debilitating period of decline, and 1988 saw a boom in productivity, demand and production.

The Textile and Clothing Industry

The textile and clothing industry is an old staple industry, a major exporter and one of the most important single industries in the Community. It has a workforce of over 1.8 millions, representing nearly nine per cent of total manufacturing employment. It is, however, beset by problems of foreign competition, a relatively slow growth of domestic consumption, and an out-moded structure associated with its nineteenth-century origins. Its precursor, the cottage industry, has retained a residual influence in the survival of sub-divided processes and the tendency to small-scale factory units, with a poor degree of integration. It is often, therefore, described as a declining industry and has faced great problems of adjustment during the present century.

Location

There has been a tradition of heavy regionalisation and concentration in localities originally favoured with water-power, coal and the local availability or import facility of the raw material. On a European scale there are four major textile zones (fig. 5.4), all of which are based upon old industrial regions: Flanders and the Rhineland; Lombardy and Piedmont; the sub-Pennine region of Lancashire and Yorkshire; and Catalonia in Spain. Several smaller, though significant areas, often medieval in origin, have survived by specialising in high-quality fabrics or by developing a monopoly in a particular material or process.

Flanders - North Rhineland

Astride the Franco-Belgian border lies one of the leading textile areas in Europe, important since the Middle Ages. It stretches across the Plain of Flanders from Lille into Belgium between Courtrai and Ghent. From its

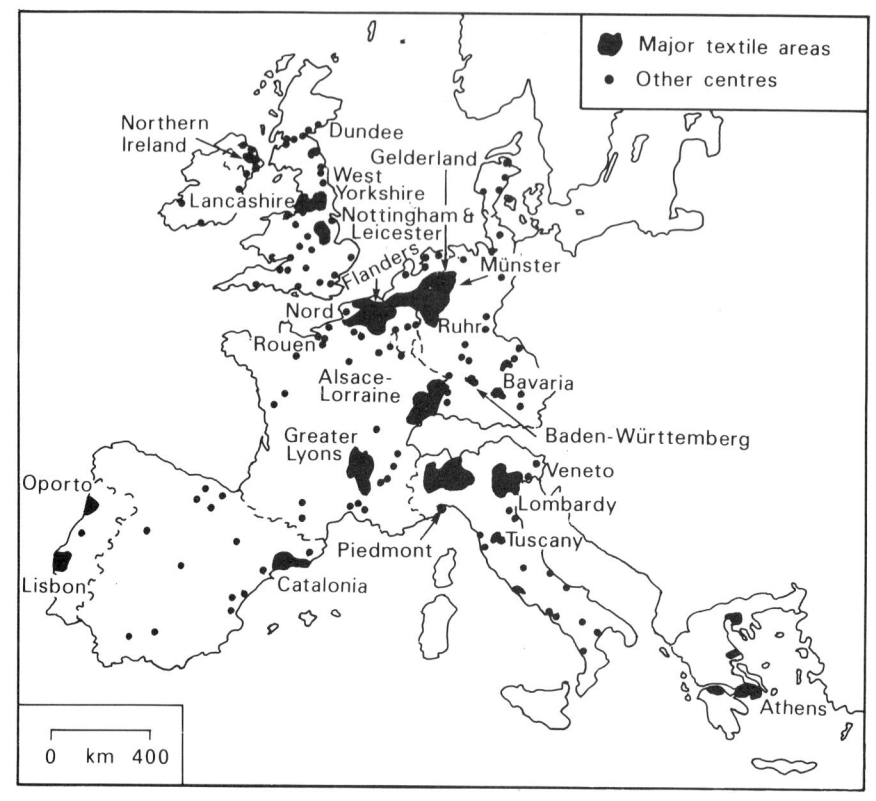

Figure 5.4 Principal textile manufacturing locations

earlier concentration on local wool from the sheep grazed on chalk escarpments, and linen using flax from the Lys Valley, it has changed into a complex modern industry. Cotton, nylon and mixed fibres are important at Lille, Roubaix-Tourcoing, Armentières and La Bassée, but fine linens, carpets, hosiery, clothing and furnishing fabrics are also manufactured, and outlying centres include Amiens, Abbeville and Cambrai. In Belgium about half of the total number of cotton spindles are in or around Ghent, but the Scheldt-Dendre Valley towns of Oudenaarde, Ranse, Tournai and Geerardsbergen are also important. Courtrai produces fine linen from Lys Valley flax, though most flax is now imported from France and Poland. The industry has moved into Brabant and Brussels, producing carpets, blankets and fashionable clothing. The original woollen industry has now largely migrated to Verviers in the Ardennes to take advantage of the soft water and local wool.

The textile areas of the Netherlands and West Germany are a natural eastward extension. In the Netherlands the industry developed from the trading activities of the Hanseatic league along the Rhine, in the towns of Breda, Tilburg, Arnhem, Nijmegen and Rotterdam/Dordrecht. During the twentieth century there has been an eastward extension to the eastern heathland towns of Hengelo, Enschede, Emmen and Twente, where nylon and rayon production is most important. In adjacent West Germany lies the

'Baumwollstrasse' of North Rhine/Westphalia, where a combination of Ruhr coal and the traditional crafts of towns such as Krefeld and Mönchen-Gladbach have produced a specialised textile industry which occupies towns around the fringes of the Ruhr. Krefeld's original speciality of silk and velvet has now been supplemented by nylon and terylene. The cotton industry is centred upon Wuppertal, Mönchen-Gladbach and Bocholt. Düsseldorf and Cologne are important too, and the area extends northwards to Munster and Bielefeld, which specialises in linen.

Lombardy and Upper Piedmont

There are over 400 000 workers in the Italian textile industry, the bulk of which is concentrated along the northern Alpine fringe of the Po Valley, stretching from Biella through Milan, Bergamo and Vicenza to Padua. The principal concentration of cotton manufacturing is within a radius of 60km from Milan, in the industrial satellite towns of Varese, Gallarate, Busto Arsizio and Legnano. There is quite intense regional specialisation, particularly in the silk manufacturing arc of Como, Varese and Treviglio, and the concentration of woollens and worsteds in the Biellese and Bergamasque sub-Alpine valleys. Immediately to the south of Milan, the towns of Vercelli, Magenta and Pavia manufacture synthetic fibres. Another major woollen and knitwear area is based upon Padua, Vicenza, Schio and Valdegno, in Veneto Province.

The Lancashire/Yorkshire sub-Pennine region

The rise of the textile areas of northern England was caused by the coincidence of several factors in a unique situation, which led to the establishment of two areas which dominated world textile production in the nineteenth century. The natural factors were the suitability of the Pennines for sheep-rearing, abundant lime-free water, numerous sites for water-power, coupled with a high relative humidity, and power from the Lancashire coalfield. These were augmented by crucial human factors. These were the energy of pragmatic non-conformists and entrepreneurs, and their ability to accumulate capital, the inventions of textile machinery by Kay, Arkwright and Crompton, and the lack of guilds in this area. The construction of port facilities, canals and railways as the expansion of these industrial areas got under way, and as imports of raw materials became necessary, was a further factor of concentration.

There was an extraordinary degree of specialisation in Lancashire, and in 1931 84 per cent of all cotton operatives in the United Kingdom were in East Lancashire and the adjoining part of Yorkshire. Manchester was the commercial centre, bank and warehouse. Spinning was localised in an arc of towns close to Manchester, and weaving more particularly in the group of towns north of the Rossendale Fells and in the Ribble Valley. The finishing trades (bleaching and dyeing) and clothing were not quite as localised, factories tending to be limited to one or other activity, which led to small production units and a lack of integration.

West Yorkshire developed an equally concentrated woollen textile region, based upon the Leeds/Bradford conurbation. Here the principal specialisation was in the type of product rather than, as in Lancashire, in processes. Long wools for worsteds predominate in Bradford and the north-west of the region, short wools for woollen products in the south-east, whilst carpets are made at Halifax, and Leeds is the ready-made clothing centre. Bradford is the financial and commercial centre of the industry.

Catalonia, Spain

Barcelona, once capital of the Catalan kingdom, is the most important manufacturing city of Spain. It is an industrial, commercial and financial centre, but its wealth was originally based upon its port, the cotton trade and the development of the textile industry. Water-power was a major factor of textile development in Catalonia and has now been replaced by modern hydro-electric plants in the Ebro valley. Cotton, wool and synthetic fibres are manufactured in Barcelona and a ring of satellite towns - Badalona, Manresa, Granollers, Mataro and Sabadell. Textile machinery is also one of the range of engineering industries in the city. Although the textile industry is widely spread throughout Spain and other cities such as Bilbao, Catalonia is the greatest single concentration, with over 3 000 small factories.

Other areas

1. **Alsace-Lorraine**: This originated as a textile area in the medieval period. The Vosges mountains provided local wool, soft water and fast-flowing streams, and the industry has survived in factories around Epinal, Mulhouse, Belfort and Colmar, chiefly with cottons, thread, fine linen and hosiery.
2. **Greater Lyons**: Silk-making originated in the fifteenth century from exiled Italian merchants and is still so important that 80 per cent of French ouput still comes from this area. Rayon and nylon has developed to supplement the natural silk.
3. **Bavaria and Baden-Wurttemberg**: The cities of South Germany have a traditional cotton textile industry at Stuttgart, Karlsruhe and Augsburg.
4. **Peninsula Italy**: There is a significant woollen area in Tuscany (Prato and Florence in the valley of the River Arno) and at Rome. Woollens have been supplemented with synthetics established by large companies like Snia-Viscosa. Local raw materials are an advantage, with mulberries supporting silkworms in the Marche and eucalyptus being grown for rayon production. In the south large factories have been set up under the auspices of the Cassa per il Mezzorgiorno at Caserto and Frosinone (Naples) and Pisticci (Taranto).
5. **The East Midlands Hosiery Belt of England**: This is dominated by Leicester and Nottingham, but stretches south in the valley of the River Soar around Hinckley, and north to Mansfield. This area expanded rapidly in the 1960s, because of the expanding market for knitwear and the ease with which synthetic fibres can be used. Around Nottingham it is

associated with the original lace industry and the early working of silk and cotton, and around Leicester with woollens.
6. **Northern Ireland**: Ulster is a world-famous linen-manufacturing region. Belfast dominates the industry, but Lurgan, Lisburn, Portadown and Ballymena are also important. Although competition from cheaper goods and the decline in demand for high quality specialist linen has taken place, the industry still has a large export market and a large development of synthetics has taken place.
7. **Dundee, Scotland**: Linen and jute are the two products of this very specialised town. Dundee is only a little less important than Belfast for linen and its jute industry employs several thousand people. Dunfermline damasks and Paisley fabrics are other famous specialities.
8. **Kidderminster in Worcestershire, England**: This town is a specialised carpet centre, based originally upon local wool and now dependent upon skilled labour and up-market products.
9. **Rouen**: This is an isolated cotton-manufacturing town originally based upon raw cotton imports through Le Havre.
10. **Portugal**: Local wool and imported cotton form the basis for the textile industry in Portugal. Lisbon, Oporto and Coimbra are the principal centres with woollens, cotton and garment manufacturing, but there are several hundred small factories spread through most of the major towns. The long coarse wool (churra) produced in the north goes mostly to make rugs and carpets, whilst the finer merino wool of the south produces woollen cloths.

Decline and adjustment

The cotton industry, particularly in Lancashire, has shown the classic symptoms of decline and the need for restructuring an old industry. In the nineteenth century Europe dominated world textile production, and as late as 1900 Great Britain accounted for fifty per cent, and the other EC nations for thirty-five per cent of world textile exports. The huge fall in British exports this century was caused by the developing nations of Asia, formerly her export markets, beginning to build up their own textile industries. Lancashire felt the loss of markets most heavily because of its high degree of specialisation in cotton fabrics which were particularly oriented towards the export trade. By the 1920s the inherent disadvantages of Lancashire had become apparent: the lack of local raw materials; static home demand; the near impossibility of competing with the cheap labour of India, China, Japan and Hong-Kong; and an increasingly obsolete industrial structure with old machinery. Production dropped and unemployment rose. There has, therefore, been no alternative for Lancashire but to rationalise the structure of the industry, to improve efficiency, and to reduce capacity and the labour force at a socially acceptable rate.

In other parts of the Community there has been a similar pattern of events. The French and Belgian textile industry is still characterised by its small production units and long domestic traditions. During the 1960s France had

	1958	1971	1981	1986
West Germany	607	499	336	258
France	518	425	293	228
Belgium/Luxembourg	172	121	68	59
Italy	481	542	476	409
Spain				217
Portugal				75
Greece				59
Netherlands	103	76	33	22
United Kingdom	815	622	395	246
Ireland	neg	42	19	11
Denmark	neg	20	15	15
Total	2696	2347	1635	1599

Figure 5.5 Changes in employment in the textile industry (thousands) (Source – *Eurostat*)

about 5 000 mills employing nearly half-a-million workers. Changes in the international division of labour were being seen within Europe, with declining production in the mature north-west European countries being matched by a shift of production to lower cost centres in southern Europe. During the period 1965 - 77 over 900 000 jobs in textiles were lost in north west Europe, whilst at the same time nearly 300 000 were created in Italy, Spain, Portugal and Greece. In Italy the production of woollens and hosiery, in particular, expanded rapidly. Since 1977, however, contraction has been particularly rapid, with all countries of the EC experiencing factory closures and falling employment. The mill towns of the Nord and Alsace-Lorraine have been badly affected, with the textile labour force of the Nord region reduced from 170 000 in 1951 to 60 000 in 1982, and below 50 000 by 1986.

The economic crisis since 1973 has particularly affected the textile industry, because a period of falling demand coincided with a sharp increase in imports of inexpensive textiles from low-cost, developing countries. In the Community as a whole, over 3 500 factories have closed and 800 000 jobs have been lost since 1971 (fig. 5.5). To allow time for modernisation and reorganisation, the Community is negotiating import limitation agreements with some thirty low-cost textile-producing countries.

There is now a definite trend towards larger companies and a greater integration of production to gain the economies of scale. In the United Kingdom in 1946 there were over 1 300 registered companies, but now there are three large groups, including Courtaulds and Coats/Viyella, third and fifth in world rank. Courtaulds have accomplished a substantial amount of horizontal integration and now control 45 per cent of cotton spinning capacity. Coats/Viyella are represented in all processes of the industry from spinning through to clothing production, and are a good example of vertical integration. In the high-income European countries, the numerous fashion changes, such as the advent of jeans and trouser suits, which require quick decisions and manufacturing changes are best taken by such large well-capitalised companies, although even they can be overtaken by very rapid changes. Although a much greater degree of horizontal and vertical integration has been achieved in the UK, many of the southern European producers, such as Italy and Spain, still have a predominance of small-scale and family firms. Portugal, in particular, has an unintegrated structure with several hundred small factories, labour intensive and domestic production techniques.

Although the picture of textiles is one of overall decline, not all branches of the industry have experienced the same rapid decline as cotton. Production of woollens has remained virtually static whilst that of synthetic fibres has dramatically increased. Woollens have had a greater ability to specialise in high-quality fabrics and to concentrate on the high-value domestic market. The woollen industry was never quite so dependent upon the export market as cotton, and thus had fewer potential contraction problems. Wool can be blended easily with synthetic fibres and has benefited from the rapid rise of the carpet and hosiery industries. The volume of knitted goods has doubled since 1960.

Man-made fibres

Man-made or synthetic fibres have exercised a decisive influence upon the character of the textile industry during the twentieth century. Their principal role has been as a supplement to, or substitute for, natural fibres. They are lower in price and are much more versatile, capable of being blended to varying degrees with natural fibres. There is a great range of synthetic fibres of chemical origin, including in the United Kingdom Nylon, Terylene, Courtelle and Acrilan, and their varients in France (Crylon) and West Germany (Dralon). Rayon is a cellulose wood-pulp product, but this is of declining importance compared with the synthetic coal- and oil-based fibres.

Production of synthetics has risen dramatically within the last thirty years. It is now much greater than that of natural fibres (fig. 5.6) and this has partially masked the fall in production and manpower which has taken place in these natural fibres. The widespread introduction of man-made fibres has had a significant effect upon the traditional textile areas. The new ranges and qualities of the synthetics have helped resuscitate the industry and have given a new lease of life to old manufacturing areas. The cotton towns of Lancashire are an example of the residual strength of the industry in the area where specialisation developed to its greatest extent. After fifty years of decline, it now looks as if the recent improvements in machinery including the new ring loom, and the new ranges of synthetic fabrics, have given the area new life. The industry has moved to new locations, often in the areas of peetrochemical production as at Teesside (Billingham and Wilton), and into development areas as at Pontypool, but in the main, it has succeeded in maintaining its heavily regionalised character. Geographical inertia has played a large part in this as the textile regions have always been heavily capitalised in plant, machinery and skilled labour.

The widespread substitution of man-made fibres has had another important effect. Mixtures of fabrics such as polyester - cotton and wool - Terylene are increasingly rendering the traditional divisions within the industry obsolete. The influence of markets has become more important as half of the textile industry's output of cloth goes directly into consumer clothing. Rapid fashion changes have thus dictated a much closer identity of interests between the branches of the industry. The old processing distinctions are breaking down, and spinning, weaving, knitwear and clothing manufacture are increasingly part of one organism, the vertically integrated company. This has involved new capitalisation, rationalisation, amalgamations and the introduction of new production techniques. Most significant of all has been the growing association of textiles with the chemical industry, particularly petrochemicals and synthetic fibres.

The large chemical and petrochemical synthetic fibre groups have become financially involved because they produce raw materials on a large scale for the textile industry. It is therefore natural that they should secure their outlets by controlling the means of production and marketing. This has led to the acquisition of textile companies. Imperial Chemical Industries, by incorporating British Nylon Spinners, and Courtaulds, who absorbed British Celanese in 1957, are now important components of the textile industry in the United Kingdom. In Italy Snia-Viscosa is in association with Montedison, the chemical group; Hoechst of West Germany are entering into arrangements

	Wool fabric	Cotton fabric	Synthetic fibres (Total)
West Germany	30	176	951
France	16	120	199
Belgium/Luxembourg	–	45	221
Netherlands	–	11	
Spain	20	84	313
Portugal	16	82	63
Italy	98	162	680
United Kingdom	25	39	333
Ireland	–	11	74
Denmark	–	–	17
Greece	25	56	27
Total EC	230	786	2878

Figure 5.6 EC textile production 1986 ('000 tonnes) (Source – *Eurostat*)

with textile companies. The EC has successfully negotiated the multi-fibre agreements which go some way towards restricting imports from cheap labour countries. A radical reorganisation of the textile industry has been taking place, particularly in the mature north-west European producers. The response to world competition is rationalisation and integration, the substitution of capital for labour, greater specialisation in up-market products, innovation in fashion and design, flexibility in production and modern marketing techniques. There are signs that during the 1980s the long decline in the textile industry, at least in north-west Europe, has been halted or reversed.

The Chemical Industry

The chemical industry has developed very rapidly in Western Europe during the last thirty years. The West German chemical industry grew by an average of twenty per cent per annum between 1950 and 1970, twice the growth rate for German industry as a whole. It occupies a key role in the complex industrial economy largely because it supplies raw materials upon which industry depends. In the United Kingdom, for example, only about twenty per cent of the products of the chemical industry enter the home consumer market directly, whereas at least 65 per cent are used by other sectors of industry. The textile industry has long depended upon chemical bleaches and dyestuffs, but more recently synthetic fibres from petrochemicals have partially replaced the traditional natural materials and have helped to resuscitate and transform the range and quality of textile products. The footwear industry depends on tanning materials, synthetic resins and rubber, and now increasingly, plastics. Fertilisers and crop production chemicals are another rapidly developing sector of the industry. The heavy chemicals division produces acids and alkalis, and pharmaceuticals (drugs, medicines, cosmetics, photographic goods, soaps and toiletries) are manufactured as high-value specialist, lighter chemicals, more specifically for the consumer market.

This great complexity of products is matched by a variety of locations (fig. 5.7). There are three broad types of location: at a raw material and energy source; at the point of import or trans-shipment of bulky raw materials; and near the market for the product. All three factors may operate at different periods of time.

Raw materials and energy

Coalfield locations

The coalfields provide a major concentration area as coal was initially both a raw material and a source of energy. Although coal is no longer the key source of energy and raw materials, the centres of heavy industry continue to provide a major inertia focus for chemical manufacturing. The major chemical manufacturing regions on the continent are the Ruhr, Saar, Sambre-Meuse, Limburg and Kempenland coalfields. The Heavy Industrial Triangle plays a significant part in the chemical industry. In the Ruhr, the main chemical centres are Duisburg, Düsseldorf, and Leverkusen near Cologne. The coke-oven plants produce heavy chemicals, including coal tar, benzene, ammonia and sulphuric acid. There is also a new development in the north of the Ruhr at Marl-Huls, where natural gas is piped from the Ems gas-field. The Dutch South Limburg coalfield produces synthetic rubber, ammonia, and petrochemicals near Maastricht. In the Sambre-Meuse Valley, Liège is the centre for heavy chemicals, and in northern France the former coalfield towns of Béthune, Lens and Douai produce aniline dyes, ammonia and acids.

Other mineral sources

A specific mineral resource may determine the location of a chemical industry: for example, sulphur extraction from natural gas at Lacq and St. Marcet in the Pyrenees; the potash deposits at Mulhouse used for fertilisers; salt deposits in Lorraine (Dombasle and Sarralbe); and gas at Cortemaggiore in the Po Valley. Perhaps the largest area of this type is in Lower Saxony around Hanover, where oil, potash and salt account for the large-scale manufacture of fertilisers. In Italy there are many dispersed locations, including the processing of sulphur at Ragusa in Sicily, and in Emilia-Romagna, and potash at Campo-Franco. In Spain the Basque coast has considerable non-ferrous metal deposits, and these have given rise to heavy chemicals at Bilbao, whilst Oviedo and Gijon have fertilisers, glass and ceramic factories. The mineral deposits of the Sierra Morena in Southern Spain have given rise to chemical industries at Linares, Seville and Huelva in the Guadalquivir and Rio Tinto valleys.

Hydro-electricity

The availability of hydro-electricity is another localising factor. In Italy there are plants at Terni in the Appenines, Crotone in Calabria, and at Bolzano in the Alto-Adige for nitrate fertilisers. In France an electrochemical industry has developed in the Durance Valley (Argentiere), at Grenoble in the Isère Valley, and in the Pyrenees, south of Lourdes.

Import and transhipment points

Oil refineries and petrochemicals

The influence of cheap transport has often led to the expansion of existing centres which were originally based upon raw materials. Oil refineries have become the principal locational factor for the petrochemical industry; cheap transport by inland waterway or pipeline is also important. In the United Kingdom the expansion of existing centres has occurred where the original

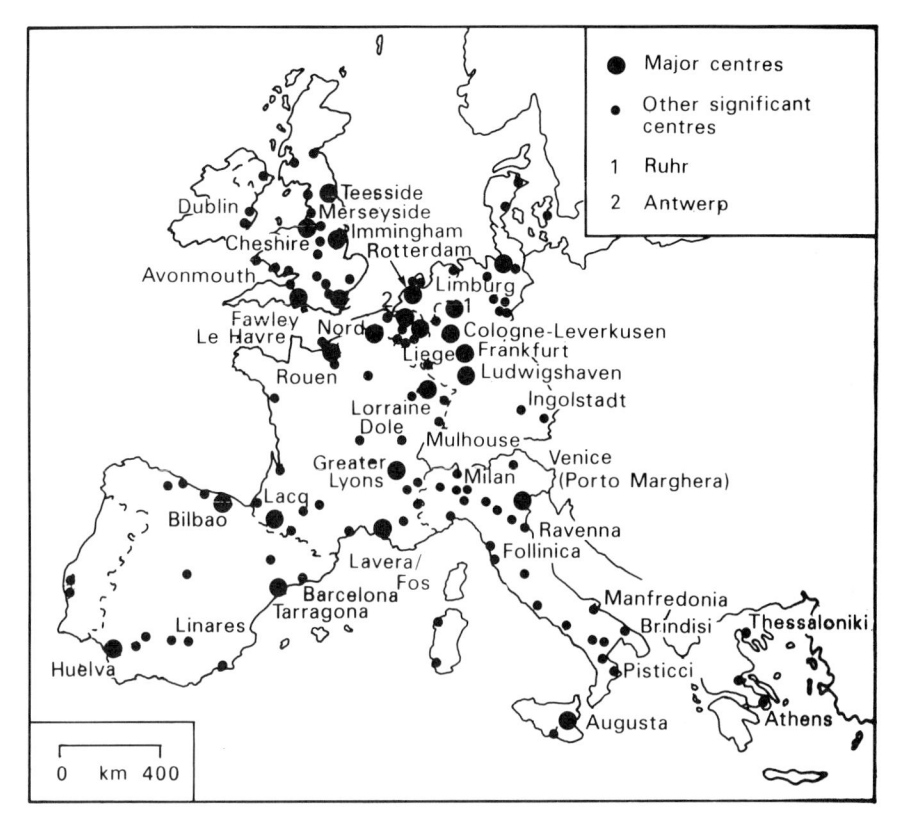

Figure 5.7 EC chemical industry locations

coal factor is complicated by others. The Merseyside chemical area stretches from St. Helens through Runcorn and Widnes into Cheshire at Northwich, and was originally based upon the saltfields of mid-Cheshire and the Lancashire coalfield. The oil refinery at Stanlow, the glass industry of St. Helens, the supplies of bleaches and dyes needed by the textile industry, the cheap import facilities of Liverpool and the Manchester Ship Canal for tropical vegetable oils, as well as limestone from Derbyshire, have compounded matters so that the industry now relates to a whole series of factors.

Teesside is a similar example where the main centres, Billingham and Wilton are linked by a pipeline under the Tees for movement of petroleum by-products. Anhydrite and salt from beneath the Tees estuary and also coal from Durham were the main raw materials at first. Much of the real impetus for the recent vast growth of the Wilton complex, making plastics, Terylene, and other synthetics, however, has been the development of Teesside as a major oil-refining and petrochemical centre.

In many cases transshipment points at deep-water estuaries or along large rivers, have become initial growth points. The oil refineries of the Rhine delta at Rotterdam and Antwerp, produce petrochemicals. The Dutch towns of Arnhem and Nijmegen along the Rhine have chemicals and rubber works. Similar developments have occurred at Marseilles, Thameside, Humberside,

116 THE NEW EUROPE

ICI Ammonia plants, Billingham complex, Teesside

Hamburg, Le Havre and Southampton Water. Italy has developed petrochemical locations at her major ports: Genoa, Naples, Augusta in Sicily, and Bari.

Inland areas

The best example of an inland transportation break-point is the mid-Rhineland, centred upon Frankfurt, Mannheim and Ludwigshaven. The Rhine axis reflects the ease of importation along a major waterway. The complex of Badische-Anilin/Soda-Fabrik AG (BASF) is at Ludwigshaven and is the largest in Europe, employing 45 000 people.

The oil and natural gas pipelines extending from Rotterdam to the Ruhr and Frankfurt, the South European pipeline from Trieste to Ingolstadt, and from Marseilles to the Rhine at Karlsruhe (fig. 2.7) are of increasing importance for the location of chemical factories.

Market Locations

Branches of the chemical industry are widely distributed in the major cities. Here are found the lighter, higher value chemical products which require a good labour supply and consumer market proximity. Paris, Brussels and London are the largest centres with pharmaceuticals and cosmetics. Lyons, Nottingham, Cologne and Manchester, are others. In Italy, the single most important area is the Milan-Turin axis. Milan employs one-third of the total chemical workers in Italy, manufacturing a very wide range of products for the very large consumer market in Northern Italy.

In addition to the consumer market, in many cases the market for associated products is an important factor. Examples are the crop-protection chemicals and fertilisers made at Hanover and Brunswick, close to the intensive agriculture of the Borde of Saxony, and cities such as Ghent, Turin and Lyons which produce dyestuffs for the textile industry.

Structure of the chemical industry

In this rapid growth industry there is scope for extensive research and development, automation, and capital investment. Certain characteristics therefore emerge. It tends to be organised in very large units which are capital-intensive and owned by a few giant companies. There has been a spectacular increase in the size of the factory units, particularly in the field of petrochemicals. In 1960 the ethylene catalytic crackers had an average capacity of 50 000 tonnes, but now Imperial Chemical Industries on Teesside have a 450 000-tonne plant of this type.

Each EC country has at least one major chemical group which exercises a partial monopoly, but the scale of operations does vary considerably. In West Germany, the three largest firms (Hoechst, Bayer and BASF) share most of the industry; in France there is much less concentration with over sixty firms sharing at least half the market; in Italy, Montedison accounts for 75 per cent of the total sales. The United Kingdom is dominated by ICI which ranks as one of the largest companies in the western world and is a multi-national with many subsidiaries in Western Europe and the rest of the world (fig. 3.9). Another feature is the growing influence of the major oil companies such as Shell BP and Esso which exercise control over the raw material sources for petrochemicals. In Italy ENI, the State oil holding company, now has major interests within the chemical industry.

The chemical industry benefited enormously by the dismantling of tariffs and increase in intra-community trade in plastics, synthetic rubber and artificial fibres. European production of chemicals is one-third of total world output, although demand was cut sharply during the oil crisis of 1973. High value added goods are increasingly concentrated in the EC, whilst lower value production is transferred by the multi-nationals to Third World locations. The importance of the industry can be measured by the fact that American investment is very high. Probably a quarter of the total United States investment in Western Europe is in the EC chemical industry. Mergers at community level are now needed to reach the American size-level. Little has been accomplished so far, except in the field of photographic chemicals, by the merger of Agfa of West Germany and Gevaert of Belgium.

6

Agriculture: the Common Agricultural Policy

A highly productive farming region

The European Community is one of the most productive agricultural areas in the world and large parts of its landscape have become almost totally humanised after centuries of continuous cultivation. Sections in the regional chapters which follow are devoted to the variety of farming landscapes and products, for instance: the productivity of the 'pays' of the Paris Basin; the efficiency of the dairy farming and specialised horticulture of Denmark and the Netherlands; the rich croplands of Northern Italy; the Rhineland vineyards; and the subsistence farming of the Mezzogiorno.

Scale and diversity

The largest single advantage of the European Community is its scale. The large market of over 320 million people is a tremendous incentive to farmers, but of even greater significance is latitudinal extent. The European Community stretches from latitude 35 to latitude 58 degrees north and covers an area capable of producing most foodstuffs apart from those which require tropical conditions. France is a major producer of wheat and maize, Denmark of pork and dairy products, the UK and Spain have the largest numbers of sheep, and Italy provides a large proportion of the Community's rice. These are just a few indications of the range of products within these latitudinal limits.

There are, however, great variations between the member states in the importance attached to agriculture (fig. 6.1). In peripheral countries such as Greece, agriculture accounts for over sixteen per cent of GDP and there is a large proportion of the labour force employed in agriculture. In the UK agriculture occupies only 2.8 per cent of the working population, but UK farmers supply about 55 per cent of the country's food. Agriculture is therefore a valuable and very cost-effective part of the UK economy, playing a part out of all proportion to its manpower. In West Germany and Belgium agriculture provides only 1.7 and 2.4 per cent of GDP respectively for, like the UK, they are heavily industrialised and all three are importers of food. Portugal and Italy are also net importers of food. Denmark, Ireland, Greece, Spain and the Netherlands are net exporters of food. France has the largest agricultural production in the Community, emphasising her large areas of farmland (fig. 6.2) and position as 'The Granary of Europe' (fig. 6.3).

	Agriculture as a % of gross value added GDP	Food and agriculture imports as a % of total imports	Food and agriculture exports as a % of total exports	Net importer or exporter of food
West Germany	1.7	11.4	4.8	Importer
France	3.9	11.1	15.3	Exporter
Italy	4.9	13.9	6.4	Importer
Netherlands	4.3	13.1	19.4	Exporter
Belgium/Luxembourg	2.4	10.5	9.8	Importer
UK	1.4	11.8	7.5	Importer
Ireland	10.6	12.6	25.8	Exporter
Denmark	5.5	10.9	27.9	Exporter
Greece	17.1	15.7	27.2	Exporter
Spain	6.0	10.2	14.3	Exporter
Portugal	7.7	11.0	8.2	Importer

Figure 6.1 Agriculture: the importance to each member country, 1985 (Source – *Eurostat*)

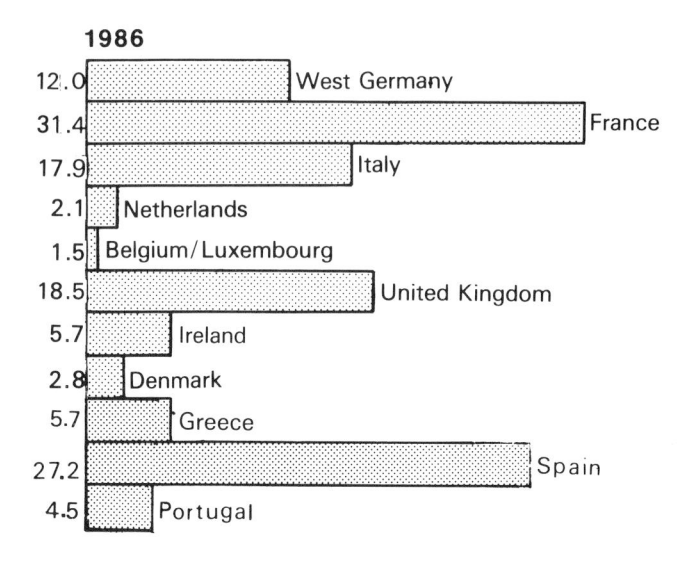

1986

West Germany	12.0
France	31.4
Italy	17.9
Netherlands	2.1
Belgium/Luxembourg	1.5
United Kingdom	18.5
Ireland	5.7
Denmark	2.8
Greece	5.7
Spain	27.2
Portugal	4.5

Figure 6.2 The total agricultural area 1986 (million hectares)

Land utilisation figures (fig. 6.4) show that Denmark and Portugal have the largest proportion of arable land, followed by West Germany, France and

	Crop Production (1000 tonnes)					Livestock numbers (1000)		
	Wheat	Barley	Maize	Rice	Sugar beet	Total cattle	Sheep and goats	Pigs
France	29473	11025	11526	53	28190	20108	11560	12063
West Germany	10165	9784	1177	–	20378	14648	1386	24180
United Kingdom	13643	10273	–	–	8283	11902	26024	7955
Italy	9196	1599	6425	1072	12005	8866	12856	9274
Denmark	2198	5486	–	–	3441	2490	69	9422
Netherlands	974	217	–	–	6999	4912	1024	14063
Belgium/Luxembourg	1290	846	–	–	5867	3146	136	5836
Ireland	507	1564	–	–	1426	5626	2917	980
Greece	2165	713	2034	106	2247	761	17632	1130
Spain	5258	9639	3122	465	7487	4954	20102	15780
Portugal	455	82	560	145	43	–	3800	–

Figure 6.3 Selected figures of agricultural production 1986

Spain. The United Kingdom, Ireland and the Netherlands are predominantly pastoral reflecting their maritime climatic characteristics. The proportion of tree crops (vines and olives) illustrates the characteristic production of the Mediterranean countries. This adds yet another element to the scale, complexity and diversity of the agricultural scene, which will now be examined in greater detail.

Climate and agricultural regions

Climatic and physical factors have the effect of creating four main agricultural zones (fig. 6.5).
1. **North-west Europe** is exposed to westerly winds from the Atlantic, and the normal climatic regime is therefore wet and variable throughout the year, with mild winters and cool summers. This maritime climate is characterised by a mixed farming regime with a bias towards a grassland and stock-rearing economy and specialised dairy farming. This is common on the coasts and lowlands of Ireland and in the United Kingdom, Normandy, Brittany, the polders of the Netherlands, and Schleswig-Holstein in the North German lowlands. Denmark is a special case: only nine per cent of the cultivable land is under grass, although it is within the coastal maritime belt; 45 per cent is under mixed cereals grown as stockfeed. This intensive method of feeding cattle indoors on grain is a more cost-effective way of producing dairy products.

Arable	Permanent pasture	Permanent crops (fruit, olives, vines, etc.)	
54.2	44.7		Belgium/Luxembourg
91.8		7.7	Denmark
60.6	37.8		West Germany
57.3	38.5		France
18.8	81.1		Ireland
52.4	28.3	19.2	Italy
43.5	54.7		Netherlands
37.4	62.2		United Kingdom
50.9	31.2	18.1	Greece
57.5	24.4	18.1	Spain
64.1	16.8	19.0	Portugal

Figure 6.4 Land-use - percentage of agricultural area

Rotation grassland. Friesian cattle, silos and modern farmhouse buildings in Flevoland, Netherlands.

2. **Towards the interior of the continent** the transitional continental climate has colder, though fairly short winters, but has sunnier, hotter summers than the coastlands. With a lower overall rainfall, arable farming is much more important and cereals tend to dominate. This climate is combined with the presence of extensive lowlands, covered with fertile loess (limon), in areas such as the Paris Basin, Flanders, Picardy and the Börde of Westphalia and Saxony, giving rise to the community's grain and sugar-beet producing area. East Anglia belongs to this agricultural type as it has the lowest rainfall in the United Kingdom. The additional advantage of an extensive chalky boulder-clay lowland has made it the United Kingdom's principal area of cereal cultivation. The Basin of Aquitaine has a long growing season and is climatically almost part of the French Midi. The plain of Lombardy is transitional rather than truly mediterranean in climate, and both Aquitaine and Lombardy have a rich and varied pattern of agriculture with market gardens, orchards, vineyards and cattle pastures. More than half the land is, however, under cereals, particularly wheat and maize and, therefore, they merit inclusion as part of the community's major arable farming regions.

Arable land on the Belgian High Plain, viewed from the site of the Battle of Waterloo

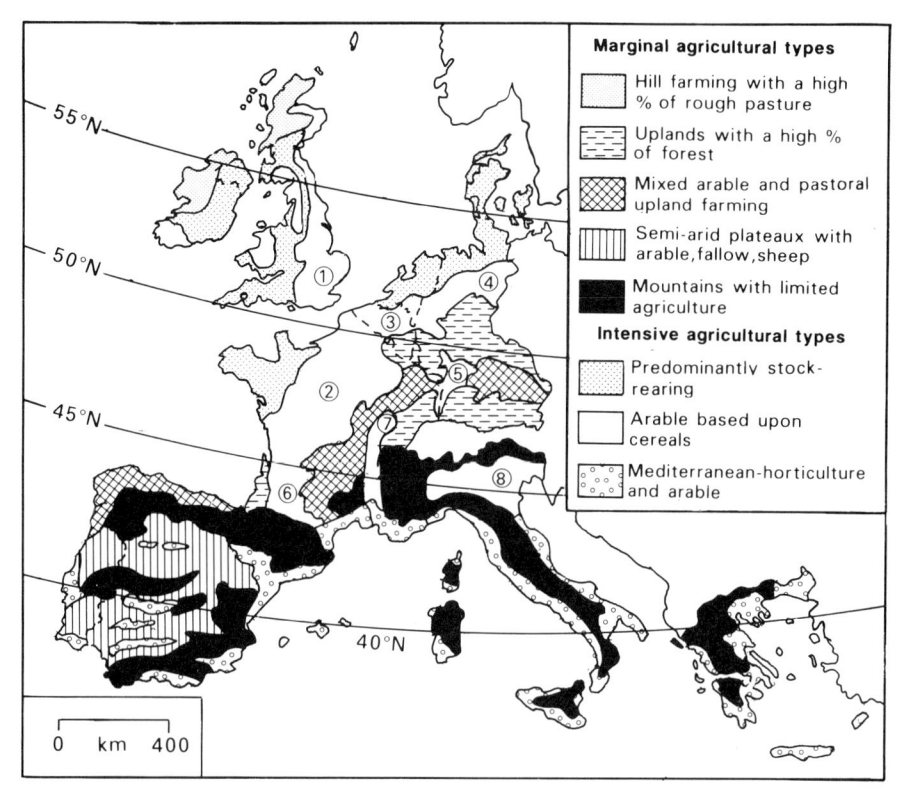

Figure 6.5 Principal agricultural regions (based upon macro-climatic and relief criteria)
(1) East Anglia and the lowlands of eastern England (2) Paris Basin and scarplands
(3) Flanders and High plain of Belgium (4) Börde of Westphalia and Saxony (5) Rhine-Main valley (6) Basin of Aquitaine (7) Rhône-Saone valley (8) Plain of Lombardy

3. **The dissected plateaux and Alpine mountain zones** stretch across much of the interior of Europe. Agriculturally, these are marginal farming areas lying at an altitude of between 300 and 2000 metres and characterised by a cool damp climate, considerable rainfall and winter snow, together with exposed conditions and thin soils. The hill-farming and stock-breeding of the Pennines, Lake District and Welsh mountains, is a valuable element in the UK stock-rearing economy. The Ardennes, Black Forest and Rhine Highlands have extensive forests which provide valuable timber. There are large areas of pasture and moorland with low rural population densities. The French Massif Central is more varied with areas of forest, interspersed with areas of rye, oats and buckwheat cultivation, whilst on the south-western margins are the limestone 'Causses' which traditionally have provided grazing for sheep, producing the famous Roquefort cheese. The Iberian peninsula of Spain and Portugal is the most extensive semi-arid zone in Western Europe with large areas of mountain and plateau. One-third of Spain is rough grazing or waste, with large parts of the Meseta plateau and the central Sierras given over to sheep-grazing and goats. Cattle rearing is more common in the humid west and north, in

Galicia and the Cantabrian mountains.

The Alpine fold mountains provide another variation on this theme. The high western regions, with heavy precipitation, have fine stands of timber, but the main form of livelihood is usually stock-rearing based upon the alternate use of alpine and valley pastures in the classic transhumance system. In sheltered valleys, as for example north of Grenoble in the Isère valley (the Gresivaudan) are vineyards and orchards, whilst in more remote areas subsistence farming and depopulation is the usual pattern. A large part of the Italian Appenines are basically suited to tree crops such as the olive and vine, but forty per cent of peninsular Italy is too steep for cultivation anyway, and soil erosion in the past has seriously damaged its capacity for agriculture of any kind.

4. **The southern coastal fringes of the community**, including the Midi of France, the huertas of southern Spain, the Greek coastal plains and the Italian lowlands, have the traditional summer drought of the 'mediterranean' regime, but areas of lowland are very limited, and usually protected by high backing mountains.

The agriculture varies so much with local conditions that the classic mediterranean complex of olives, vine and cereals is only a very basic picture, and increasingly intensification is occurring where irrigation is available. The region of Northern Italy (chapter 16) which stretches from the Ligurian Riviera into Lombardy, illustrates this variation. Intensive horticulture, viticulture, fruit and floriculture is common on irrigated and terraced coastlands, with intensive crop and cereal production and cattle rearing in the interiors.

Local variations due to micro-climatic effects, aspect and market demand

Whilst the description of these four basic climatic divisions and their agricultural responses are useful in describing the latitudinal range of agricultural production available in the community, nevertheless there is a mosaic of diverse farm types within this broad picture. Specialised agriculture depends upon locally favourable circumstances and is found in restricted areas (fig. 6.6).

Viticulture

Although the large-scale areas of viticulture lie in the Mediterranean regions including Languedoc, southern Spain, Provence and much of Italy, nevertheless there are locally favoured environments which allow the vine to flourish much further north.

The northern limit of the vine lies approximately from the river Loire to Koblenz on the Rhine. The sheltered slopes of the Rhine Gorge and Rift Valley, the warm soils and southward facing scarps of the Champagne Pouilleuse near Reims, and the southward-facing Cote d'Or in Burgundy are included in these specialised high-quality wine-producing areas.

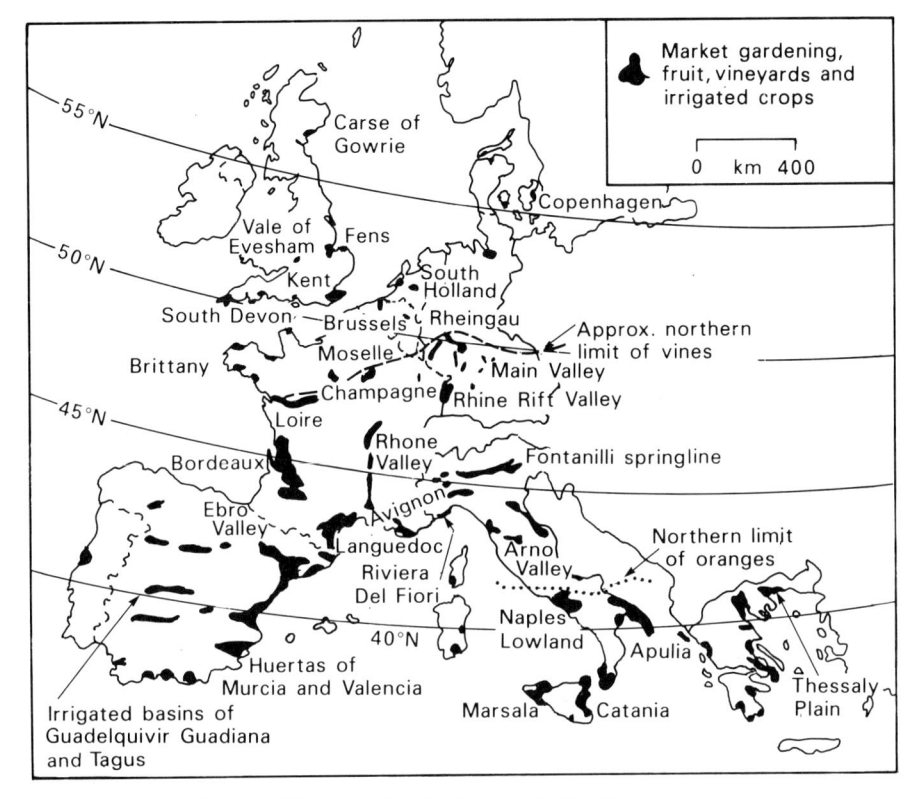

Figure 6.6 Specialised and intensive horticulture and viticulture

The Vale of Evesham

The Vale of Evesham in south Worcestershire experiences a warming effect with mild winters and early springs because of the funnelling effect of the Bristol channel upon the westerly winds. As a result it is an important area in the UK for the intensive cultivation of fruit and vegetables.

Brittany

In Brittany the cultivation of primeurs (early vegetables) is made possible around the coast in sheltered bays such as St. Malo, Roscoff and Quimper. mildness of the winters and early springs are associated with the Westerlies and North Atlantic Drift.

The Netherlands

Much of the intensive horticulture of Randstad, Holland, with its vegetables, glasshouses, and bulb cultivation, is on mixed soils where sand has been blown inland from coastal dunes over the peat to create a fertile, easily worked soil. The nearby urban markets in the Netherlands itself, West Germany and the UK have also been a major stimulation.

Urban markets and transport

In addition, economic factors have become increasingly predominant. The food requirements of the large city populations have created 'Von Thunen' type conditions, and land is intensively farmed immediately around the city. The perishable and highly-priced fruit and vegetables are freed from transport costs of any magnitude and are in close proximity to their urban market. The environs of Paris, London, Hamburg and the Randstad illustrate this. A modification of this occurs where transport provides an easy and inexpensive route to the market. The cultivation of primeurs (early fruit and vegetables) in the Rhône delta near Avignon is not only to early springs and intensive irrigation, but also to the development of fast access routes to Paris.

Self-sufficiency

The diversity of land-use across the member countries gives a picture of considerable integration. The Common Market has combined twelve countries into a unit capable of producing the majority of its own food. These countries, together, have a broad self-sufficiency in food, adding great strength to each member country (fig. 6.7). For example, the grain-producing area of East Anglia is not sufficient to feed the UK's large population, but the European Community as a whole can produce enough grain. The large areas of agricultural land and the range of climate types stretching from northern Scotland to Sicily and Malaga, make the European Community product greater than the sum of its parts. The law of comparative economic advantage operates so that the UK as a traditional grass and livestock region is a producer of beef, pigmeat, and dairy produce. The Mediterranean areas such as Spain, Greece and Italy (fig. 6.8) have a climatic monopoly within the

Vineyards of the Côte D'Or near Nuits St. Georges, south of Dijon

The Westland glasshouse horticultural region, near Rotterdam

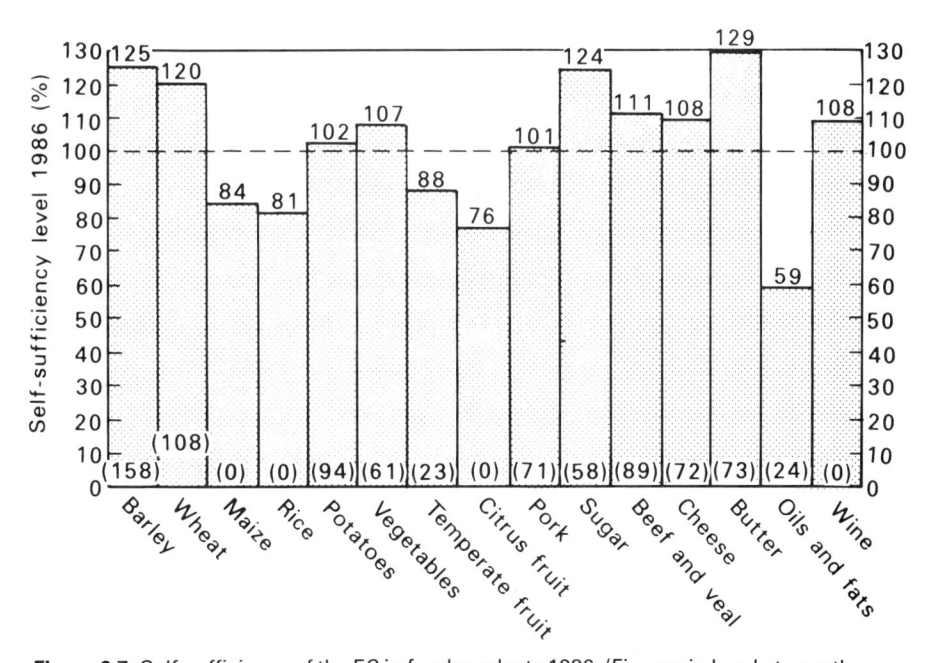

Figure 6.7 Self-sufficiency of the EC in food products 1986. (Figures in brackets are the comparative levels for the UK alone)

128 THE NEW EUROPE

European Community in the production of sub-tropical crops such as vines, olive oil, rice and citrus fruits, and advantages in tomatoes and early vegetables. Denmark is a major supplier of dairy produce. The potential self-sufficiency of this large and productive area is the key to an understanding of the Common Agricultural Policy and is the reason for the early importance attached by the Community to a policy for control and restructuring of farming and food production.

Farm structure and land tenure: the traditional picture

There are very large differences in the cultural traditions, social systems and farming techniques of the member states, and it is these variations which illustrate many of the problems inherent in the Community's farm policy.

Land tenure and farm size varies considerably. At one end of the scale large parts of Spain and the south of Italy have tiny peasant holdings (minifundia) alternating with vast estates owned by absentee landlords

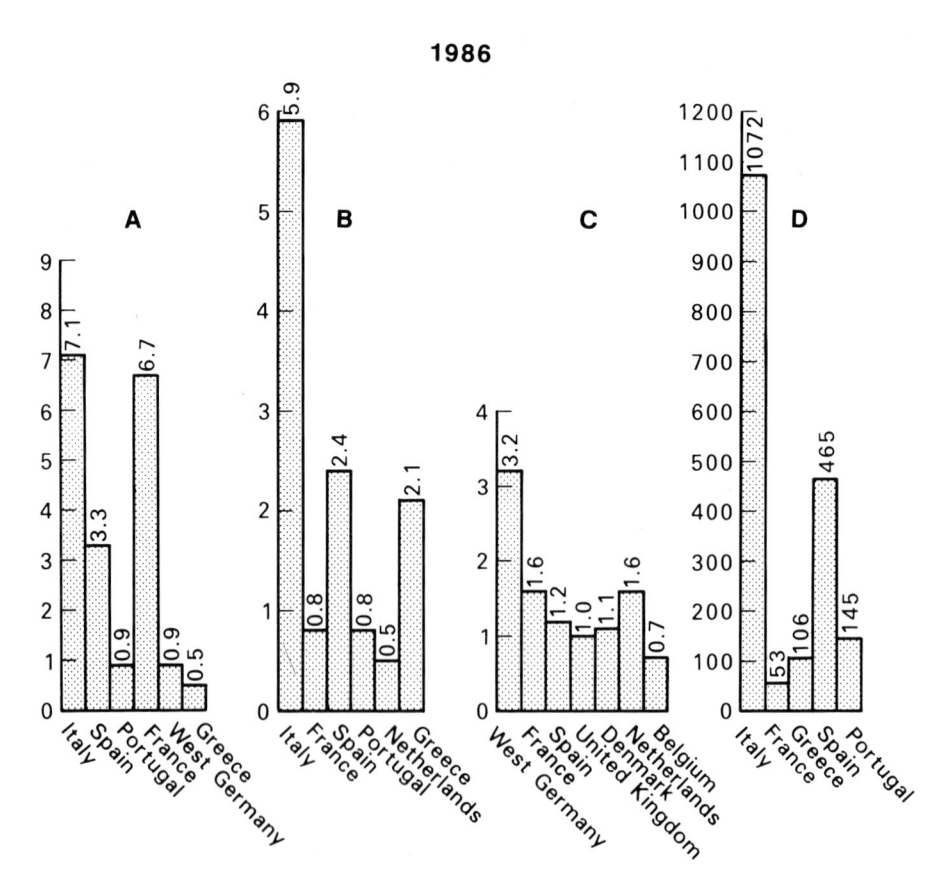

Figure 6.8 Production of selected specialised crops, 1986 **(A)** Wine (million tonnes) **(B)** Tomatoes (million tonnes) **(C)** Pork, ham and bacon (million tonnes) **(D)** Rice (1000 tonnes)

(latifundia). The underdeveloped latifundia, with their typical extensive monoculture of wheat and day-labour system, exist alongside peasant smallholdings of under one hectare on two or three widely separated patches of land. Even today about 70 per cent of the total agricultural holdings of Italy and Greece are under five hectares in size (fig. 6.9(b)). Of course, particularly in the traditional economies of the mediterranean countries, the peasant farm is the very significant supporter of the family unit in rural areas. In the modernised and wealthier farming areas of Lombardy a normal system of tenant occupation on farms nearer the average size occurs. France, on the other hand, has a tradition of owner-occupation and the average farm size is twenty-five hectares, but there are real differences between regions. In the north-east most farms are over thirty hectares, whilst Brittany, Aquitaine and much of the Midi have a majority of farms under twelve hectares in size. The tendency to small farms, which are difficult to work for profit, is compounded by the system of widely scattered holdings which are a remnant of the medieval open-field system and a legacy of the European code of equal inheritance. West Germany is also characterised by a proportion of small farms, and in Bavaria and Swabia particularly these are often fragmented. An additional complicating factor is that many farms combine part-time agriculture with commuting or with tourism. In the Netherlands, however, many of the farms on the reclaimed polders are state-owned and rented out to the farmer in consolidated plots of land. By contrast, the UK has a more mature farm structure with the average farm size at 65 hectares and, more significantly, it has the greatest number of large farms over 400 hectares in the

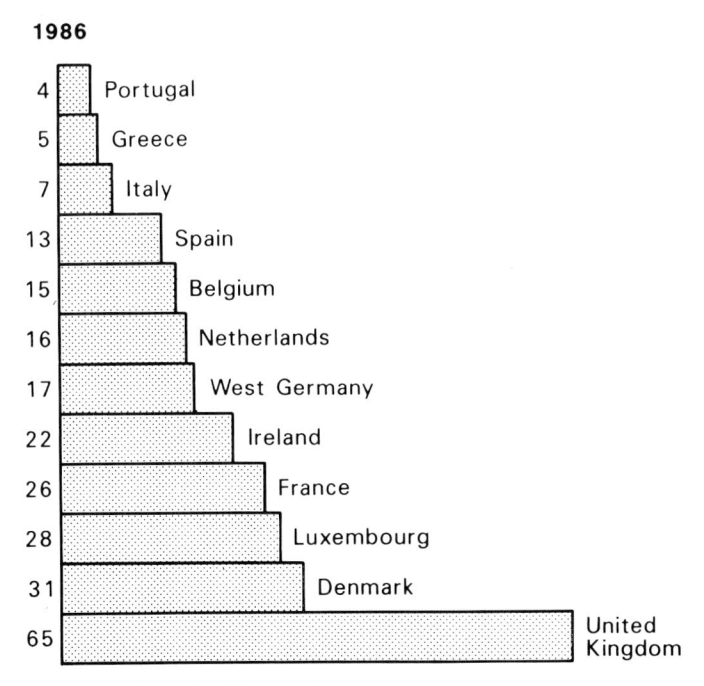

1986

Hectares	Country
4	Portugal
5	Greece
7	Italy
13	Spain
15	Belgium
16	Netherlands
17	West Germany
22	Ireland
26	France
28	Luxembourg
31	Denmark
65	United Kingdom

Figure 6.9(a) Average farm size (Hectares)

130 THE NEW EUROPE

Size groups (ha)	Denmark	West Germany	Greece	France	Italy	Netherlands	United Kingdom	EC Total
1 < 5	11	32	71	21	68	24	12	47
5 < 10	18	19	21	15	17	20	12	17
10 < 20	27	23	7	21	8	30	16	15
20 < 50	34	22	2	30	4	24	27	15
≥ 50	10	4	0	13	2	3	33	6

Figure 6.9(b) Percentages of agricultural holdings by size groups (selected)

Community, often run as a company by a manager (fig. 6.9(a)). Again, however, major regional contrasts are shown by the predominance of small family mixed farms in the north and west, such as the Pennines and central Wales, and the large cereal farms of eastern England owned as capitalist estates by finance houses or insurance companies.

Many areas of the European Community had too large a farm population for efficiency. During the 1950s the original Six had a very significant problem with Italy having 6.5 million people or 41 per cent of her labour force on the land (fig. 6.10). This was a remarkable picture of regional contrasts in a rapidly developing industrial country which yet retained a large rural peasant economy. Even the Netherlands, a much more efficient agricultural country, had fourteen per cent. By contrast the United Kingdom had five per cent of its working population employed in farming. Many of the older peasant farmers of France and Italy had an innate conservatism and lack of enthusiasm for change, and had difficulty raising and investing the capital required to mechanise efficiently.

The tremendous variations in yields from one part of the European Community to another have been a major problem (figs. 6.11). Traditionally, the Netherlands had a consistently high yield of commodities such as wheat, a reflection of intensive efficient farming, and of orientation to the food requirements of large urban populations. Denmark developed the most efficient farming systems in the world and the United Kingdom's yields, particularly in its favoured lowlands areas, are very high. By contrast the yields for major crops in France and Italy reached only half the levels of those in the Netherlands. A good indication of productivity levels is made by comparing the British farmer who produces food for twenty people; the Dane for seventeen; the Frenchman and the German farmer for nine people. A model summarising these regional differences which also introduces the concept of the core and the periphery, is that taken from the ideas of Van Valkenburg (fig. 6.13). An index of high production is based on the following criteria; average production values for arable land; milk yields; and yields for eight selected crops. Figure 6.13 shows a high production area across the north European plain focussed upon the Rhine delta and adjacent areas. Conversely, there are peripheral regions in which agriculture is much less favoured, including some of the Mediterranean region, south-western France and the Atlantic fringes.

The problems of size and fragmentation, together with an excessive farm labour force, inadequate mechanisation, backward farming techniques and low yields, were widespread, causing considerable imbalance in the rural sector. When the European Community was formed in 1958 half its farmland was in need of consolidation. It was to take advantage of the benefits of scale to produce sufficient food and to improve the structure of farming throughout the whole community, that the Common Agricultural Policy was formulated.

The Common Agricultural Policy

Protectionist policies in agriculture are not new: most governments in Western Europe have always protected their farmers for a range of social, economic, strategic and political reasons. Farming tends to be high cost, and protection of the home producer against cheaper world competition, thus maintaining a certain level of self-sufficiency, has always been deemed to be necessary. Structural reorganisation and land reform have also been a long-term aim of governments who have also felt it politically necessary to retain the support and votes of the rural community. World markets for agricultural products are notoriously unstable and unpredictable, depending upon the weather and fluctuations in supply and demand. In addition, rural farm incomes are traditionally lower than those in industry and the service sector, and farming is the dominant occupation in remote and physically difficult regions, which therefore need considerable assistance.

One of the earliest examples of Community policy-making was the establishment of the Common Agricultural Policy in 1962. From the end of World War II there was a paramount need for the security of food supply to be maintained and the original six in the European Community saw the CAP as a means of achieving early success in integration policies. France and Germany saw it as a 'quid-pro-quo' for French food surpluses set against German industrial products in the new common market. The CAP is a system of protection in which the European Community food market is regulated. Its objectives, defined in Article 39 of the Treaty of Rome, are to increase agricultural productivity, to ensure a fair standard of living for those people in the agricultural sector, to stabilise markets in a high risk business, and to ensure reasonable consumer prices. It thus establishes a zone of Community preference in one of the world's major food-producing areas, in which its farmers and growers can produce ninety per cent of the food needed by the 320 million consumers.

The policy is administered by the European Agricultural Guidance and Guarantee Fund (EAGGF), the words of which spell out its two major practical functions: the guidance and improvement of farm production and the guarantee or protection of the farmer's income by means of a common price support system. The central aim of the policy is to establish a single market for all farm products. This means that national, fiscal and physical barriers to trade must be abolished, with unrestricted intra-community free trade. There must be stabilised and guaranteed prices at the same level throughout. Easy access to all parts of the European Community by fast transport routes would thus reduce to a minimum the costs of transport. This ideally means success for the most efficient producer and the effect of favourable environmental factors should make itself increasingly felt. Under the law of comparative economic advantage this should mean that regional specialisation in agriculture will be the norm. United Kingdom farmers would specialise in dairy products, lamb, beef, pork and poultry as well as cereals, because the combination of arable and grassland and a livestock economy is particularly fitted for the prevailing climatic and physical conditions. Conversely, Italy and southern France have great climatic advantages in horticulture and a monopoly in vines, olive oil and citrus fruits. Thus, each region produces at its own level of advantage, giving regional interdependence in farm products. The potential self-sufficiency which has already been mentioned for the nine countries is enormously strengthened by this specialisation in each region.

During the early years of the European Community, there is no doubt that the CAP was a potent force and a success story for an integration policy. Protection between the member states was swept away and the basic features of a single market in farm products was established. With high financial returns to farmers under the common price support system, production increased substantially through the 1960s.

The price support system

The protection of the farmer's income

The protection of the farmer's income has always been one of the major purposes of the Common Agricultural Policy. It is recognised that it is more difficult to maintain farm incomes at a level comparable with those of the service and industrial sector. The reason lies partially with the composition of the original Six and the fact that France and Italy in particular have always had a large and electorally powerful farming section in their population (fig. 6.10). France and Italy relied upon the CAP to act as a compensatory factor in return for opening their markets to West German industrial products. The protection of the farmer's income, the guarantee policy, works in the following way. A common pricing system is applied to each product and a target price is fixed each year which estimates a fair return for an efficient farmer in open competition. Every farmer can therefore sell on the open market for a price related to the target price, but if the price slumps there is an intervention price at which the local intervention organisation, run by the Community, will step in and buy the product. This intervention price is about eight per cent below market price and is a type of 'floor' or reserve price. In addition, if a foreign producer wishes to sell food within the market, he has to pay a levy at the frontier which raises his prices to the level of the target price, thus excluding any low cost imports from undercutting the Community farmers' products. So the farmers' incomes are protected at an artificial level, the target price, but there is one safeguard for the consumer. If the market price for a product goes over the target price competition from imports will bring it down to the target level again, thus providing a measure of price regulation.

To pay for this price support, the member countries pay to the Fund the levies which they have received from food imports, customs duty receipts and a proportion of their receipts from value added tax. In effect, the indirect taxes paid by the consumer are the source of the farmers' guaranteed protection.

Agricultural guidance and reform

The other positive function of the Common Agricultural Policy is guidance and reform.

Since 1950 there has generally been a dramatic improvement in the agricultural structure. Several factors are involved and it is difficult to identify relative importance. Initially, there can be no doubt that the stimulating effects of the large internal market had a major effect, particularly upon France, whose export of agricultural produce doubled between 1966 and 1971. Two-thirds of these exports are to member countries. Equally, the policies of national governments have had a considerable effect. The 'Cassa per il Mezzogiorno', the comprehensive plan designed to aid the Italian south with industrial development has also carried out substantial measures of farm

Packing Danish blue cheese at a dairy near Copenhagen

improvement and modernisation. The French 'Remembrement' policy was officially introduced as early as 1941, with the object of consolidating severely fragmented farmland which covered much of the country. It has been only partially successful, largely in areas like the Paris Basin, with its open-field system. In the more traditionalist west and south, such as Brittany, Aquitaine and the Massif Central, where it was needed much more, there was considerable resistance to change. Official agencies called SAFER and FASASA were set up in 1960 and 1962. These buy land which is used to consolidate fragmented holdings and to enlarge smallholdings into viable units. They also encourage farmers in western France to retire at age 60, and retrain men leaving the land for other employment. Finally, a national network of agricultural markets was set up to improve distribution throughout France. In addition a third very significant factor has been rural-urban migration, particularly in France, West Germany and Italy. During the 1960s this massive rural exodus was a major factor in slimming the agricultural workforce. Some mountain areas of marginal farming like the Jura, Vosges and Massif Central are in danger of becoming depopulated. French government policy changed to promote a balanced agriculture with different stimuli, depending on regional needs. Hill farmers are now given bonuses to increase their cattle herds, and grants are made to young farmers who will settle in areas with population below the minimum desirable level (11 people per square kilometre). On the other hand, in areas like Brittany where there are

Co-operative wine producers in Provence

Percentage of labour force employed in agriculture

	1950	1958	1965	1971	1972	1974	1976	1981	1986
Belgium	11.3	9.0	6.1	4.4	4.2	3.7	3.4	3.0	2.9
France	28.3	23.3	17.0	13.2	12.9	12.0	10.9	8.6	7.3
West Germany	24.7	15.0	11.0	8.3	7.8	7.3	7.1	5.9	5.3
Italy	41.0	33.0	24.7	18.9	18.2	16.6	15.5	13.3	10.9
Luxembourg	24.0	17.0	13.5	10.1	9.3	6.6	6.1	5.6	4.0
Netherlands	14.1	12.0	8.0	6.9	6.9	6.6	6.5	5.0	4.8
United Kingdom	5.1	4.0	3.2	3.0	3.0	2.8	2.7	2.8	2.6
Denmark	–	22.0	–	10.9	9.8	9.6	9.3	8.5	6.2
Ireland	–	38.0	–	26.9	25.7	24.3	23.8	19.2	15.8
Greece								30.3	28.5
Spain									16.1
Portugal									21.9

Figure 6.10 Agricultural labour force 1986

too many farmers, there are inducements to retire or amalgamate. There is room for the small family farm giving a living to three or four people. The large, highly mechanised grain farms of the Paris Basin needs entirely different conditions from those required by the smaller, highly efficient dairy farm of fifty cattle, typical of Holland and Denmark. In addition funds have been allocated by the European Investment Bank and the EAGGF for dairy processing, fruit and vegetable processing plant, and retirement pensions for older farmers. The combined effects of economic forces, national policies and the guidance policies of the Common Agricultural Policy have seen a transformation in the farming structure over the past twenty years.

The percentage of the population employed on the land has been substantially reduced (fig. 6.10). The total labour force fell from 19 million in 1950 to 11.5 million in 1966, and to 7.3 million in 1981, although differences are still marked. In terms of productivity, cereal yields have doubled (fig. 6.11). Large increases in mechanisation and the use of fertilisers are also recorded (fig. 6.12). Irrigation, land drainage and intensification of production techniques, with new capital investment in livestock breeds, crop strains and fertilisers, have resulted in a marked reduction in the amount of land and labour needed to produce a given amount of food.

Problems for the CAP

Since 1968 a whole range of problems have emerged for the CAP.

Surpluses

The price support system, by giving high financial returns to efficient and inefficient farmers alike, gave them a continual impetus to increase output, and raised production to surplus levels by the late 1960s. Dairy products in particular are very prone to these, and the intervention cost of buying and storing large quantities of food, became very substantial. The 'butter mountain' was the first major sign of this, but has been followed by 'cereal mountains' and 'wine lakes'. The perception of the CAP became one of a high-cost and wasteful system. The problem is essentially that the regulated commodity price which protects a small-holder in the Massif Central, encourages efficient farmers in the Paris Basin to raise output and profits, thus creating a spiralling problem. The Mansholt plan, published in 1969 by the then Commissioner for agriculture, was the first attempt to solve these problems. It recommended that in the medium-term up to 1980 up to five million workers should be withdrawn from farming and either receive pensions or be retrained for other employment. Also, five million hectares of marginal land were to be taken out of agriculture and used for afforestation, leisure parks and nature reserves. The aim was to further cut the labour force, eliminate the marginal high-cost producer, raise efficiency, lower target prices, and eliminate surpluses. Its real significance was to indicate that the CAP needed to evolve into a social and economic policy for poorer rural regions. However, Dr. Mansholt was really ahead of his time, and very little was done to control the spiralling costs of farm support.

The CAP and Trade Agreements

The Community is substantially self-sufficient in many major foodstuffs (fig. 6.7). Nevertheless, imports of food have increased substantially owing to a considerable rise in living standards. Tropical fruit and vegetable oils are the

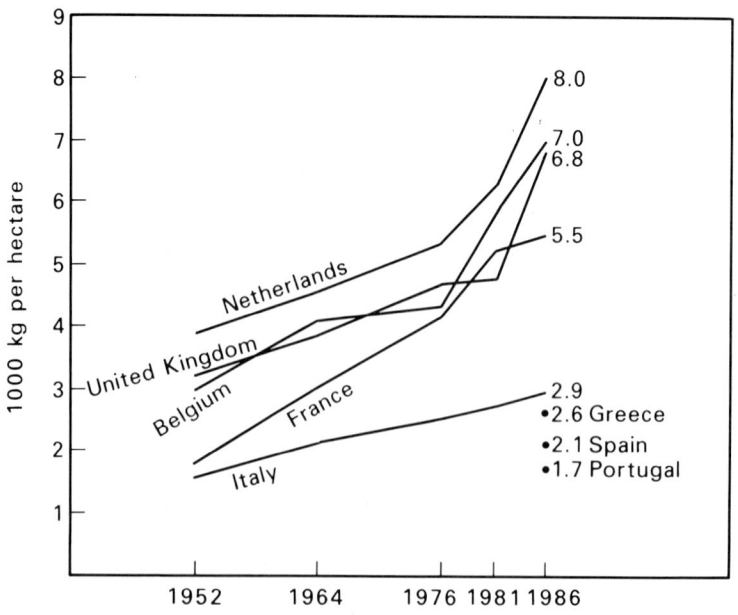

Figure 6.11 Wheat yields

	Use of fertilisers (kg per hectare)		Tractors per 100 hectares	Use of fertilisers: regional variations
	1965	1986	1986	
France	91	181	5.1	Over 200 kg per hectare in the Paris Basin
Netherlands	239	346	8.5	Over 450 kg per hectare in the market gardening areas
Italy	49	111	5.3	
Spain		58	2.3	

Figure 6.12 Use of fertilisers and machinery

main deficiencies. Cereal imports, particularly, have soared since 1958 because of their use as feed-stocks for animals. The position is complicated further by the export of soft wheat, of which the European Community has a surplus, and the larger quantity of imports of hard wheat. The protective nature of the CAP does cause disagreements with countries such as the USA, over the different standards of imported goods, on the 'dumping' of surplus products for a cheap sale, where these affect the internal producers. There is periodic raising of tariffs or exclusive orders. It even happens within the European Community, as for example, in the 'lamb war' when France refused to accept UK lamb imports in an attempt to protect her sheep farmers.

A logical extension of the Common Agricultural Policy is the augmentation of the Community's temperate products with those tropical foodstuffs which it cannot produce for climatic reasons. Many ex-colonies, particularly in Africa and the Caribbean are associated with the European Community. In 1963 eighteen African states, most of them ex-colonies of France, became associates at the Yaoundé Convention. Since then, however, many of the Great Britain's former colonies have become associated, culminating in the Lomé Agreements of 1975 and 1984. The associates obtain tariff-free entry into the Community for their staple commodities, and they also obtain aid for development projects through the European Development Fund. The Community has an assured supply of primary products and tropical food-stuffs, and a preferential market for its manufactured goods (Ch.7 Trade).

The United Kingdom and the 1973 enlargement

Within the Community a clash of interests developed with regard to the agricultural policy and there are many underlying tensions. West Germany and the United Kingdom are heavily industrialised and urbanised, and population has outstripped food production since the nineteenth century. They are thus large-scale importers of food, and by means of import levies, they both make large contributions to the Common Agricultural Policy funds. The United Kingdom, in particular, because of her very efficient farming system, obtains few development grants from the Common Agricultural Policy for structural reforms. France and Italy on the other hand are large-scale food producers and France is a large food exporter. Ireland, though a small volume producer, has a small population and is therefore a major exporter of dairy produce. These last three members have acute structural problems and gain immensely from the guaranteed prices, export subsidies, large internal market and grant aid implicit in the Common Agricultural Policy. The Netherlands and Denmark are two members whose agricultural exports exceed imports and this reflects their traditional specialisation and efficiency in dairying and horticulture, combined with a small population.

The United Kingdom, in general, finds the Common Agricultural Policy unfavourable to her own interests. The circumstances of her unique position with early industrialisation, limited land resources and a large urbanised population led to the adoption of a cheap food policy from the 1840s onwards. Under a policy of free trade, cheap imports from anywhere in the world were allowed to enter the country and most of these were from territories originally under British control. United Kingdom farmers (high cost producers, relative to those of Canada and New Zealand) were compensated with deficiency payments which made up the difference between their high costs and prices, and the price levels of the lower cost imports. The farmer stayed in business with a subsidy, the imports entered freely at a low price level, the tax-payer paid the difference, but the consumer enjoyed very low retail prices for food. This was a fine arrangement for a country with a small farming population, needing essential food imports and also controlling vast agricultural areas of the world. This situation was transformed during the Second World War and since. The need to increase production and self-sufficiency during the wartime period 1939 and 1945, the current 'agricultural revolution', and the acceptance that governments have an obligation to manage and maintain an efficient national food production machine, had led to a much greater contribution by British farmers to an overall national self-sufficiency rate of 55 to 60 per cent and to virtual self-sufficiency in eggs, milk, potatoes, barley, pork and poultry.

The United Kingdom has abandoned her former position as a free-trade importer from the Commonwealth and the rest of the world, and entered one of European protection. The United Kingdom farmer is generally in a very good position. With guaranteed markets and prices his efficiency has meant a great stimulation of production upon integration into the large European market which is aiming for substantial self-sufficiency. Specialised groups such as the horticulturalists have problems because of climatic disadvantages, but in general the United Kingdom farmer has experienced increased demand for, and high production of cereals and livestock products. However, there is the problem of late entry to a system designed for different needs. The ultimate aim of the Common Agricultural Policy is that the agricultural strength of the whole shall be the basis for the specialisation and reform of all the parts, and that the integration of twelve countries will create a strong food and agriculture system which can supply all the needs of 320 million people. However, for the United Kingdom at present, the advantages are finely balanced. The consumer is within a managed price system where prices are high, but supplies of food are plentiful, stable and of good quality. The UK obtains only small benefits from the guidance fund because her agriculture, with the exception of hill-farming, needs little assistance or restructuring. The contributions which the UK pays into the budget because she is a major importer of food have been the cause of continuing tensions and negotiation during the 1980s.

Inflation and currency problems

The situation has deteriorated since 1973. The Mansholt reform proposals for medium-term improvements were overtaken by events, with dairy products, cereals and sugar in structural surplus, and the guaranteed prices to farmers out of control. Rapid inflation since 1973 was largely responsible for the increasing cost of the Common Agricultural Policy price support mechanism. The second problem relates to the accounting mechanism of the CAP. The common price system required a set of internal exchange rates to convert national currencies to a common denominator. These were 'Green Currencies' or agricultural money, and examples are the 'Green Pound' and 'Green Franc'. The ECU (European Currency Unit) has developed from this. With the currency chaos of the 1970s and floating exchange rates, the national currencies have been revalued. The periodic changes in value of the pound, French franc and the German mark have not been accompanied by similar changes in the value of the 'green rates'. The 'Green pound' has therefore been different in value from the real pound and the 'common prices' throughout the Community diverged considerably. As a result, the Community had to introduce a system of MCAs (Monetary Compensation Amounts) which offset the difference between the 'green rate' and the real exchange rate. For the future operation of the CAP the 'green currencies' must be part of a stabilised and adjustable exchange system. The introduction of the European Monetary System (EMS) is likely to provide the basis for this future stability.

The Mediterranean enlargement

The agricultural problem regions, Italy and south-western France, continue to have physical difficulties, fragmented holdings, remoteness, inadequate infrastructure and low productivity. Their special needs were considered in the Mediterranean aid package of 1978, with a range of measures for irrigation, advisory and cooperative services and upland forestry measures.

However, since the entry of Greece in 1981 and Spain and Portugal in 1986, the agricultural situation has deteriorated again. Community agricultural output has been raised by one quarter, and the agricultural workforce has now risen from 7.3 million in 1981 to 10 million in 1986. Their mediterranean products add to existing imbalances in wine, fruit, olive oil and tomatoes.

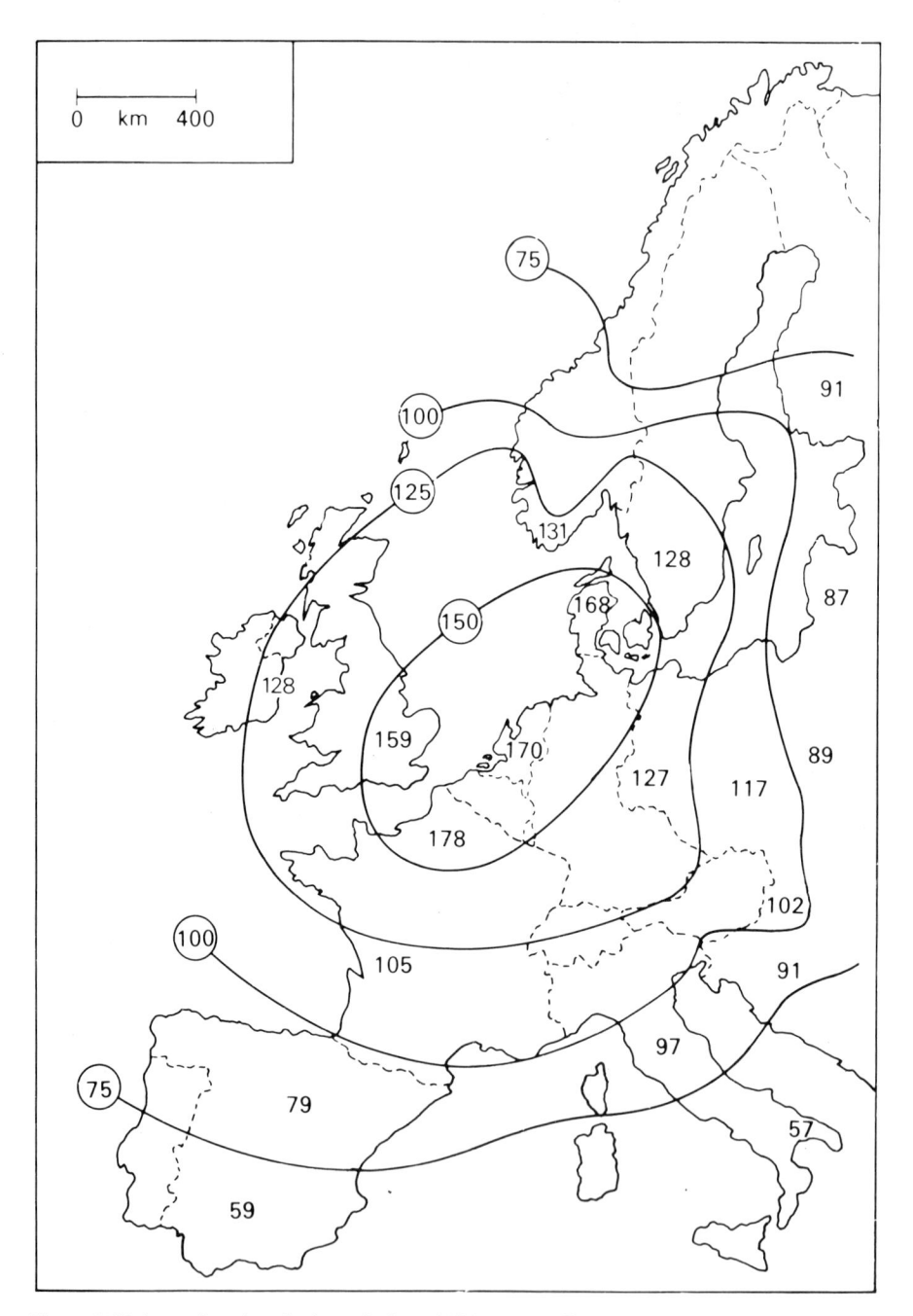

Figure 6.13 Intensity of agriculture. Index of 100 average European yield per acre (Source – Van Valkenburg)

Conflict has arisen with existing producers in France and Italy, adding to political tensions. Structural problems are enormous, with little or no land reform having taken place in Spain and Portugal. The economic divide between rich and poor farmers is now even more extreme than before.

Since 1983 the Integrated Mediterranean Programmes (IMP) have recognised the problem by coordinating aid to the whole mediterranean region from the EAGGF, and the Social and Regional funds for agricultural modernisation and diversification.

Conclusion

Contemporary Issues

The problems of the 1980s and 1990s are distinctly different from those of the 1950s, and yet the CAP has only just begun to adapt to changing circumstances.

The CAP still takes over 60 per cent of the Community budget, and these considerable resources need to be transferred to other funds which take account of industrial decline and urban problems. Food production is in structural surplus in many sectors and the intervention and price support system has tended to continue these costly surpluses. However, there are major regional imbalances between the Community's farming regions. The CAP illustrates some inherent contradictions. It is basically a strategic food policy. Whilst its economics may easily be questioned, it is increasingly taking on the role of a social and regional policy.

Alongside this have emerged a range of crucial environmental questions. Damage to delicate eco-systems through use of fertilisers, irrigation and land drainage, and by changes in farm structure leading to the cereal mega-farms with the attendant dangers of soil-erosion, merely illustrate the dramatic changes which have taken place. It is now questioned as to whether the medium term benefits of expanded food production, at high cost, are worth the long-term costs to the environment.

Clear contrasts have emerged between the remote marginal rural problem zones with negative pressures, and the dynamic lowland rural areas where there are very positive pressures for economic development, exurban residences and alternative rural land-uses. The original aims of the CAP have largely been achieved with the creation of a large internal market with specialised commercial farming and virtual self-sufficiency. The contradictions and rigid mechanisms of the system were embedded in the political interests and structure of the Community and have proved difficult to change. The beginnings of reform have been in place since the Mediterranean packages of 1978 and the IMP of 1983 were announced.

In 1983 a series of revisions were proposed by Commissioner P Dalsager, who said that the CAP was a luxury the Community could no longer afford. They may be summarised as 'a policy of lower market-oriented prices . . . helping small and medium-sized farmers to modernise and adapt . . . without stimulating the output of products in structural surplus'. His proposals were taken further by Commissioner F Andriessen in 1985 who stressed that farmers were stewards of the landscape, wild-life and rural resources. This indicates the changing perceptions of the role of farmers and farming.

In February 1988 at the Summit meeting in Brussels, final agreement was made on a comprehensive and binding reform of the CAP. A legal limit is

142 THE NEW EUROPE

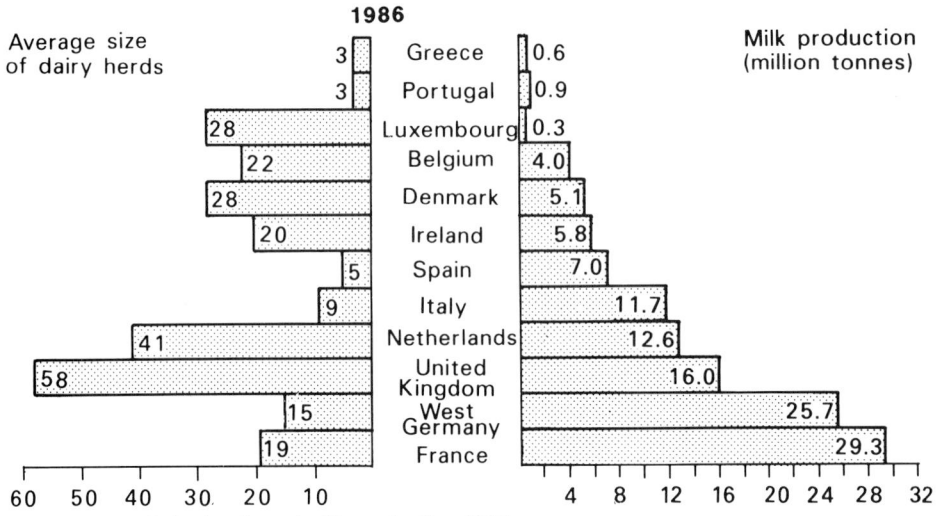

Figure 6.14 Dairy herds and milk production 1986

Average size of dairy herds	Country	Milk production (million tonnes)
3	Greece	0.6
3	Portugal	0.9
28	Luxembourg	0.3
22	Belgium	4.0
28	Denmark	5.1
20	Ireland	5.8
5	Spain	7.0
9	Italy	11.7
41	Netherlands	12.6
58	United Kingdom	16.0
15	West Germany	25.7
19	France	29.3

placed upon farm price support with controlled annual growth so that it will reach no more than £20.7 billion in 1992. Thus, expenditure in agriculture as a percentage of the Community budget will fall from 70 per cent to 55 per cent by 1992. There are quotas and production ceilings called 'stabilisers' for all products which if breached, trigger price cuts and taxes. This included, for instance, a 10 per cent cut in milk quotas, and 13 per cent reduction in guaranteed prices for beef. Alternative land-use such as forestry, receives subsidies. There is also a 'set-aside' scheme under which arable farmers are paid to leave 20 per cent of their land fallow, or convert it to grazing or non-surplus crops. The agreement was combined with a doubling of aid to declining and inner-city regions.

Free trade and trading groups.

The view of the classical economists such as John Stuart Mill was that international trade between nations led to a more efficient deployment of world resources by assisting countries to specialise in the production of those goods in which they possessed a comparative economic advantage. The ideal was therefore a global multi-lateral free trade situation, but in practice this has rarely existed for very long.

In the nineteenth century, largely under the influence of Great Britain, there was an extension of 'free trade' throughout the world. This meant that food and raw material supplies from large-scale producers in the newly discovered continents could be purchased at low cost, largely for the benefit of the rapidly industrialising Western World. At the same time, manufactured goods from Europe were exported on a large scale. During the twentieth century there has been a move towards protectionism as all newly independent countries have sought to protect their own developing industries by placing protective tariffs on goods formerly bought in Europe. Equally, they have looked elsewhere when freed from the necessity of trade with the former colonial power. This is shown in the decline in the proportion of United Kingdom exports going to the 'sterling area' (largely synonymous with the Commonwealth) (fig. 7.5). As high as 80 per cent during the nineteenth century, these exports declined to 38 per cent (1960), and by 1986 to 15 per cent.

Here is the essential basis for combining the resources of the small, highly industrialised nations of Europe on a continental scale. The EC gives them a guaranteed home market of over 320 million people which is a huge internal trading unit with an industrial base and capability for vigorous competition in export markets. It is a 'regional economic grouping' which depends upon a substantial and uninterrupted flow of goods and services. It is therefore in the common interest that barriers to international trade are as low as possible and this has been partially achieved through GATT (General Agreement on Tariffs and Trade) and the international conferences such as the Kennedy Round.

The European Community operates as a trading unit on two planes: internal free trade in a common market of the member states; a customs union in external trade.

The future landscape and the CAP

With the advent of the single market in 1992, the principal geographical patterns will remain very much as now. Regional specialisation should intensify with each country producing what it does best., The UK will continue to remain a mixed livestock-cereal economy, with specialist dairy and meat products according to the law of comparative economic advantage.

Although less land is needed for efficient food production, its concentration in fertile lowlands such as the Paris Basin and lowland England, will be tempered by other pressures upon these areas such as urban expansion, roads and airports. There is a case for supporting types of production which pay greater respect to ecological principles, such as medium technology farming in which less stress is given to maximum output, and the landscape is conserved in its natural state. A major reduction in imports of feedstock for dairy farms would be a result of this.

The marginal lands will see rather more change. Forestry is one obvious alternative with conifers for commercial use but broad-leaved species also have an important part to play. Forestry is not only associated with timber production, but also water supply, recreation space, and tourism. Regional and National Parks such as the Camargue in France are multiple land-use zones. Conservation of mountains, distinct environments, forests, heathland and wetlands, requires careful control of the ecological balance.

The CAP has to reflect a compromise between food importers such as the UK and West Germany, and food exporters such as France and Greece. Its pricing policies must take account of the needs of the farmer, the consumer and the landscape. In the remote mountainous and rural regions of the Community, the impact of the current changes will be hard. The maintenance of the rural population and the service infrastructure requires the CAP to keep a major guidance role as a social and regional policy.

7

Trade: the world's largest trading group

The EC is now the world's most significant trading group and in 1986 the enlarged community was responsible for nearly 35 per cent of the value of world trade (fig. 7.1).

There are considerable variations in the relative importance of trade to each country. In 1986 the Netherlands and Belgium/Luxembourg, with their commercial traditions, had an export trade which was equal to 48 per cent and 59 per cent respectively of gross domestic product (GDP). This is far higher than the equivalent figure of 20 per cent for the UK and 27 per cent for West Germany, usually considered as essentially exporting countries (fig. 7.1). France and Spain have lower proportions with trade accounting for seventeen and twelve per cent of GDP respectively reflecting a greater self-sufficiency in

	Exports (million ECU)	*Imports* (million ECU)	*Exports approximately % of GDP*
West Germany	247517	194368	27.2
France	121377	130551	16.5
Italy	99401	101947	16.2
Netherlands	85851	81279	48.0
Belgium – Luxembourg	71068	70401	58.7
United Kingdom	131621	127524	19.4
Ireland	13621	11821	51.4
Denmark	22126	23639	26.3
Greece	5950	11550	14.2
Spain	26982	33299	11.6
Portugal	7319	9608	25.3
EC Total	806958	796005	22.9
World	2085000	2684000	

Figure 7.1 EC and world trade, 1985/86

food production in France, and the long period of isolation of Spain from European and world markets. The distinction between trade surplus countries like West Germany, and deficit countries such as Greece may be seen.

Why such a large involvement?

The reasons for this large involvement in foreign trade are complex, but may be identified under three headings:

1. **Widespread industrialisation** was based in the nineteenth century upon mineral resources which have become depleted to the point of exhaustion. The mineral resources of the Community, though extensive in some commodities such as coal, are incomplete, with major deficiencies in certain crucial areas, such as petroleum, high-grade iron-ores and non-ferrous metals (fig. 7.2). Many of the more accessible coal-seams have been worked out, so the coal is now difficult to exploit and is high-cost, and the remaining French iron-ores have an iron content of only thirty per cent. There are adequate reserves of salt and potash and the reserves of natural gas and petroleum in the North Sea will redress the balance considerably for a time. Nevertheless the basic picture is one of deficiency in raw materials, with a corresponding reliance on large-scale imports (eighteen per cent in 1986).

2. **A heavily-urbanised population with a high standard of living.** Resources are used at an ever-increasing rate in industrialised societies, and there is

	European Community production	*Main producers*	*% of World production*
Coal	227.9	United Kingdom, West Germany	7.0
Crude Oil	143.9	United Kingdom	5.4
Copper	0.07	Spain	0.8
Iron-ore	22.2	France, Spain	0.1
Bauxite	3.8	France, Greece	4.2
Salt	29.3	West Germany, United Kingdom, France, Netherlands, Italy	18.4
Potash	5.5	West Germany, France, Spain	18.8
Lead	0.2	Ireland, West Germany, Spain, Greece	5.3

Figure 7.2 Selected resources of the EC, 1985/86 (million tonnes)

	Imports (percentage)				Exports (percentage)			
	EEC 6	UK	EC		EEC 6	UK	EC	
Commodity group	1958	1973	1977	1986	1958	1973	1977	1986
Foodstuffs	25	30	14	12	10	7	10	10
Fuel and raw materials	46	25	28	18	10	8	9	9
Manufactured goods	29	45	58	70	80	85	81	81

Figure 7.3 External trade - main commodity groups (by value)

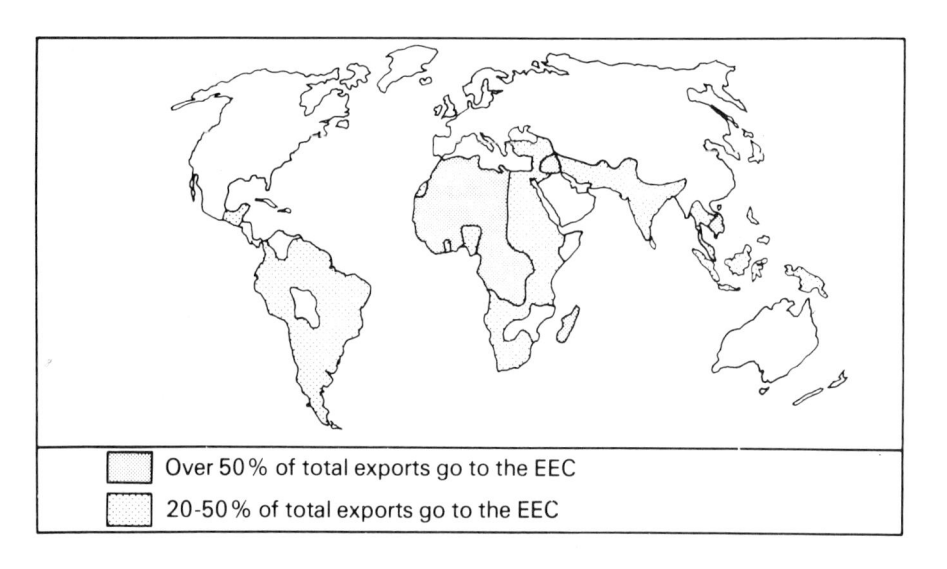

Figure 7.4 The Community's share in the exports of the developing world

an underlying propensity to import semi-finished materials for further processing, together with finished manufactures, which add to the range of choice open to the consumer (fig. 7.3). Wealth generates imports and the richer the society the greater is the reliance upon a complex movement of imports and exports of all categories.

3. **Food is an important element in the community's import trade**, despite a potential self-sufficiency in temperate foods. Ten per cent of imports in 1986 were of foodstuffs, including temperate foods from the USA and Australia, and tropical foods from Africa and Asia (fig. 7.4). This

illustrates the importance of the former colonial territories, and explains their status as associate trading partners of the EC. The Yaoundé Agreement, and the Lomé Convention of 1975/1984 were both preferential trade and aid treaties.

The Common Market and intra-community trade: trade creation

The Common Market theoretically allows goods, services, people and capital to circulate freely throughout the member countries, the basic prerequisite for this being free trade, the abolition of all internal customs duties, tariffs and quotas. This has effected a very rapid expansion of trade within the single market. The original Six completed the abolition of all internal tariffs by 1969, but new members have adjusted gradually to the Community levels via transition periods. A single market of 320 million people now extends from Scotland to Sicily. The real expansion of trade relating to this is in intra-community trade, and the period from 1958 to 1972 saw a substantial increase between the member states, with trade growing in value during this period at an annual rate of over fifteen per cent. Although the recession slowed this process down for a time, since 1975 the value of trade has continued to grow at an annual rate of sixteen per cent. The individual members raised their proportion of trade with each other from 25 per cent to 58 per cent (fig. 7.6). Those with the highest levels of integrated trade have either a limited base to their domestic economies such as Ireland, or have a tradition of European trade such as the three Benelux countries. All have benefited: Italy has become an exporter of fashion clothing and consumer durables; France and the Netherlands have greatly expanded their exports of agricultural produce; and West Germany's persistently favourable balance of trade is due principally to the opening of EC markets to her efficient industry. The UK has a vastly increased market for the products of its specialised engineering, chemical, vehicle, electronics, telecommunications and aerospace industries. West Germany is now Britain's single largest export market, and the EC now takes 49 per cent of Britain's exports. The European market is one in which Britain must succeed if she is to thrive as a major industrial and trading nation.

The direction of British trade was already swingingly strongly towards the industrialised markets of Western Europe during the 1960s (fig. 7.5). Since the enlargement of the Community in 1963 the United Kingdom, Ireland and Denmark have rapidly increased their interdependence with the other EEC countries (fig. 7.6). The Mediterranean enlargement of the 1981-6 period involves a lengthy transition period whilst the tariff barriers behind which Spanish industry has sheltered during its development are gradually removed. The three mediterranean countries import about eighty per cent of their energy, and add to EC overproduction of mediterranean food products. Thus major imbalances in some internal trade patterns have emerged whilst absorption is taking place. Intra-Community trade is the means by which the special expertise and comparative advantages of each member country may

148 THE NEW EUROPE

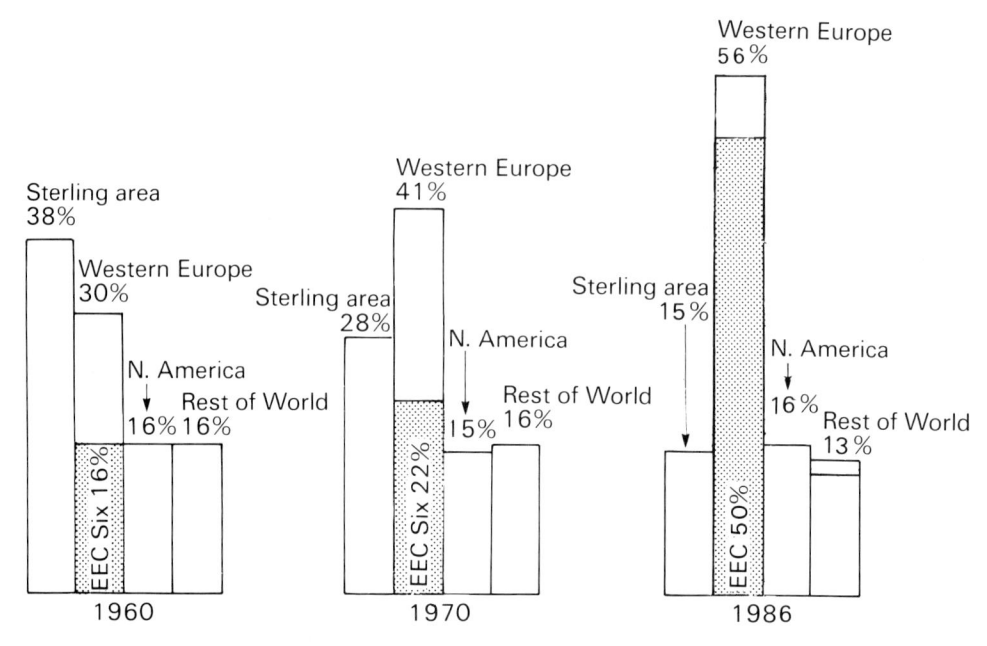

Figure 7.5 Direction of British exports

be used with greater efficiency to achieve interdependence and convergent economies.

The Customs Union and external trade: trade diversion

At the same time as the Treaty of Rome provided for the reduction to zero of internal tariffs, it also required all members to conform to a common external tariff (CET). This was accomplished in 1968 for the original six members. New members have been absorbed gradually through transition stages over a number of years. This is a Customs Union and is the distinguishing feature of the EC. This may be compared with the purpose of EFTA, which was to promote free trade between its members, with each country keeping its trade relations with third countries quite separate. The EC has formed a single trading unit without internal barriers and with a common external tariff wall. Essentially, therefore, it is creating a single economic unit from twelve separate ones. The common external tariff deals with external trade relationships, that is, with the rest of the world. Although in global terms the CET is set at low rates, the average for industrial goods being six per cent and there are trade arrangements with some associated countries at zero rates, this does involve some *trade diversion* from other countries. The CET tends to replace imports from non-member countries with suppliers within the Customs Union. Thus the benefits of *trade creation* in the enlarged markets have to be set against *trade diversion* from non-members.

	1957	1971	1986
Belgium/Luxembourg	44	63	70
France	21	50	64
West Germany	24	47	54
Italy	21	42	55
Netherlands	41	55	61
United Kingdom	16	22	50
Ireland	–	69	73
Denmark	–	32	53
Greece	–	–	58
Spain	–	–	51
Portugal	–	–	59
EC 12 (average)			58

Figure 7.6 Intra-Community trade. Percentage of total imports coming from other member countries

The character and direction of external trade is summed up as follows. Thirty per cent of imports are of raw materials and foodstuffs, whilst seventy per cent are manufactured goods. Eighty per cent of exports are of manufactures, and nineteen per cent are primary products (fig. 7.3). There is thus a large, complex, and increasingly multi-lateral movement of manufactured goods throughout the developed world. About eighty per cent of total trade by value is with the industrialised and wealthy countries within the EC and the rest of Western Europe, and with North America and Japan (fig. 7.7). There is a major imbalance of trade in imports from Japan. Trade with the underdeveloped countries overall accounts for fourteen per cent of the total, but a large proportion of that includes a heavy reliance upon imports of petroleum from OPEC. Although rather less oil is now imported because of North Sea supplies, this will change as the North Sea supplies diminish.

The real value of a trading group such as the EC is that because of its high rate of growth and development, it has the ability to create and sustain a large volume of world trade and to provide the aid for development which is needed. This is the very significant trade creation factor. In fact, many studies have shown that the generally low levels of the CET have not led to trade diversion on the scale which might have been thought. The Treaty of Rome stated that 'member states aim to contribute . . . to the harmonious development of world trade, the progressive abolition of restrictions on international trade, and the lowering of customs barriers'. The enlargement of the EC has been a major factor in creating and maintaining a more outward-looking trade and external policy. Historically, Europe has had traditionally privileged links with the developing countries, and the accession of the UK with her

150   THE NEW EUROPE

traditional involvement in world affairs has done much to strengthen those ties.

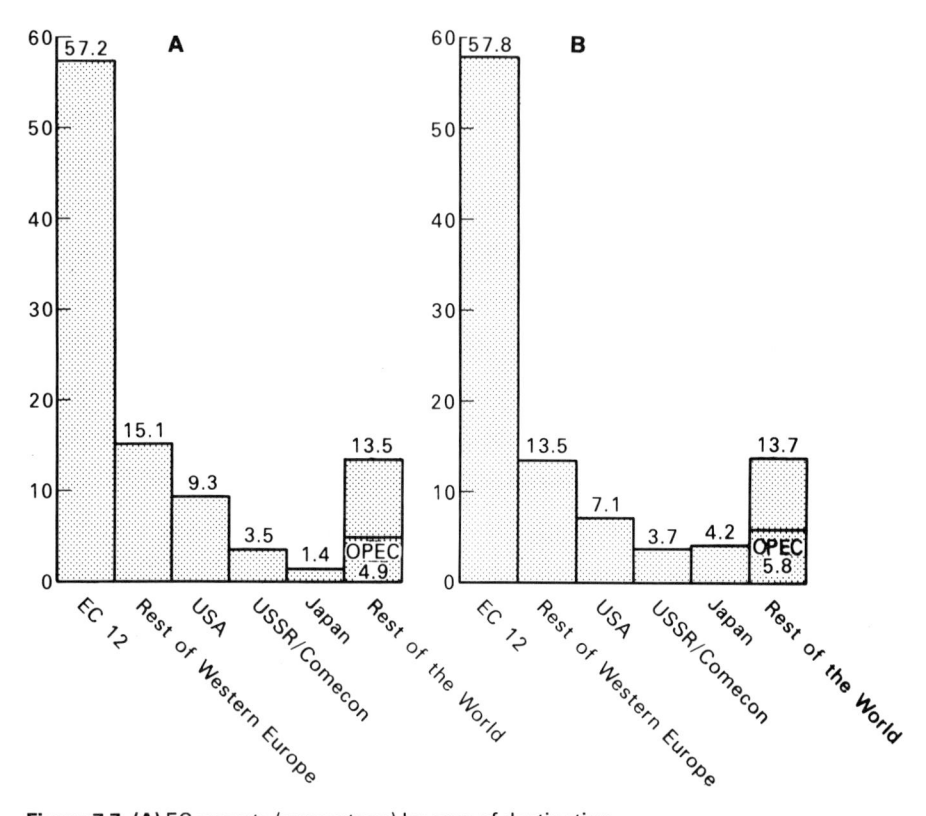

Figure 7.7 (A) EC exports (percentage) by area of destination
(B) EC imports (percentage) by area of origin

Trade and external relations

The development of a world trading interest is based upon the European special relationship with former dependent territories. In 1964 the Yaoundé Agreement came into operation with eighteen African countries setting up a free trade zone and arrangements for financial aid. At the same time, the early Association agreements with Greece and Turkey were signed. The very important 'Kennedy Round' on tariff reductions in 1968 marked the emergence of the EC as a trading unit and a negotiating entity on the world scene. The EC has now embraced a commercial role and external relations. It has developed an extensive system of generalised preferences for many types of products, and has a complex network of association and trading agreements with many groups of countries. One of the most important is the Lomé Convention, with sixty-six African, Caribbean and Pacific ex-colonial territories, better known as the ACP countries. This was first signed in 1975 and renewed in 1981. Lomé III dates from 1984. The convention covers free trade for virtually all products from ACP countries, co-operation and development aid. Ninety-five per cent of exports from the ACP countries enter the EC free of duty and there are special arrangements for products such as rum, bananas and sugar. The STABEX scheme worked like an insurance scheme against bad years. A minimum income is guaranteed to ACP countries for their exports of 44 products including cocoa, coffee, tea, peanuts and sisal. In 1984 this was supplemented by MINEX, a similar stabilisation scheme covering a range of ACP exports including manganese, phosphate and bauxite, tin and iron-ore. There are also agreements with the EFTA countries, the Maghreb countries of Tunisia, Algeria and Morocco, and with countries in the eastern Mediterranean Basin such as Cyprus, Israel, Lebanbon and Egypt. The main provisions of these agreements are unrestricted access to Community markets for industrial goods, customs preferences for certain agricultural products, and financial aid. Significantly, the Community's developing spheres of economic interest lie principally with its former dependent territories, with the other industrial countries of Western Europe, and within the Mediterranean Basin. Eastern Europe too could well have a growing role.

The Mediterranean Basin is rapidly becoming part of the Community's economic sphere of interest. It is a major supplier of primary products and, lying as it does on the southern flank of the Community, is clearly an area of strategic interest to Western Europe. 'Geopolitical reasons in themselves make an impressive case for the necessity of a coherent European Community policy on the Mediterranean. A glance at the map proves it. Look first at the Balkans and then at the edge of the Atlantic. Take in the Dardanelles and the petrol-producing region of the Near East; remember too that the Mediterranean is the inescapable north-south axis for links between Europe and Africa. We must question whether the Community could survive a serious disturbance in the Mediterranean region . . .'. (Lorenzo Natali, European Commissioner with responsibility for Mediterranean affairs, 1981).

Other barriers to trade

Although the Customs Union and the dismantling of all internal tariffs has been the most significant single act in the formation of the Common Market, nevertheless there are many technical and physical barriers to trade which must be eliminated before a unified economic system comes into being. Industrial standards, metrication, and public health and safety regulations are areas where harmonisation is needed. Monopolies are another problem. The commission has to approve monopolistic trading agreements, but generally these are not discouraged as they have to be seen in a world context rather than a national one. Cross-frontier Community mergers are actively encouraged, but the harmonisation of company law, financial liability, tax laws and mercantile practice is necessary before long-term intra-Community mergers are feasible. Public purchasing policy is another area for agreement, as in a single market each government should purchase the cheapest and best quality product available, even if this involves a cross-frontier product. As yet, each country tends to purchase from within its own industries, quite naturally, but this is an indication that economic union is still far from complete. Finally, and perhaps most importantly, a fast integrated transport system is the best way of ensuring unrestricted movement of goods throughout the Community.

8

Transport: movement on a continental scale

The historical legacy

Transport in the EC is a complex network of systems which have evolved over a long period of history, under national conditions of development, and as a response to market forces, generally with a lack of strategic planning. The primary axes were dependent upon historic trade routes such as the east-west Hellweg, and the Rivers Rhine and Rhône beside which great cities such as Cologne and Lyons developed. During the industrial revolution the rapid development of waterway and railway networks was most marked in the coalfields and heavy industrial triangle. The national basis of operations meant that unique systems developed within national frontiers, and the transport infra-structure between countries was poorly developed and fragmented up to World War II.

The Treaty of Rome and subsequent enlargements of the Community have increased to a continental scale the planning of transport operations and highlighted the urgent need for integration of the transport networks. Increased speeds have reduced the time needed to cross Europe and it is often better to think of travel in terms of time rather than distance. The Italian Riviera is two hours away from London by air or twenty-four hours by road or rail and sea. Economic, political, cultural and tourist contacts have expanded rapidly.

At the same time there is a growing complexity of transport systems and an associated need for efficient linkage and integration. Canals and railways, for long dominant in long-distance movement, are now supplemented or supplanted by motorways. Air travel is very significant for medium and long-distance passenger and tourist transport, and the hovercraft is adding to cross-channel and other short-distance sea routes. The integration of all these is illustrated by the container revolution, long-distance coach services, channel ferry services, railway car sleepers, by technological feats like the Channel Tunnel project and the trans-Alpine tunnels and the development of the Euro-route systems of motorways.

There are also other specialised elements which form part of the totality of movement in the Community. Movement of capital, goods, liquid fuel, and labour are considered in other chapters.

154   THE NEW EUROPE

	Inland waterways		Railways		Merchant fleet	Civil aviation	Road freight	Principal road network
	Length in use (km)	Tonne-kilometres (millions)	Length of line operated (km)	Freight Tonne-kilometres (millions)	1000 tonnes gross	Passenger-kilometres (millions)	tonne-kilometres (millions)	(km)
West Germany	4461	41472	27709	57941	6177	24273	98615	493000
France	6324	8394	34678	50828	8237	32858	97094	802000
Italy	2237	neg	16185	16584	8843	13638	75000	300000
Netherlands	4832	27308	2824	3275	4301	16859	18831	97000
Belgium/Luxembourg	1554	4888	3982	7874	2400	6061	10587	136000
United Kingdom	1631	78	16729	17190	14344	43580	99067	370000
Ireland	neg	neg	1944	601	194	2190	3727	92000
Denmark	neg	neg	2471	1216	4942	14865	8342	70000
Greece	neg	neg	2461	687	31032	6300	10352	40000
Spain	neg	neg	13466	11645	6256	16547	91500	167000
Portugal	124	neg	3613	1265	1437	4275	na	52000

na = not available   neg = negligible

Figure 8.1 (a) Comparative transport data 1985. Compare the relative densities of the transport networks with population and area (fig.1.6).

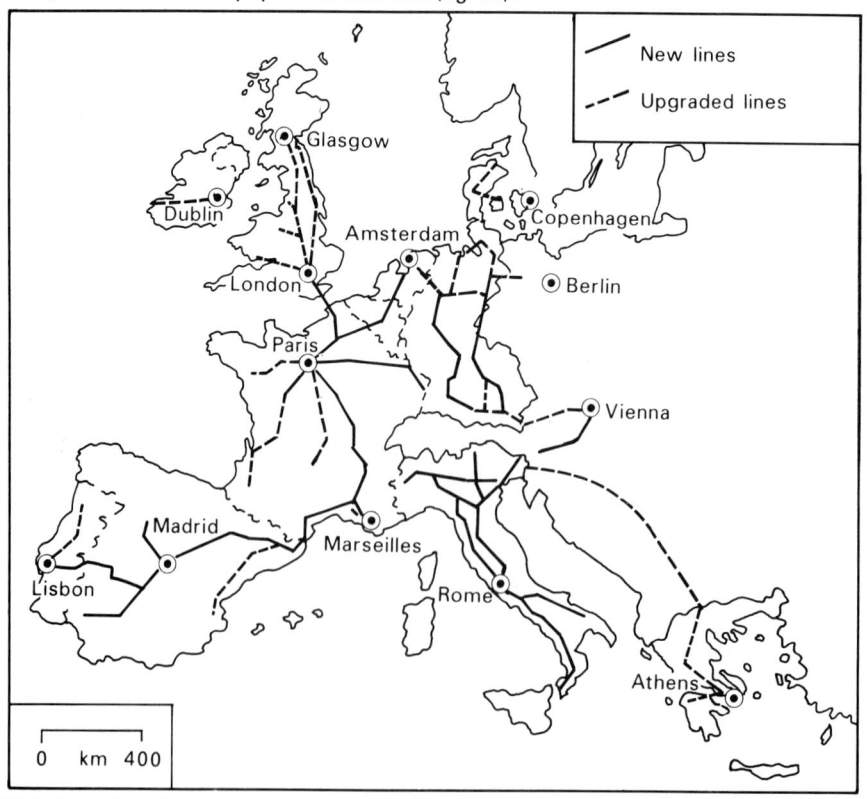

(b) Railway Development: high speed lines developed and planned for the 1990s

Railways

The railways were built largely in the nineteenth century industrial period, and often reflect national patterns with the main lines radiating from the capital city to the frontiers. Very few were built with an international viewpoint in mind, but a standard gauge was adopted across Europe, which has proved beneficial because of the twentieth century growth of inter-city cross-frontier services.

Despite their inflexibility, high capital costs, and substantial competition from other forms of transport, railways still play a major role, although there are wide variations between the member states (figs. 8.1 (a) and 8.3). The highest density is in north-west Europe with its industrial and urban concentrations. France has a very comprehensive network of railways and they carry the highest number of passengers and greatest tonnage of freight. The relatively large area of France and the long distances benefit rail haulage. Government support has been substantial since the railways were completely rebuilt after 1945. French Railways (SNCF) is a public service with competition from road haulage reduced by restrictive licensing. West Germany's railways carry similar amounts of freight and passengers, and have a very dense network relative to area, but are less important relative to other forms of transport, specifically waterways. Italy has a small network of railways compared to her area, partially a result of the retarded growth of the system up to the 1930s. The great length (1000 km) of peninsular Italy, the rugged nature of the relief and the inability of much of the area south of Lombardy to sustain a high level of traffic were all factors in this retardation. Although the railways are now state-operated, they still constitute a minimal system compared with the northern Community countries.

The United Kingdom railway system has experienced considerable reduction. A relatively small land area, high densities of population and a well-developed network of roads combined to emphasise the disadvantage of the railways as a relatively inflexible form of transport. The railways have had to operate in open competition with, and have lost ground to, the roads both in passengers and haulage. The United Kingdom was over-endowed with railways owing to the 'railway mania' of the nineteenth century, when much duplication of railway lines occurred. This very dense network has been reduced during three periods of rationalisation. By the 1930s these were four large companies: London, Midland and Scottish; London and North Eastern Railway; Great Western Railway; and Southern Railway. In 1948 the system was nationalised as British Rail; and during the 1960s the Beeching Report and subsequent pruning reduced the track by about one-third to 17 000 km. The United Kingdom network has a greater density than France, but it is considerably less used, particularly for freight.

In other EC countries the use of rail systems varies according to other factors. Belgian railways are a high density system and are well used for freight which is indicative of the general level of industrialisation; by contrast that of the Netherlands is a smaller system because of the competition from its extensive waterway system. Spain, Portugal, Ireland and Greece have a low-density network indicative of retarded industrial and urban development. All EC countries have experienced a marked relative decline in the amount of freight carried by rail since the late 1960s. Nevertheless, in volume-distance terms, the railways are still very significant, particularly in France and West Germany (figs 8.1 (a) and 8.3). Their continuing importance depends on the movement of bulk freight, inter-city routes and suburban commuter services.

However, there are signs of a revival of the railway systems as environmental concern increases and as road and air traffic approach saturation point. It has been shown that high-speed railway lines between major city regions over medium-distances compete very well economically with other forms of transport. In France, the TGV has been operating from Paris to Lyons and Marseilles since 1981 and has proved a great success. It is to be extended to Bordeaux and possibly, Strasbourg. Other countries such as Spain (Madrid to Seville and Barcelona), Italy (Naples - Rome - Florence - Milan), and West Germany (Hanover - Frankfurt and Cologne - Frankfurt), are developing high-speed lines (fig 8.1b).

One of the most significant developments involves several countries. The French TGV is to be extended north to Lille, Belgium and the Dutch network at Amsterdam. It will also link with the Channel Tunnel. British Rail are planning a new high-speed link from the Tunnel to London through Kent, and are also considering links to the north of England and Scotland, by-passing London.

Major investment in new technology and substantial government subsidies are needed for these developments. Many lines will be upgraded rather than being entirely new. Nevertheless, the railways are increasingly being seen as a stimulus to growth in the peripheral regions. It may be that the railway system is about to enter another stage of its development in the 1990s.

Inland waterways

The importance of inland waterways is related to the occurrence of large navigable rivers and the lack of physical barriers. Denmark and Ireland have no commercial waterways of any significance. In Italy only the River Po and its tributaries are of any value for freight as the rest of the peninsula has rugged terrain with short mountain torrents and few rivers of any size. The Iberian peninsula offers few possibilities for navigation. The UK depends even less upon inland water transport for a variety of reasons which include topography, the limited size of barges characteristic of the early narrow canal system, and the disuse into which it fell after its virtual destruction by railway competition.

Those countries of the North European plain drained by the Rhine and its tributaries, however, have the natural basis for an extensive canal system (fig. 8.2). Eastern France, Belgium, the Netherlands and West Germany depend to a large extent upon inland waterways for the transport of bulk cargoes such as oil, iron-ore, coal, grain and petroleum products. West Germany's economic axis trends north-south along the length of the Rhine and movement is primarily between the industrial areas along its banks. Duisburg,

Frankfurt and Cologne are the most important inland ports. The importance of the Rhine can be seen in the total West German tonnage carried by waterways (fig. 8.1a). In the Netherlands half of all freight is carried by water (fig. 8.3), reflecting the traditional character of the country, but also underlining its position at the mouth of the Rhine, with Rotterdam as the great entrepôt of Europe. In Belgium, the Scheldt and Meuse rivers, and the Albert and Juliana Canals form a very important link between Antwerp, the Rhine and north-eastern France. To France the main value of inland waterways is the linkage of the Paris Basin via the Seine to the coast, via the Oise/Nord Canal to Lille, and via the Marne and Moselle system with

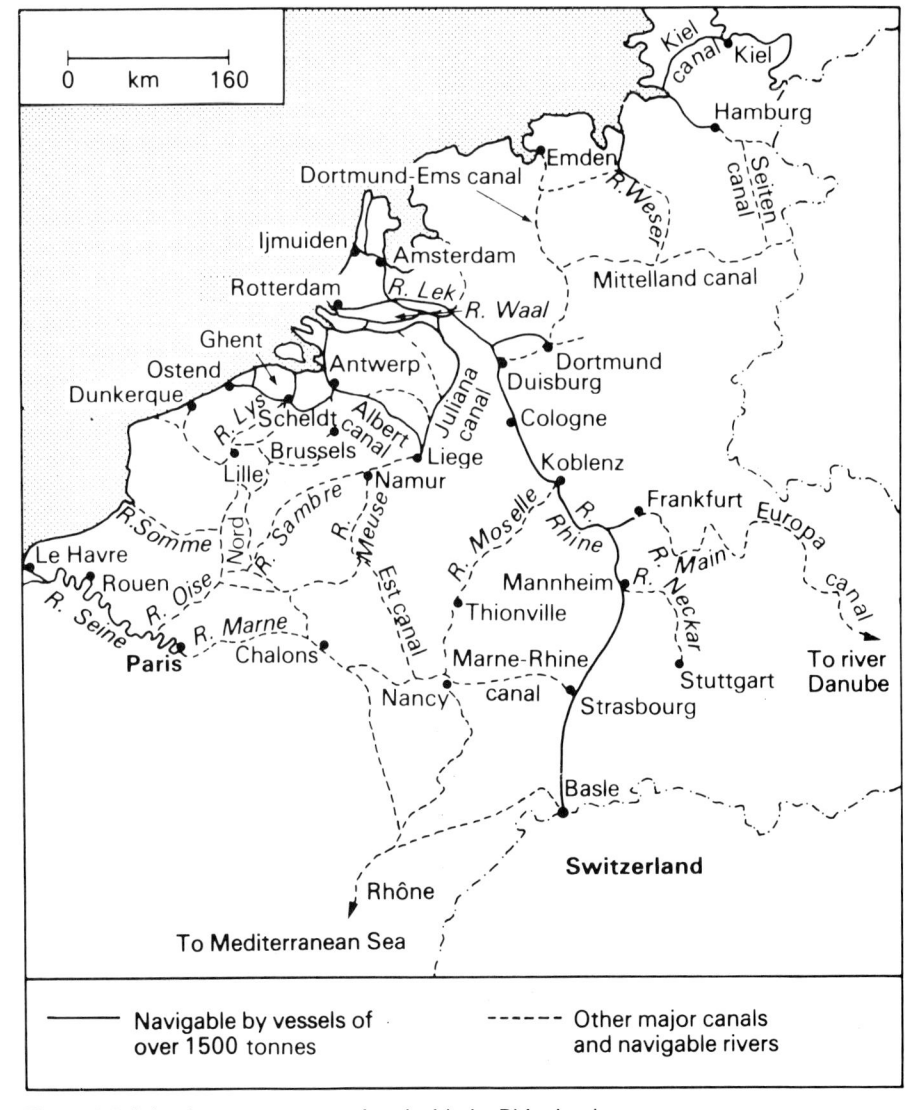

Figure 8.2 Inland waterways associated with the Rhinelands

The Rhine Gorge with passenger and barge traffic

Lorraine, Strasbourg and the Rhine. Although the French canal system is very extensive, much of it is outdated, confined to the north and east of France, and by comparison with railways is under utilised.

Road transport

All countries of the EC have a well developed network of roads but there is a marked difference between the much higher densities in the 'core' countries of the north-west and peripheral countries such as Spain, Portugal and Greece (figs. 8.1(a), 8.5, 8.6). West Germany and Belgium possess the nearest approach to a balanced transport system with the three major forms of inland transport: road, rail, and waterway, each carrying substantial proportions of freight (fig. 8.3). The United Kingdom and Italy are proportionately the greatest users of road transport with over seventy per cent of all freight being carried by road. Since the 1960s the creation of fast motorway links within and between the EC countries has done most to make road transport the principal means of transporting freight. In Italy the reasons are largely historical, since until the 1930s there was a very poorly integrated system of railways, with no through routes over the 1000 km from north to south, and very great difficulties east-west across the Apennines. This minimal network was associated with late unification and weak industrialisation of Italy up to the 1930s. To cope with her rapidly expanding economy since 1945, and aided by modern technology, the Italians have constructed an excellent Autostrada system with some superb feats of engineering. Perhaps the most famous is the Autostrada Del Sole, which links Milan to Naples with many bridges, tunnels and dramatic views. The long north-south distances have been overcome by motorways, but even more significant is the construction of motorways and tunnels through the Alps which have done much to end the relative isolation of Italy from the trunk of Europe (fig. 8.4). Along the Riviera to France from Genoa, from Turin up the Val D'Aosta and through the Mont Blanc Tunnel, from Milan northwards to the Simplon and

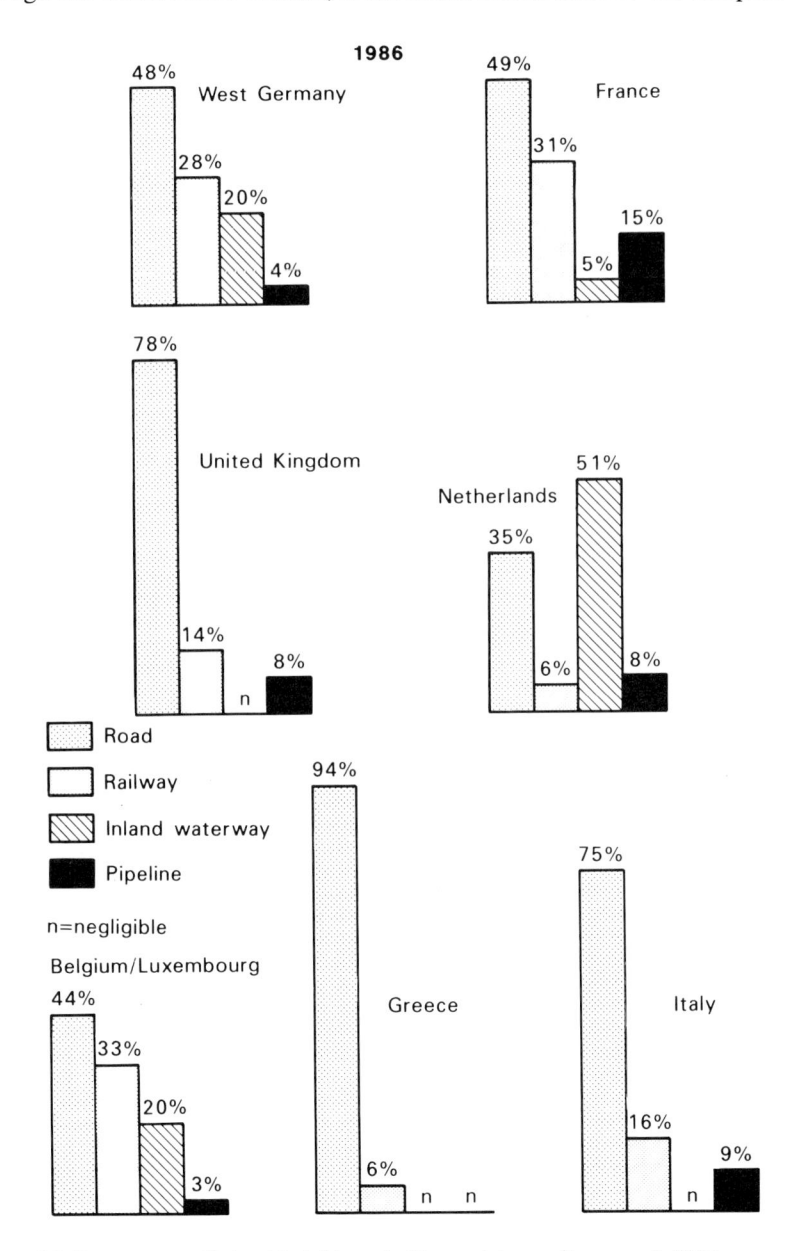

Figure 8.3 Percentage of inland freight carried by each type of transport, 1986 (data based upon volume-distance and measured in tonne - kilometres)

St. Gotthard Passes, and from Verona northwards to the Brenner Pass, the Italian Autostrada are linked by fast routes to the rest of the EC.

Although Italy's network is a critical element in her economy, the West Germany autobahn system is more extensive and was originally constructed as part of the economic recovery programme before 1939. Following the post-war division of Germany, the main trend has become north-south based essentially upon two lines of movement (fig. 8.6b). The Rhineland is the primary artery, linking the Ruhr with Bonn, the Saar, Frankfurt, Mannheim and Stuttgart. A second line runs from the North Sea ports of Hamburg and Bremen south to Hanover, Brunswick and Kassel, and thence to Nuremburg and Munich. East-west routes link up these two axes at major urban nodes.

The other EC countries have motorway systems which reflect their axes of greatest activity. The importance of London and Paris is illustrated by the concentration of roads and motorways upon these two cities. In the United Kingdom the radial pattern based upon London is supplemented by a figure 'H' which is centred upon the industrial Midlands. This links the industrial areas of Northern England and Scotland with London, South Wales and the Bristol area. The French system is based essentially upon links from Paris to the Lower Seine, Lille, Dunkerque and the Nord, and south to Bordeaux, Lyons, Marseilles and the Côte D'Azur (fig. 8.5).

Belgium and the Netherlands have a greater mileage of motorways in proportion to area than any other part of the EC. This reflects three

Autobahn through the Spessart uplands near Frankfurt, West Germany

significant geographical facts: a high level of economic activity; a high density of population; and their position as the focal zone of the Community (fig. 8.6a).

The motorway system of the Community reflects several important geographical relationships - the new mobility and affluence of the population with over 90 million private cars, nearly one car for three people. It is one of the ideal ways in which the EC can develop a modern integrated cross-frontier system of transport. Most significant of all, however, is their influence in changing industrial location. Those areas with accessibility have become favoured zones for footloose industrial growth. The Frankfurt-Mannheim-Stuttgart corridor, the Autostrada Del Sole and the M4 corridor in the London city region have become excellent examples of industrial and service sector growth. These are the new 'sunrise' areas and 'industrial boulevards' of the EC.

Air transport

Air transport is concerned principally with the international movement of passengers, but at the same time there is considerable domestic and intra-Community movement. Freight traffic is increasing rapidly, and London

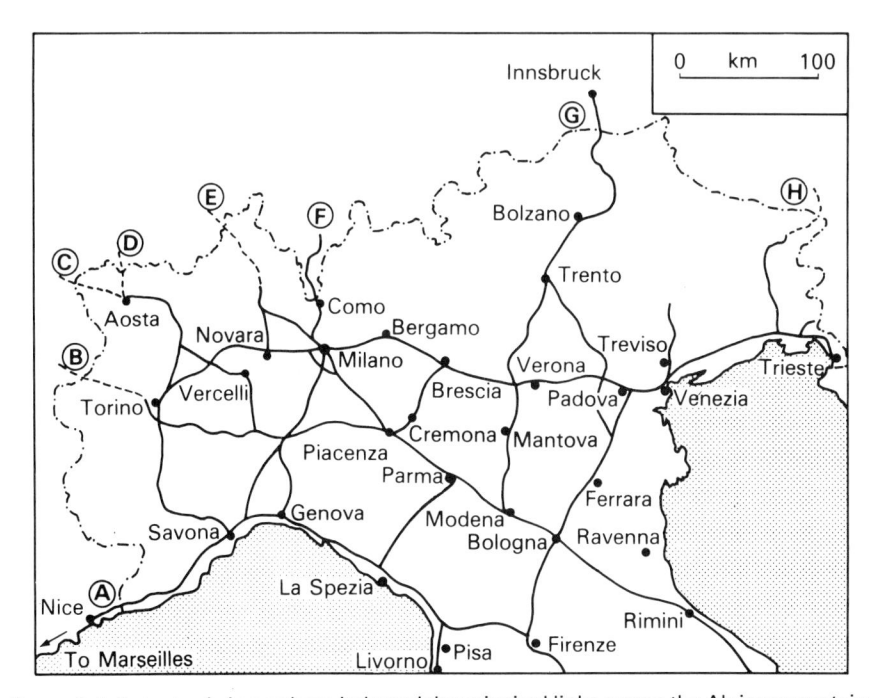

Figure 8.4 Autostrada in northern Italy and the principal links across the Alpine mountain system (A) Riviera coastal motorway to Marseilles and the Rhône valley (B) Mt. Cenis pass to Chambery and Lyon (C) Mt. Blanc tunnel to Chamonix and upper Rhone valley (D) Grand St. Bernard pass to Lake Geneva (E) Simplon pass to Switzerland (F) St. Gotthard pass to Zurich (G) Brenner pass to Innsbruck and Munich (H) Villach (Austria) Salzburg-Munich route

162   THE NEW EUROPE

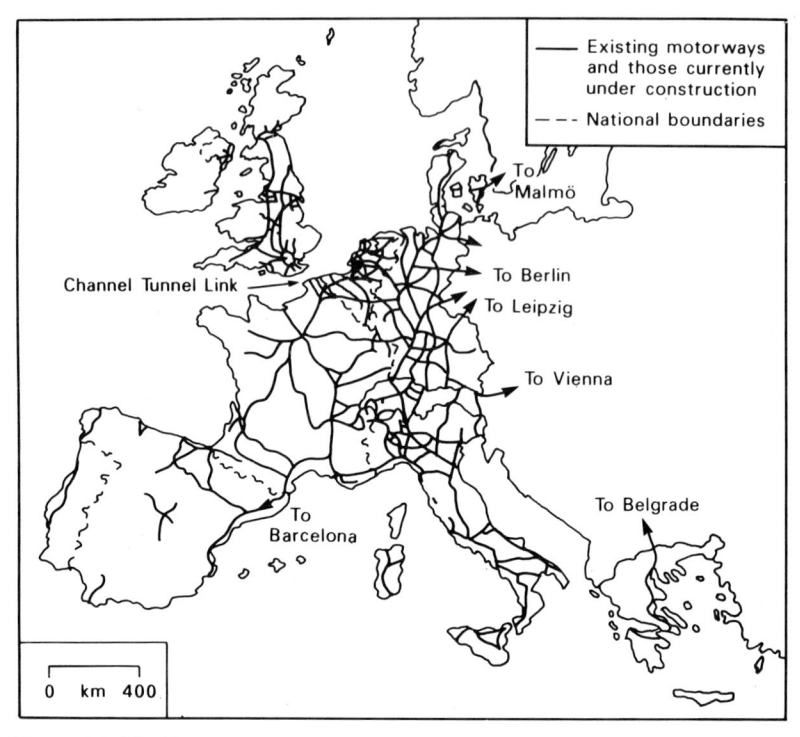

Figure 8.5 The E-route motorway system

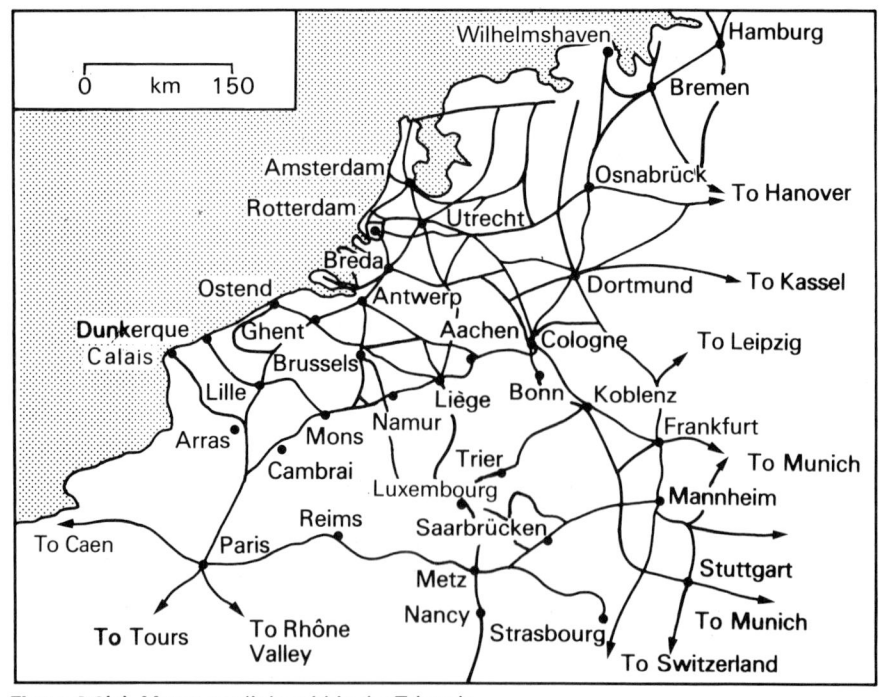

Figure 8.6(a) Motorway links within the Triangle

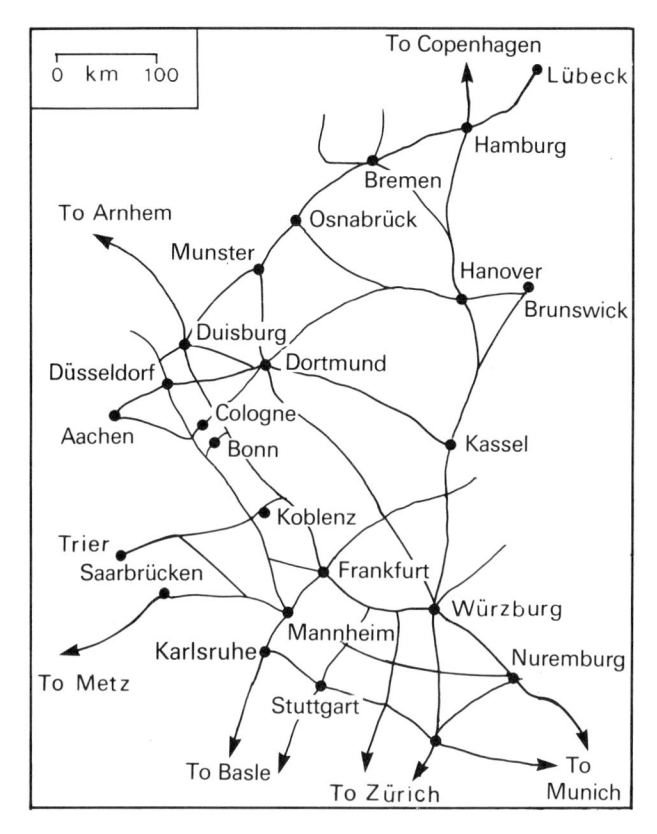

Figure 8.6(b) The West German autobahn network with major cities

Heathrow is now one of the busiest freight transit points in the UK. Some indication is given (figs 8.1(a) and 8.7) of the relative importance of national air-fleets, passenger density and principal airports. Although there are over a hundred major airports in the EC, five city regions dominate air traffic. London Heathrow, supplemented by Gatwick, Luton and the planned expansion at Stansted, is the busiest passenger airport in the world. Its major problems, common to all large airports, is the congestion when several large passenger jet aircraft arrive in close proximity, the continual need for expansion of terminals, the position peripheral to the city with the need for transfer of passengers, and the environmental problems of large airports in residential areas. The significance of London is related to the United Kingdom's traditional international relationships, whilst Paris, Amsterdam, Frankfurt and Rome reflect a combination of the nodality of capital cities and core regions, together with their position as transit points between Europe and the rest of the world. A large number of regional airports such as Munich, Brussels and Milan have developed, because of vastly increased air traffic, to supplement the major airports. One of the most important and fast-growing airline operations during the 1980s has been tourist charter flights, largely to the Mediterranean.

164   THE NEW EUROPE

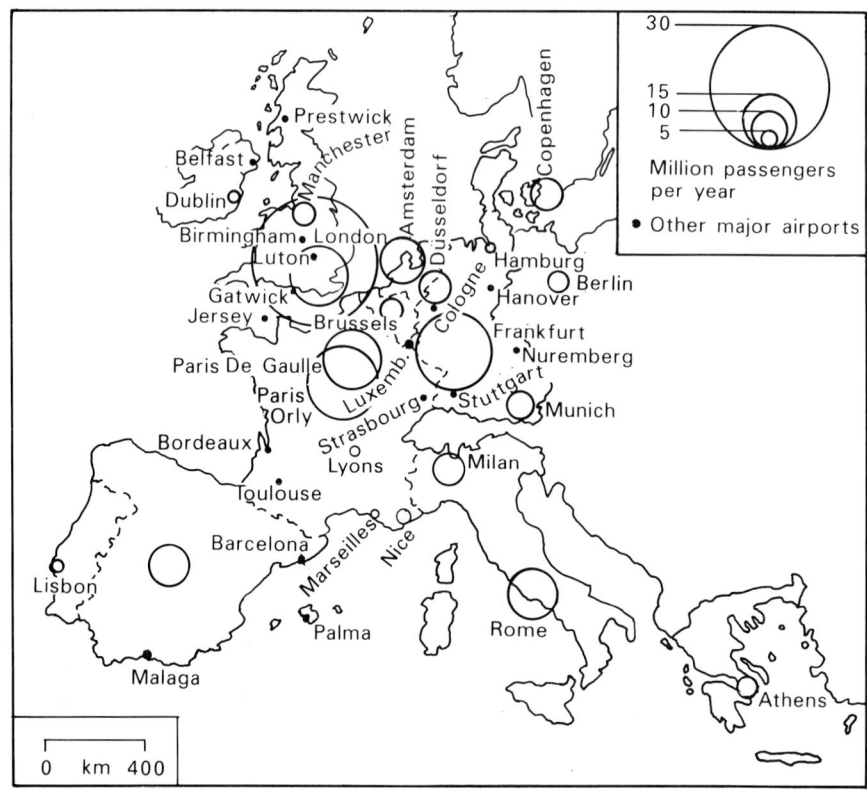

Figure 8.7(a) Major EC airports (relative importance, 1986)

Schiphol airport, Amsterdam, one of the busy airports of Europe

London Heathrow	31.3	Munich	7.6
Frankfurt	19.5	Manchester	6.1
Paris Orly	17.7	Brussels	5.6
Gatwick	14.9	Athens	5.4
Paris De Gaulle	14.6	Hamburg	4.7
Rome	12.9	Berlin	4.5
Amsterdam	11.4	Nice	3.2
Madrid	9.9	Lisbon	3.0
Copenhagen	9.4	Dublin	2.6
Milan	7.9	Lyons	2.6
Düsseldorf	9.9	Marseilles	1.9

Figure 8.7(b) EC airports, 1985. Total number of passengers (millions)

Sea transport

Shipping is related mainly to trade, and here, although the original Six had a substantial merchant shipping fleet, the accession of the UK and Greece to the Community has completely changed its character. The United Kingdom's fleet is half as large again as the original Six put together (fig. 8.1(a)) and one of the largest merchant carrying fleets in the world flies the Greek flag. It is a function of international trading traditions that this significance is so marked, and it contrasts with the mainly continental character of the original Six. In addition the accession of the United Kingdom, Denmark, Greece and Ireland has meant that intra-Community sea traffic, hitherto largely coastal, is now an essential part of movement between member states.

The map of ports and sea traffic (fig. 8.8(a)) shows the area of greatest significance. The North Sea coast and English Channel, from the Tyne to Southampton and from Le Havre to Hamburg, have the greatest number of large ports and carry most passengers and freight. These are amongst the busiest sea-lanes in the world, and are another significant reflection of the high level of economic activity in the northern part of the Community. In addition, there are major ports on the Atlantic coasts of the United Kingdom and France, and the Mediterranean ports act as input points to the Community from the south.

A most significant change in port geography has taken place during the last forty years. Changes in the patterns of sea-borne trade, the growth of channel passenger ferry services and advances in technology are all reasons for this. Increase in vessel size and the development of containerisation and roll-on, roll-off facilities (fig 8.8(b)) have led to a decline of traditional port cities based upon estuary or lowest bridging point sites, such as Amsterdam and London, and the development of and concentration at downstream deep-water terminals. London-Tilbury and Marseilles-Fos are examples of these large deep-water Euroports.

There are a number of 'Euroports' which are international in character,

166 THE NEW EUROPE

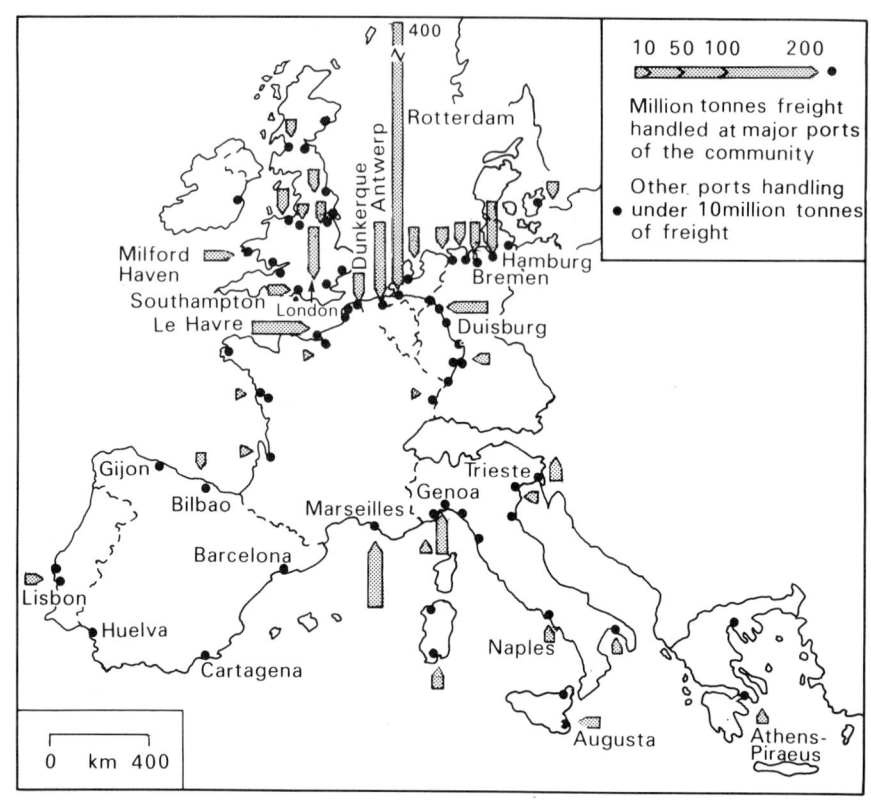

Figure 8.8(a) Seaports and inland ports of the EC (Average amounts of freight handled, 1985)

Container wharf at Pernis - Eemhaven, Rotterdam

	Total	Imports	Exports
Hamburg	10033	4598	5435
Bremerhaven (Bremen)	8074	3067	5007
Le Havre	5415 4356	2381	3034
Marseilles-Fos		1760	2596
Genoa	1988	–	–
Rotterdam	22242	9392	12850
Antwerp	10719	3648	7071
Zeebrugge	2155	953	1202
Other UK ports	10485	5576	4909
London (Tilbury)	3872	1964	1908
Southampton	1811	892	919
Felixstowe	6839	3602	3237
Hull	1687	1012	675
Piraeus	1533	946	587

Figure 8.8(b) Container traffic at major ports, 1985 (thousand tonnes gross)

UK	5.8
Netherlands	5.4
West Germany	4.0
Italy	3.0
France	2.9
Belgium	2.4
Spain	2.1
Others	2.5
Total EC	28.1%

Figure 8.8(c) World container traffic, 1985 (percentage)

handle a large tonnage and serve a substantial part of the whole Community. Rotterdam is by far the largest, although its tonnage figures are relatively distorted by the very large amounts of petroleum which it receives. Other major Euroports are Antwerp, Bremen, Le Havre, Hamburg, London, Tilbury, Marseilles-Fos, Genoa, Barcelona and Lisbon. A feature of most EC countries is that trade is dominated by one or two of these major Euroports.

168   THE NEW EUROPE

In the case of the United Kingdom, there is a great difference. As well as London (Tilbury) and Felixstowe, there are over seventy small and medium-sized ports (fig. 8.8(b)) (related to the maritime traditions and favourable estuary locations around the UK shores) which handle the country's trade. The volume of container traffic (fig. 8.8(b)) gives a good illustration of the dominant ports, and also shows those ports which have grown rapidly in recent years, to take advantage of the cross-channel growth in trade. These include Dover and Felixstowe, but ports in the south and east have all generally benefited from the growth in European trade and transhipment. By contrast, ports in the west and north have declined with the reduction in relative importance of deep-sea cargo trade. Liverpool is one such example.

Principal lines of movement: core and periphery

The transport system as a whole within the Community is dominated by several corridors in which the rate and density of movement is at an extremely high level fig. 8.9). There are two primary north-south axes: the French corridor from Le Havre to Paris via the Seine, and thence to Lyons and Marseilles via the Rhône Valley; and the Rhine corridor from Rotterdam via the Ruhr to south Germany. These are supplemented by the Stuttgart-

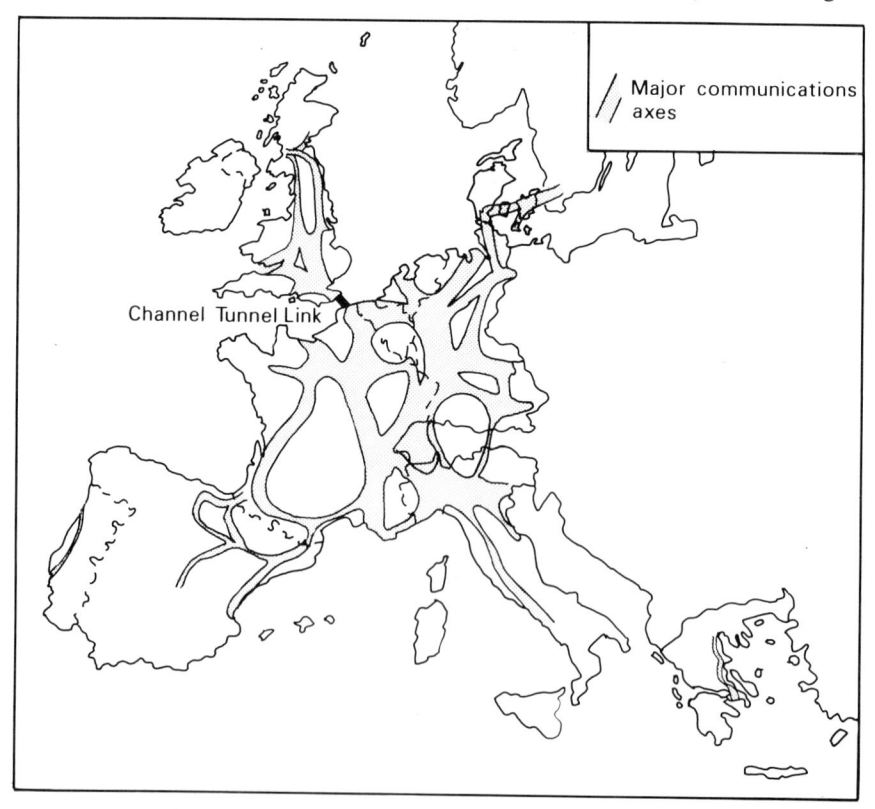

Figure 8.9 Principal lines of movement within the EC

Munich, Frankfurt-Main Valley and Hamburg-Bremem-Ruhr axes, and are linked to each other via the Belfort corridor and the Nord-Brussels Sambre-Meuse Valley. To the north the main UK axis from Lancashire to the south-east region is linked to the continent by the Channel sea lanes, whilst to the south the Italian line from Milan-Turin-Genoa is integrated into the main stream by means of trans-Alpine routes. The concept of a Community core will be developed subsequently, by showing that the main stream of movement is determined within a zone joining Manchester, Southampton, Paris, Marseilles, Milan, Frankfurt, Rotterdam and Leeds.

By contrast, the peripheral regions such as Ireland, Iberia, southern Italy and Greece tend to have localised or isolated movement patterns, for a range of physical, spatial and historical reasons.

The populous regions of the Euro-core with their access to markets, labour, finance, etc. gained immensely from the strengthening during the post-war period of these major European axes of movement or 'Euroroutes'. Evidence suggests that the effect of transport integration has been to reinforce access to the core regions at the expense of the periphery Economic activity is stimulated and reinforced in the Eurocore, and although new sub-axes emerge such as the Lyons-Grenoble-Marseilles axis, generally the improvement of transport facilities has had a limited effect upon peripheral regional growth. In effect the core regions have been enhanced whilst the periphery has been further peripheralised.

The Treaty of Rome and transport policy

The divergences that have emerged in the transport systems of the EC indicate the legacy of a number of separate national transport systems, which, though connected, do not operate as a single integrated system. There is a compelling need for Community action in the formulation of a coherent transport policy. It is set out in the Treaty of Rome that 'the community should establish a common transport policy to enable the free movement of people and goods over national boundaries'. A Common Transport Policy (CTP) was formulated as early as 1961 but in practical terms has accomplished little owing to the reluctance of national governments to allow real and practical progress. The two most important objectives are the establishment of fair competition and a regulated transport market. To promote fair competition there is the need to formulate a uniform system of transport taxation, with licensing rules which will allow lorries to make journeys throughout all member countries, and a standardisation of working hours and conditions. This involves the use of the tachograph, the automatic meter which monitors the lorry's journey. One of the most difficult problems has been the standardisation of overall lorry weights. In the UK the 50 tonne maximum weight is considered to be environmentally unacceptable. It is intended to achieve a regulated transport market by controlling freight rates; by maintenance and stabilisation of the railway system; and by a common sense of containerised freight to integrate the road, rail, inland waterway and

170 THE NEW EUROPE

sea transport network. It is likely that future policies will also attempt to look at the co-ordination of air and sea transport, because the enlarged Community is no longer linked by a continuous land surface and distances are much greater. Rhine shipping is such a major feature in the economy of five of the member states that some form of integration will be necessary. The co-ordination of air traffic control is now a serious issue.

The most important practical step, however, towards fast, uninterrupted movement and the completion of the Customs Union amongst the member countries has been the adaptation of separate motorway systems into a European classification. The 'E' road system (fig. 8.5) links the major trunk roads and motorways across national frontiers. Some intra-national routes have already been carefully co-ordinated, for example, the Dutch motorway from Rotterdam to Utrecht and Arnhem which connects with the German autobahn system in the Ruhr and Rhineland (fig. 8.6). The South Belgian motorway from Liège to Namur, Charleroi and Mons connects directly with the French system either south to Paris or north to Lille and Dunkerque. The Antwerp-Liège motorway joins the Germany system at Aachen. The EC Commission encourages consultation procedures which allow for this cross-frontier integration. The European Investment Bank has also contributed to many projects of this type such as the Brussels-Paris motorway, the Val d'Aosta-Mont Blanc and Grand St. Bernard tunnels and motorways, and the Brenner Pass route linking the Italian Autostrada system at Verona with the German system south of Munich. Paris to Cologne, Basle to Milan across the St. Gotthard and Simplon Passes, and Turin to Chambery across the Mt. Cenis pass are others which are either under construction or planned.

The partial implementation by national governments of the CTP has been criticised by the EC institutions, but the policy has been thrust into focus by the moves towards the single market in 1992. There is no doubt that the improvement of international links such as between France and West Germany, the improvement of links to the peripheral regions, particularly to the Mediterranean regions and islands, the reducing of bottlenecks in and around major cities such as London and Paris and the overcoming of natural obstacles by means of the Channel Tunnel project to improve access (figs 8.9 and 8.10) for the British Isles, are foremost on the agenda.

Journey time from London (Hours)

To	1989	1993
Paris	$5\frac{1}{2}$	3
Brussels	5	$2\frac{3}{4}$
Amsterdam	$10\frac{1}{2}$	$5\frac{1}{2}$
Luxembourg	$8\frac{3}{4}$	6
Bonn	9	$6\frac{3}{4}$
Copenhagen	$19\frac{3}{4}$	17
Rome	$24\frac{3}{4}$	$21\frac{1}{2}$
Madrid	$25\frac{1}{2}$	21

Figure 8.10 The Channel Tunnel: estimated effect upon accessibility between the UK and the continent

9
Population: the axis of city development

Diversity

The EC has a combined population of 320 million people, and includes four large countries each with populations of nearly 60 million, Spain, with 38 million, six smaller states, and the tiny Grand Duchy of Luxembourg (fig. 9.1). This economic grouping, stretching from Scotland to Sicily, and from Ireland to Greece, includes a great number of ethnic and linguistic types. The peoples of Italy and France, traditionally known as Latin or Mediterranean, speak languages belonging to the Romance group, whilst the northern part of the community is dominated by the Teutonic group of languages (German, English, Scandinavian and Dutch). The Roman Catholic Church in the south is broadly associated with the Romance language groups and more specifically with Spain, France and Italy, whilst there is a similar link between Protestantism and the areas around the North Sea. There is, therefore, a broad distinction between the northern and southern sections of the community on the basis of religion, language and culture.

In general, the language divisions are those of national frontiers, but there are several complicating factors. Minority languages exist alongside major national languages as in the case of the Welsh, Gaelic, Breton and Basque areas. Along the Franco-German frontier zone in Alsace and Lorraine is a German-speaking minority, and the whole Rhineland area in the past has been a zone of contention between the two major groups, French and Germanic. In Belgium there is a division along linguistic lines between the French-speaking Walloons in the south and the Flemings in the north. However, these distinctions have begun to lose their former psychological importance, and the economic pressures of the twentieth century, combined with political maturity and a desire for peaceful cooperation and federation, have created 'unity within diversity'. There is of necessity a system of official languages in the Community, French, English, German, Italian, Spanish, Portuguese, Greek, Dutch and Danish, the first two being in most common use.

Distribution and density of population

Distribution of population tends to reflect relative regional advantages, and indicates the countries with the highest degrees of industrialisation, popula-

172   THE NEW EUROPE

	Population	Percentage in agriculture	Percentage in industry	Percentage in services	Percentage of population urbanised
Belgium	9 853 000	3.0	29.3	67.8	96
France	55 170 000	7.3	31.3	61.3	73
West Germany	61 024 000	5.3	40.9	53.7	86
Italy	57 141 000	10.9	33.1	56.0	69
Luxembourg	370 000	4.0	33.0	63.0	69
Netherlands	14 492 000	4.8	26.8	68.4	88
United Kingdom	56 617 000	2.6	31.1	66.4	92
Denmark	5 114 000	6.2	26.9	66.1	86
Ireland	3 540 000	15.8	28.3	55.8	57
Greece	9 934 000	28.5	28.1	43.4	65
Spain	38 505 000	16.1	32.1	51.8	77
Portugal	10 157 000	21.9	34.1	44.0	31

Figure 9.1 Population, employment and urbanisation, 1986

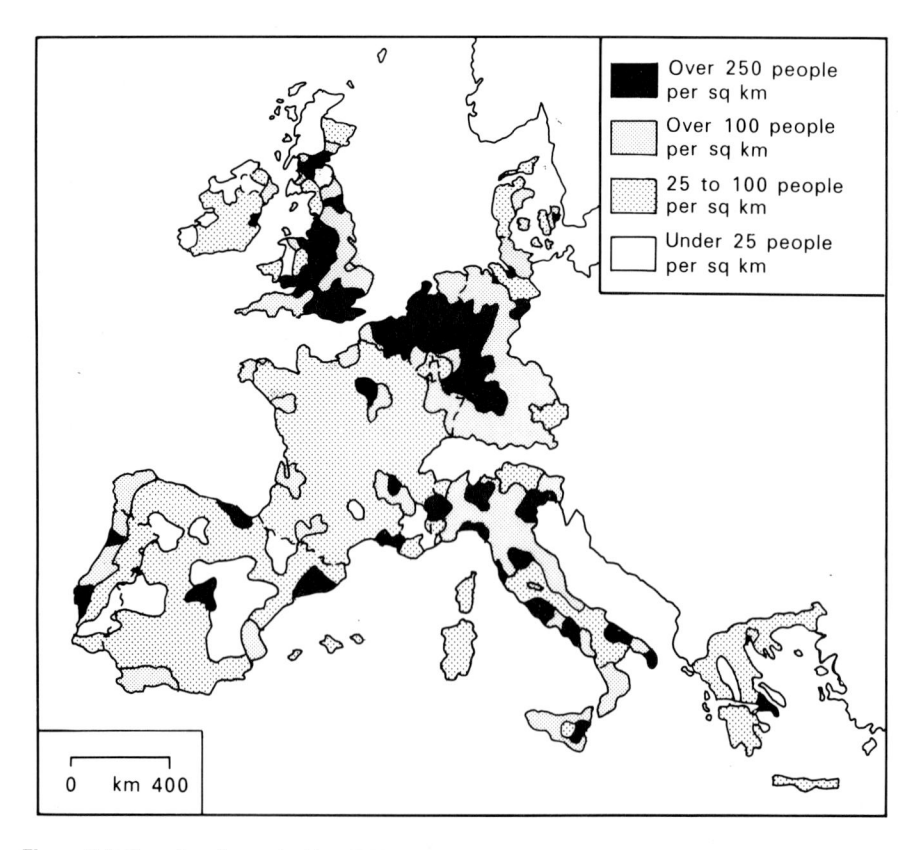

Figure 9.2 Density of population, 1986

tion, wealth and urbanisation (fig. 9.2). In summary, there is a densely populated core and sparsely settled periphery.

Using national statistics, the member states may be divided into four groups (fig. 9.3). The Netherlands and Belgium have a very high population density with over 300 people per square kilometre. The Netherlands has the highest density of all with its population concentrated in the Randstad conurbation. Belgium's population is highest around the Brussels-Antwerp area, with similar concentrations throughout the commercial cities of Flanders, and the Sambre-Meuse coalfield.

High densities of population are recorded in the UK, West Germany, and Italy, but there are considerable regional variations. The Italian population concentrations in the Lombardy lowlands, Ligurian coast, and the Rome-Naples area contrast with the sparsely populated Alps and Central Apennines. West Germany and the UK have major concentrations in their industrial zones and city regions. These favoured zones of population concentration, the German Rhinelands and the English lowland (fig. 9.2), contrast significantly with many upland areas in both countries which have extremely low densities.

France, Spain, Portugal, Denmark, Luxembourg and Greece have relatively low overall densities of under 140 people per square kilometre, reflecting their greater agricultural and rural character. France has a remarkably even distribution of population, and apart from Paris there are few concentrations of any great extent. Even the Nord and Lorraine heavy industrial areas never reached the same degree of population density as comparable areas in West Germany or the UK. In the Iberian peninsula the well-watered Atlantic coastal lowlands of Portugal have favoured a relatively dense agricultural population, whilst in Spain there are large stretches of underpopulated arid mountain plateaux. Three urban industrial concentrations exist around Madrid, Barcelona and Bilbao. Luxembourg has the thinly-populated Ardennes in the north, but Luxembourg City and the steel-making district in the south have a moderately high density (250 per square kilometre), whilst Copenhagen is a significant exception to the rule in Denmark. Ireland has the lowest density of all, with a long tradition of rural depopulation, lack of mineral resources and a peripheral position in relation to Europe as a whole. The only centre of any size is the capital, Dublin.

Economic structure and population patterns

Population density is one indication of the rate and scale of economic activity, but equally valid indices are employment characteristics and the degree of urbanisation. The proportions of the population employed in agriculture, industry and services measure fairly accurately the stage of evolution which any country has reached (fig. 9.1). The percentage of the population which is urbanised indicates a sophisticated life-style associated with a mature economy. The relative proportions for each sector of the economy in the EC countries are compared with those for the USA which has a declining proportion of employment in industry, and a high proportion of employment

Country	Population per square kilometre
Netherlands	347
Belgium	323
West Germany	245
United Kingdom	232
Italy	190
Luxembourg	141
Denmark	119
Portugal	100
France	50
Ireland	75
Greece	110
Spain	76

Figure 9.3 Density of population, 1986

in the tertiary sector. The UK closely approaches the American situation, with a well-developed tertiary sector, efficient agriculture, large industrial sector and high degree of urbanisation. The Netherlands, Belgium, Denmark and West Germany have largely similar characteristics except that West Germany has a larger industrial sector. In all these north-west European countries there is an advanced social and economic infrastructure associated with a high standard of living and urban life-style. France and Italy have developed rapidly during the last forty years, but still have an enlarged agricultural sector. There is a different emphasis when examining Portugal, Ireland and Greece, which have much larger agricultural sectors, an underdeveloped tertiary sector and often a lower degree of urbanisation. Indeed, an examination of primacy (the percentage of urban population in the largest city) shows major differences across the Mediterranean basin. Italy and Spain have a balanced, mature urban hierarchy, with only eight per cent and ten per cent respectively of the urban population in the largest city. In Greece (Athens, 46 per cent) and Portugal (Lisbon, 52 per cent) the percentage is much higher, illustrating the rapid contemporary rural and urban migration which has been concentrated heavily into the capital city region creating an unbalanced spatial structure. Athens is a rather unfortunate example of the rapid transformation without planning constraints of what forty years ago was a relatively small city of half a million people. Now there is a considerable unattractive urban sprawl which has surrounded the ancient city. These differences illustrate a complex gradient of economic types ranging from the core countries of the north-west to the less developed peripheral countries of the Community.

National averages tend to blur the differences between contrasting regions

	Total population	Agriculture	Industry	Services	Population Density (km²)
Ile De France	10 227 000	0.4	30.4	69.2	851
Piedmont	4 403 000	3.7	46.9	49.4	173
Basilicata	618 000	11.4	39.6	49.0	62
Liguria	1 775 000	2.4	28.9	68.7	328
Normandy (Lower)	1 370 000	7.8	37.0	55.2	78
N. Rhine/Westphalia	16 686 000	1.2	42.9	55.8	490
Brabant (Brussels)	2 218 000	1.0	24.4	74.6	661
Limousin	736 000	7.1	33.9	59.0	44
North Holland	2 317 000	2.6	29.5	67.9	783
Lower Bavaria	1 012 000	6.6	40.6	52.8	98
S.E. England	17 192 000	0.8	31.2	68.0	632
Crete	513 000	33.5	20.6	45.9	62
Alentejo	558 000	27.2	26.8	46.0	21
Castilla La Mancha	1 680 000	14.4	34.9	50.7	21

Figure 9.4 Regional characteristics (gross value) 1985

of economic sophistication or underdevelopment, and some idea of these extremes may be gathered from the figures which compare sample EC regions (fig. 9.4). Concentrations of population in south-east England and Greater Paris, and North-Rhine Westphalia contrast with the very low populations of Limousin, Crete, Alentejo and Basilicata. The service employment levels in Paris, Brabant (Brussels) and Liguria, the well populated regions of Lower Normandy and Lower Bavaria, and the industrialised Piedmont and North-Rhine Westphalia are of equal significance. Here the essential contrasts between regions can be seen: wealthy urbanised regions with a high level of service-sector employment; industrial conurbations with a high percentage of employment in industry; rich farming areas with high rural populations, and isolated marginal farming areas barely able to sustain even low population levels.

Areas of population decline

The rate of decline of rural, peripheral, or upland regions varies considerably. In the UK rural depopulation has been going on since the sixteenth century and has produced a farming population which is less then three per cent of the total labour force, as well as a high level of urban concentration. Peripheral mountainous areas are still losing people in spite of Government schemes to introduce employment and improve facilities. Since the 1930s the principal losses have been in the Scottish Highlands, and the uplands of the Cheviots, Pennines, the Lake District and Wales. Smaller losses are also recorded in the

176 THE NEW EUROPE

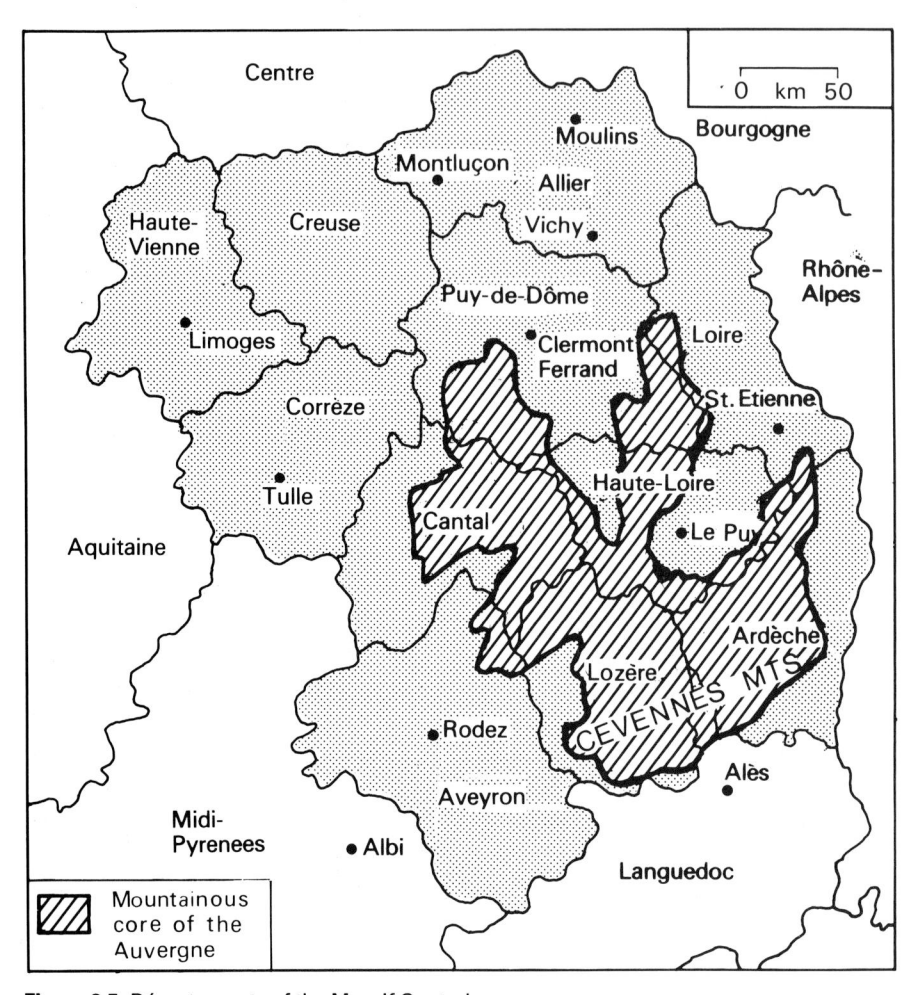

Figure 9.5 Départements of the Massif Central

remoter rural parts of East Anglia, Cornwall and the Welsh border counties.

In France, regions like the Massif Central have a complex internal population movement. Although population is lost from the central areas, there is often a concentration into lowland fringes and valleys within the zones as a whole. The Massif Central (fig. 9.5) is still losing people from its mountainous spine, particularly the départements of Haute-Loire, Cantal, Aveyron, Lozère, Corrèze and Creuse. The valley of the Allier in the north, containing Moulins and Clermont Ferrand, has seen considerable growth. Clermont Ferrand is practically a 'company town' of the Michelin Tyre Group, and also gains by being in easy communication with Paris. Limoges and St. Etienne are other old industrial towns on the margin which are minor growth areas.

In Belgium, the Ardennes are within commuting range of the Liège conurbation to the north, and towns such as Dinant, Bastogne and Spa have experienced growth in a new dormitory role. In addition, there is the development of 'second homes' in the Ourthe Valley, and upland regions like the Ardennes may well become re-populated, as they are becoming a major leisure area (fig. 13.6).

There are great differences between northern and southern Italy (fig. 9.6). The out-migration from the Mezzogiorno is a sustained response to the overpopulation, low standards of living and poor employment opportunities. Much of the movement is to Piedmont, Lombardy and Liguria, but in addition there has been a major movement into West Germany and other EC countries, a result of the free movement of labour allowed within the Community. In the decade up to 1980 the Mezzogiorno lost about half a million people.

In Greece, the rapid development of the Athens-Piraeus lowland, which now has one third of the total population of the country is an example of the excessive 'pull' of primate capital cities to the detriment of peripheral areas of the country.

Old and exhausted coal-mining areas are also responsible for pockets of either population decline, relative stagnation, or unemployment and dislocation. Commuting is a partial answer. The Belgian Borinage is an example where many ex-miners travel daily to Charleroi and Brussels. The small high-cost coal-fields of the margins of the Massif Central, Montluçon, Decazeville and Commentry are now under a complete closure plan which will intensify migration from the whole area. In the former mining vallyes of South Wales, migration and commuting to Cardiff and the coastal area has taken place. In the north-east of England there is heavy migration from the old Durham mining villages to new towns such as Washington and Peterlee. The Nord region of France has similar problems, losing over 200 000 people by migration during the last 25 years (fig. 9.11).

Areas of population growth

Population growth areas lie around the major nodes of industry, services and transport, the city regions, and in new residential and leisure areas.

Low-cost assembly points on major estuaries and transport nodes such as Rotterdam, Antwerp, Hamburg, Marseilles and Teesside have become the growth points for such industries as oil refining, petrochemicals and other basic industries. The West Midlands conurbation, at the central point of convergence of the UK motorway system, possesses a varied and regenerating range of industry and services.

The capital cities of core regions of the member states have experienced continuous population growth since the mid-nineteenth century. Their attraction lies in their functions as administrative, cultural, service and prestige centres. They lie at the focal point of the national transport network, and have two significant resources - a pool of skilled labour and a huge market. They are examples of the ultimate factor behind multi-million city growth: the non-basic or self-sustaining capacity of the large population concentration.

178 THE NEW EUROPE

Mont Gerbier in the Massif Central of France: a rugged volcanic upland of marginal farming and rural depopulation

London, the Randstad, Brussels and Paris are good examples.

The Rhine Valley of West Germany has experienced sustained population growth (fig. 9.6) but illustrates a variation on the capital city concentration theme. West Germany has a political history of many independent states each with its capital city, and with the loss of Berlin resulting from the division of Germany, is a country with a large number of almost equal regional centres. No one city dominates completely. Thus various cities in the Rhine valley enjoy specialised, yet inter-related, functions; Bonn as administrative centre; Frankfurt and Düsseldorf as financial centres; Duisburg as the major port; Cologne and Mannheim as major commercial and industrial cities. In South Germany Munich, Stuttgart, and Nuremberg enjoy extensive regional status. In Italy, the principal manufacturing and core region lies in the northern triangle of Milan, Turin and Genoa. Nevertheless, the historic, cultural and administrative functions of Rome, and the regional importance of Naples, ensure their continuing growth.

Population growth in coastal and other environmentally attractive areas is associated with wealth, holidays, retirement, mobility and the phenomenon of the 'second home'. Large areas of the English south coast have become continuous suburban developments as towns like Bournemouth and Poole, and the Bognor Regis and Brighton group, coalesce. They are the new leisure service centres. On a larger scale is the pattern of linear growth along the

Region	Net migration 1960–1975	Net migration 1982
North–Rhine Westphalia	+897 998	−29 764
Baden–Württemberg (South Rhineland)	+732 301	+5 781
Hessen (Middle Rhineland)	+704 578	+7 093
Piedmont	+644 369	−8 222
Lombardy	+866 846	+3 757
Liguria	+184 761	+288
Brabant	+188 486	−549
Lower Saxony	+251 401	+6 227
Limburg (Holland)	+686	−991
Campania	−430 992	−12 627
Calabria	−425 382	−4 939
Mezzogiorno (total)	−2 355 725	−22 168

Figure 9.6 Migration flows in sample regions

Cote D'Azur from Cannes to Monte Carlo and along the coast of the Ligurian Riviera (fig. 9.11).

The conurbation and city region

The most characteristic feature of contemporary Europe is the extent and scale of city growth (figs. 9.8 and 9.9). The percentage of the population which is urbanised is above 70 per cent in all EC countries except Portugal, Ireland, Greece and Italy.

The pattern of small nucleated cities of the medieval period was succeeded in the nineteenth century by the conurbation, a sprawling structure composed of a number of coalescing industrial towns. The city of Stoke-on-Trent comprises six towns based originally upon the North Staffordshire coalfield and the pottery industry, and is perhaps the best example of a polycentric agglomeration of industrial towns. Rhine-Ruhr, Lille-Roubaix-Tourcoing and Greater Manchester (fig. 9.7) are much larger examples. The twentieth century has witnessed such an increase in mobility that people often now live up to 60 km out of the city and commute daily to work. The life-style of the city is a great attraction, and its retail, commercial, administrative and cultural services expand so that congestion becomes the great problem in the city centre. Replacement of obsolescent property and changes in residential and employment patterns has meant displacement of people to the city margins, with new housing estates and a vast expansion in the total city area.

The term 'city' is an insufficient description. Paris city has 2.6 million people; the Paris region includes the city and four départements. It extends over nearly 2000 km^2 and has a population of 9.9 million. The London planning region, bounded approximately by a radius of 60 km from Charing

180 THE NEW EUROPE

Monte Carlo on the Cote D'Azur: centre of the French sun-belt and rapid population growth

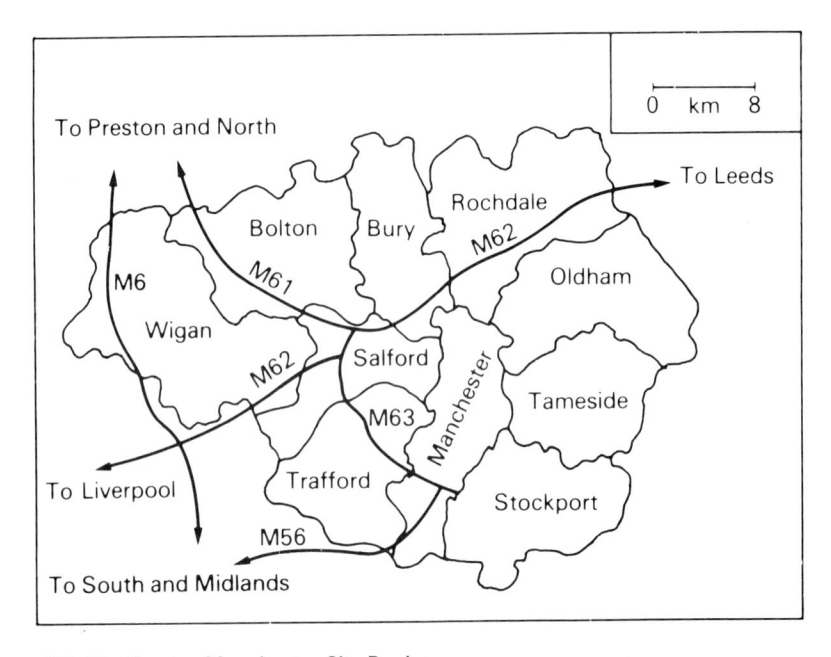

Figure 9.7 The Greater Manchester City Region

Cross station, has a population of 12.5 million people, and is really three concentric zones. The Greater London conurbation, with nearly eight million people, forms the core and inner ring, and is surrounded by the green belt, a successful device which has substantially arrested continuous growth. The outer ring consists of old centres such as Aylesbury and Guildford, and new towns such as Stevenage and Crawley, established after the 1947 New Towns

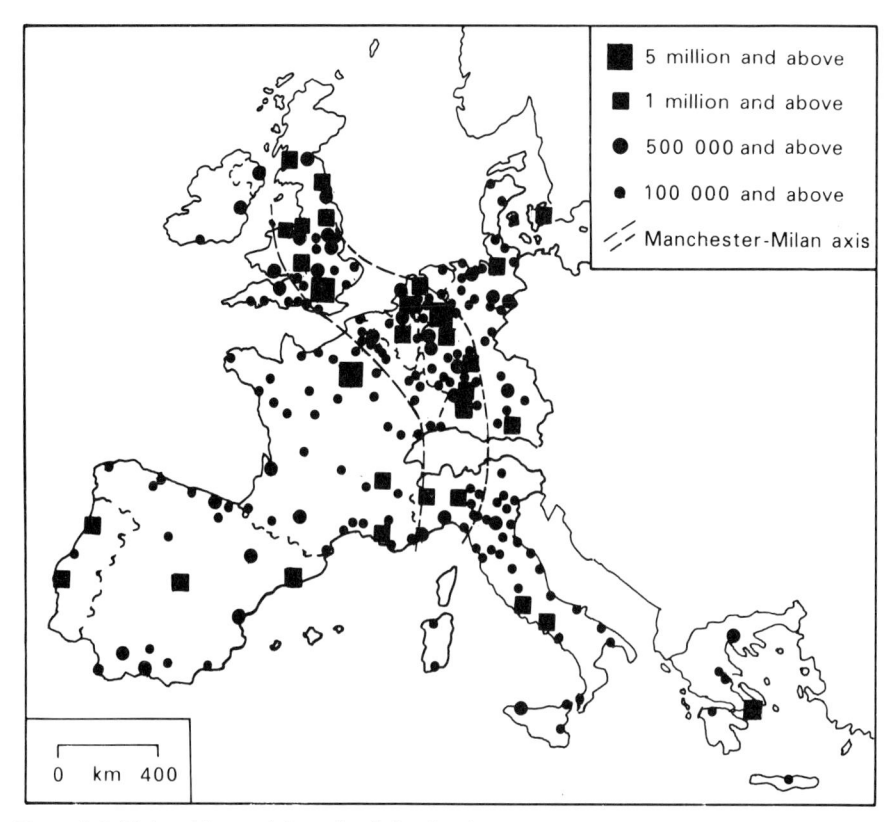

Figure 9.8 Major cities and the axis of city development

Act. The rapid suburbanisation of this outer ring has proceeded apace, so that this is now a part of London's 'commuter belt'. The effects of city growth now reach far beyond the London planning region, however. The whole of south-east England from the Wash to the Solent has two-fifths of the total population of the United Kingdom.

Such regions of dynamic growth are 'city regions' dominated by the central city and surrounded by industry and housing, leisure areas and open space. On the fringe there are old-established towns, metropolitan villages and non-contiguous residential areas and new towns. These are all linked to the services, employment facilities and life style of the city by fast suburban communications. The four largest city regions in the EC are Greater London, Rhine-Ruhr, Greater Paris and Randstad. These are primate cities of world rank and complex metropolitan regions. They are followed by a group each with populations around the 2.5 - 3 million mark; Madrid, Athens, Rome, Greater Manchester (fig. 9.9), the West Midlands and Hamburg. The very great number of medium-sized cities with populations of over 500 000 demonstrates fully the substantial degree of urbanisation which now exists.

The axis of population concentration

There are two central areas of concentration (fig. 9.2). In the UK there is a quadrilateral stretching from Manchester and Leeds to the south coast. On the continent there is a triangle best described by joining lines from Stuttgart to the coast at Dunkerque and Hamburg. These are not exclusive, and areas external to these such as Greater Paris, Lyons, Northern Italy and Rome are important subsidiary nodes. If a high rate of migration is taken as the most significant factor, the Mediterranean coastlands of France show sustained growth over a long period.

The main axis of city development thus stretches from South Lancashire to Northern Italy, often referred to as the Manchester - Milan axis (fig. 9.8). This is a wide belt which essentially crosses the English lowlands and extends to the heavily urbanised North Sea coastlands stretching from Lille through Belgium to the Randstad. The River Rhine has three great city agglomerations. Rhine-Ruhr, Rhine-Main, and Rhine-Neckar, which embrace the great regional and industrial centres of West Germany. Though broken by the Alpine mountain chain it re-emerges in the Plain of Lombardy, with its traditions of city life and the economic core area of Italy. Some sixty per cent of all the million cities in Europe are found in the United Kingdom, West Germany, Italy and the Netherlands.

Contemporary Population change: a summary

In Western Europe this may be understood by reference to three principal themes. After the traditional or pre-modern stage up to the late eighteenth century there was the demographic transformation of the Industrial Revolution in which rapid growth of population and an expanding labour force was absorbed by rapidly expanding manufacturing industries in the cities and industrial areas. This was a period of population concentration and rural depopulation. The rapid changes in population have slackened since the 1960s and during a modern phase during the 1970s and onwards there has been a population state much closer to equilibrium, with birth rates falling, the population ageing and migration patterns becoming much more complex with a particular difference emerging between north-western Europe and southern, or peripheral, Europe.

The European Community enjoyed rapid growth in both population and economic terms during the period from 1950 to 1970. All countries of the EC experienced this growth in varying degrees. The German economic miracle was a term associated with the growth in population in West Germany to 61 million people. During this period the German economy gained some seven million labour migrants as refugees from the Eastern European countries poured in and as workers from Yugoslavia, Turkey and Southern Italy also entered the country, attracted by its booming economy. France gained large numbers of people from French North Africa and in the case of the United Kingdom there were migrants from Ireland and the Commonwealth.

In addition to this international migration there was a large scale natural

City	Population 1985 (1000s)	EC rank
London Region	12 762	1
Rhine-Ruhr	10 419	2
Paris Region	9 906	3
Randstad	5 093	4
Madrid	3 218	5
Athens	3 027	6
West Berlin	2 985	7
West Midlands conurbation	2 981	8
Rome	2 828	9
Greater Manchester	2 541	10
Hamburg	2 407	11
Glasgow (Clydeside)	2 008	12
Leeds-Bradford (W. Yorkshire)	1 945	13
Stuttgart	1 935	14
Liverpool (Merseyside)	1 823	15
Barcelona	1 757	16
Milan	1 750	17
Lisbon	1 612	18
Mannheim–Ludwigshaven	1 578	19
Frankfurt-on-Main conurbation	1 520	20
Munich	1 502	21
Copenhagen	1 380	22
Oporto	1 315	23
Naples	1 233	24
Turin	1 202	25
Lyons	1 170	26
Brussels	1 075	27
Marseilles	1 080	28
Lille-Roubaix	995	29
Cologne	916	30
Dublin	915	31
Genoa	842	32
Newcastle-on-Tyne	804	33
Valencia	764	34
Palermo	714	35
Thessaloniki	706	36
Seville	674	37
Antwerp	663	38
Bordeaux	628	39
Bremen	606	40
Zaragoza	593	41
Hanover	554	42
Sheffield	538	43
Toulouse	523	44
Nuremberg	515	45
Bologna	500	46

Figure 9.9 Rank-size of EC cities

increase. The post-war baby boom reached its second peak in the mid-1960s. One of the major economic pressures which was operating was that of rural de-population. The movement towards more efficient farming, rationalisation, modernisation and intensive production of food was associated with the growth of manufacturing, mass-production and the tertiary economy in the city regions. Consumer wealth and a rapidly growing transport system have vastly increased population mobility and intensified this trend. The movement away from isolated rural, peripheral or upland environments towards the more favoured industrial regions is known as urban concentration and rural de-population. This is complemented by the economic pull factor exerted by the metropolitan cities, core areas and expanding service infrastructure and is exemplified by the economic growth of the Rhineland cities of West Germany and by the drift to the South-East in the United Kingdom.

Since the mid-1970s the rates of population increase have slowed very markedly and from 1973 to 1985 the EC countries showed only a five per cent increase in population, much lower than the previous two decades. The rates of natural increase slackened particularly in the richer countries of the core of North-West Europe. There are many complex reasons for this including changing age structure and the position of women in society. The most specific reason was the on set of the economic recession from 1973 onwards. The decline in industrial production and consequent unemployment led to a reduced demand for labour migrants, the drying up of the process of international migration and the phenomenon of return migration where the so called 'Gastarbeiter' or guest workers in West Germany were no longer welcome. In the Netherlands there was a marked downgrading of future estimates of population growth. However, the Atlantic and Mediterranean periphery, particularly countries such as Greece, Spain and Ireland, are still recording rapid natural increases in population, and thus an interesting demographic contrast emerges. Italy illustrates an interesting contrast within her national boundaries. Northern Italy is well developed and part of the mainstream economy of Western Europe and illustrates the same population characteristics. However the Mezzogiorno still retains many demographic traits typical of the pre-modern period. It has persistently high birth rates which are associated with geographical inaccessibility, a predominance of agricultural employment, low urbanisation rates and low female economic activity rates.

Case-study: France

Up to 1975 the core regions experienced the greatest volume of population increase. In France this is illustrated by figures 9.10 and 9.11. Only three regions persistently gained population on a large scale: Paris, Provence-Cote D'Azur and Rhône Alpes. These three absorbed half the total population increases from 1945 - 1975 by which time they contained over 35 per cent of the total population of France. Net migration losses were recorded in the Massif Central, an impoverished and isolated upland region in Western

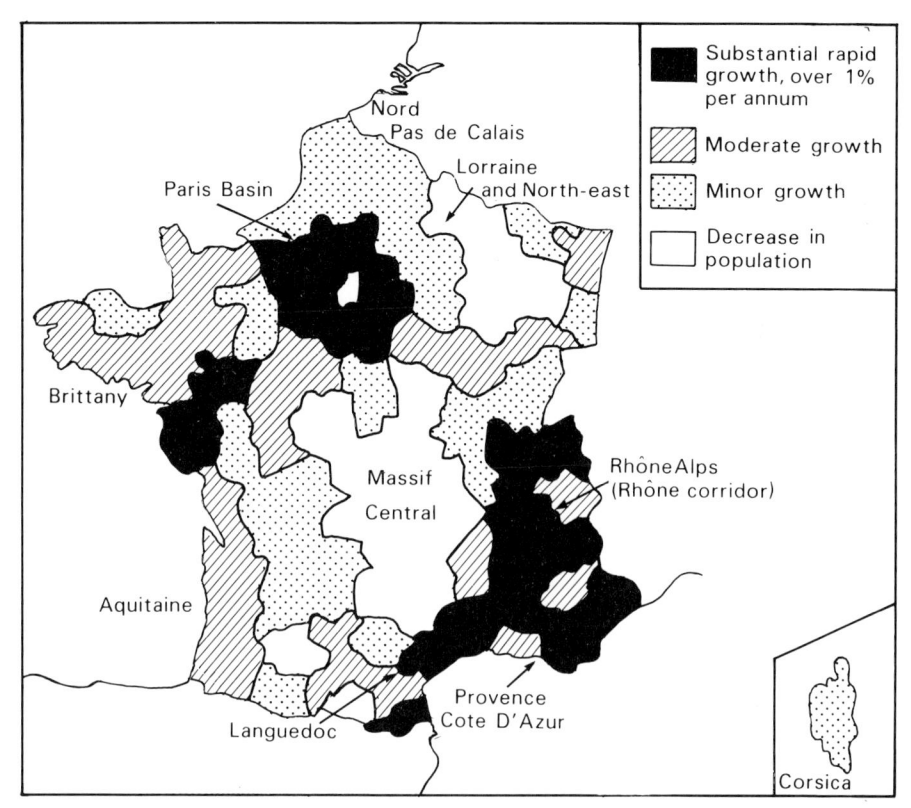

Figure 9.10 France: population change 1975–82 (source - *INSEE (census 1982)*)

France, basically an area of contracting agricultural employment with few alternative sources of employment, and in the old industrial area of the North-East where contracting employment in coal mining, heavy metallurgy and textiles was combined with a slow replacement by new industries.

From the mid 1970s onwards the spatial expression of this was considerably modified. In the first place the collapse in employment in traditional industries such as coal, steel, shipbuilding and textiles, and the recessionary pressures on the economy, dampened down buoyancy in areas which had been 'core' in the 1960s. A good example of this is Lorraine in Eastern France in which the depressed steel and engineering industry had led to considerable emigration during the late 1970s and 80s. Within the city regions a process of de-concentration and counter-urbanisation began. The process of inner-city decline and slum clearance schemes, together with increasing affluence, car ownership and the need for more space for residential purposes and the decentralisation of employment to the edge of the city region, all combined to mean decentralisation from the cities to the suburbs, away from the larger cities to small and medium sized settlements down the urban hierarchy, and from the older settled industrial areas to the accessible rural areas. Migration patterns have become much more complex.

The 1982 census in France reflects these trends. In the first place there was

	Net total migration 1954–1975	Net total migration 1982	Total population 1982 (million)
Ile De France	+1 152 000	−63 349	10.1
Lorraine	−31 939	−12 737	2.3
Nord/Pas de Calais	−202 907	−18 449	3.9
Brittany	−76 559	+7 891	2.7
Auvergne	−24 931	+723	1.3
Languedoc-Roussillon	+330 000	+17 463	1.9
Provence-Cote D'Azur	+969 568	+28 940	3.9
Rhône-Alpes	+516 105	+6 446	5.0

Figure 9.11 France: population change 1954–82

the beginning of regeneration of growth in areas which had been characterised by population decline for well over a century. These included Brittany which was now much more prosperous partly due to tourism, the Alps and parts of the Massif Central which are more accessible. There are, however, still broad areas of countryside beyond the reach of urban influence which are still in decline. Those mountainous areas of the Massif Central which are still losing population include particularly the départements of Haute Loire, Cantal, Aveyron, Lozère, Correze and Creuse (fig. 9.5). The industrial départements of the North and East are losing population, as are the inner départements of the City of Paris. It is in Paris that the most interesting changes have taken place. The Paris region now has a negative migration total and this partially reflects the success of the French authorities in checking the rate of the growth of the city. Growth has been concentrated in the five new towns on the edge of Paris and in the accessible rural areas up to 100 km out of the City. The population growth of the Mediterranean regions continues with their attraction for tourism and retirement in addition to the important industrial developments around Marseilles. Not only the Cote D'Azur but also the area of the South-East Rhône Valley around Provence and the coast, the so called 'sun belt', has seen consistent growth. Those towns and cities with the most longstanding growth include Cannes, Antibes, Aix-en-Provence, Montpellier, Avignon and Toulon.

Another important area of consistent growth is the city region of Lyons with strong growth in the Alpine - Isère - Rhône corridor, and in particular the cities of Grenoble, Annecy and Lyons itself.

Within the stages of urbanisation (urban concentration, suburbanisation and now counter-urbanisation) we may see the contrasts within the urban system of Western Europe. The population contrasts which have been shown allow us to consider the existence within the European community of a dynamic core and problem periphery.

10
Regional disparities: core and periphery

Introduction

Much of Western Europe has experienced substantial economic growth and greatly increased prosperity since 1945. More varied energy sources, increased farm production, greater industrial efficiency, greater mobility, a larger volume of trade, greater purchasing power and a wide variety of consumer goods, all create an impression of prosperity. The principal problem, however, within this wealthy society is that of 'dualism'. This is the propensity to uneven growth and the emergence of marked contrasts between natural growth areas and problem regions. Calabria in Southern Italy has less than forty per cent of the average income per head in the EC, whilst the Brussels region has 150 per cent. The basic reason for these disparities, in part at least, is the operation of the Common Market itself. Increased cross-frontier competition, and operation of the large free market amongst 320 million people, means that stronger competitors gain at the expense of weaker ones. Industry tends to move to the area which will allow it to operate at maximum efficiency and with the highest profit. There is recognition that the free market has to be distorted to a certain extent to aid the weaker regions, and that, for example, Italy's Mezzogiorno needs assistance to enable it to compete with the factories of Milan and Turin. The establishment of the principle that less intensive, inactive or declining areas require considerable aid or reconstruction, is one of the very significant innovations in the modern industrial state. It has become an important policy-making area in the EC.

The economic heartland

The preceding chapters have attempted to illustrate a major theme: the contrasts between the economic heartland and the peripheral regions of the EC.
1. In **agriculture**, the area of large-scale, intensive and efficient food production centres upon the polders, boulder clays and loess lowlands of the North-west European Plain and south-eastern England (fig. 6.13).
2. The **Home Energy Zone** (fig. 10.1) includes the coalfields of the United Kingdom and the Heavy Industrial Triangle which produce ninety per cent

188   THE NEW EUROPE

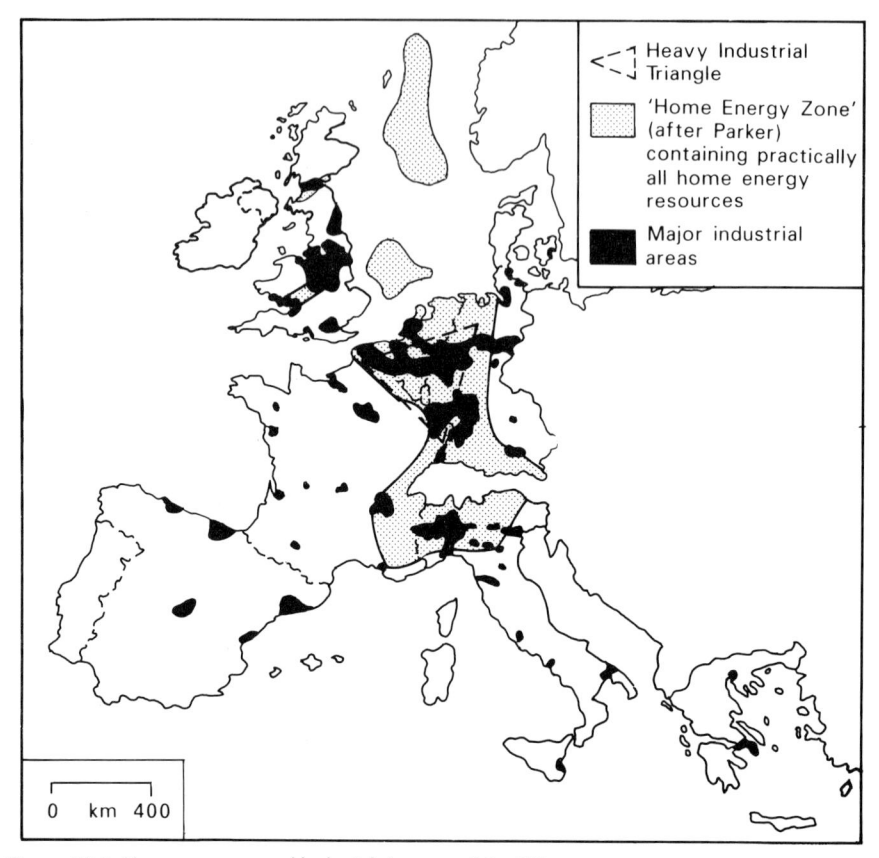

Figure 10.1 Home energy and industrial zones of the EC

of the Community's coal and lignite. Three-quarters of EC hydro-electricity comes from the Alpine zone of south-eastern France and Italy, which forms the southern portion of the Home Energy Zone. If the reserves of gas and oil in the North Sea are added, then the vast bulk of the Community's home energy supplies will continue to be found along a north-south axis which includes the North Sea littoral and the Rhinelands. This is the 'Home Energy Production Zone' of G Parker.
3. **Manufacturing Industry** (fig. 10.1) is concentrated in the German Rhinelands, Eastern France, Central Belgium and the Rhine delta, with a very large proportion of the community's coal, iron-ore, steel, refined oil, engineering, chemicals, textiles and vehicle production. To this industrial triangle must be added the Manchester-London axis, Greater Paris and the Milan-Turin core, for although physically separate, they form a logical extension of the central industrial axis.
4. **National core areas** of six major Community countries lie within this zone (fig. 10.2). Community decision-making lies between Brussels, Luxembourg and Strasbourg. London, Paris and Bonn are the capital cities of three key industrial and political powers.

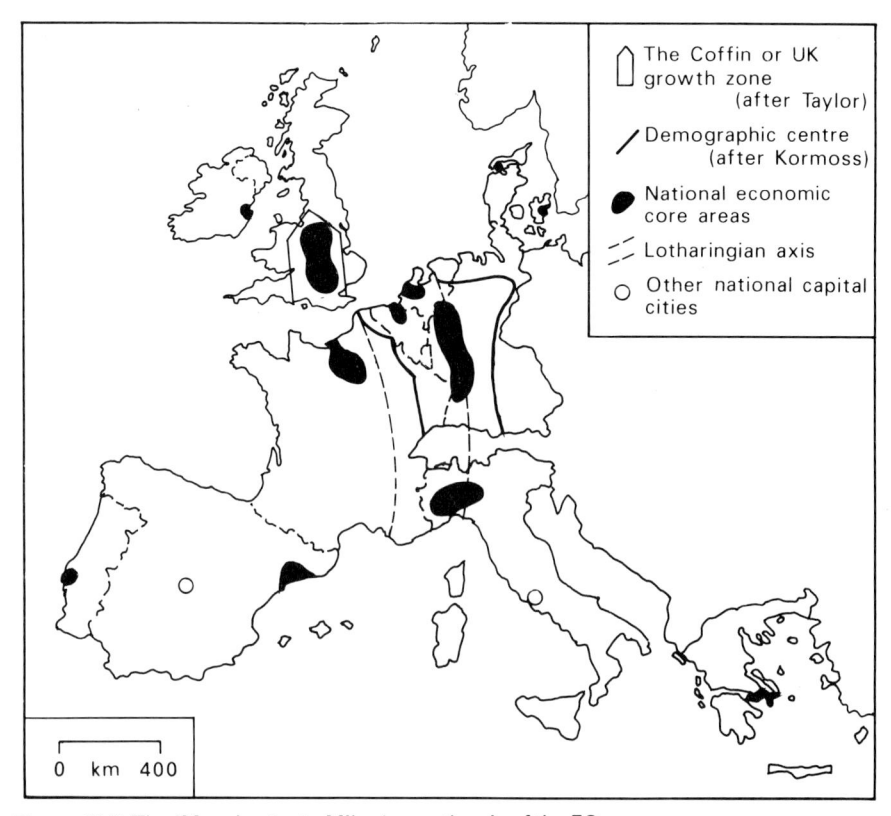

Figure 10.2 The 'Manchester to Milan' growth axis of the EC

5. **Mobility** is at its height along the Rhineland routeways (figs. 8.2, 8.5 and 8.6) and the waterway system of the Rhine and its tributaries is both physically and psychologically the central artery of the community, containing its greatest seaport, Rotterdam.
6. **The demographic heart of the community** (fig. 10.2) lies within an area delimited by Boulogne-Nancy-Stuttgart-Hanover-Amsterdam (Kormoss, 1959). In the original six-member community, this area had twenty per cent of the land area but forty per cent of the population, and contained some 80 million people. The map of population density (fig. 9.2) confirms this picture of a heavily populated core. If the UK 'coffin' (after Taylor) (fig. 10.2) is added, the existence of a population heartland in the Rhineland stretching across the North Sea into England is a reality.
7. **Wealth**, as measured by gross domestic product per capita, is another index which illustrates the relative wealth of much of the central part of the EC, and in particular the national core areas. By contrast, the peripheral areas of Iberia, Greece, southern Italy, western France and Ireland are at substantially lower levels of income (fig. 10.3). The estimates of gross domestic product, however, must be treated with some reserve as national figures are not exactly comparable.

190   THE NEW EUROPE

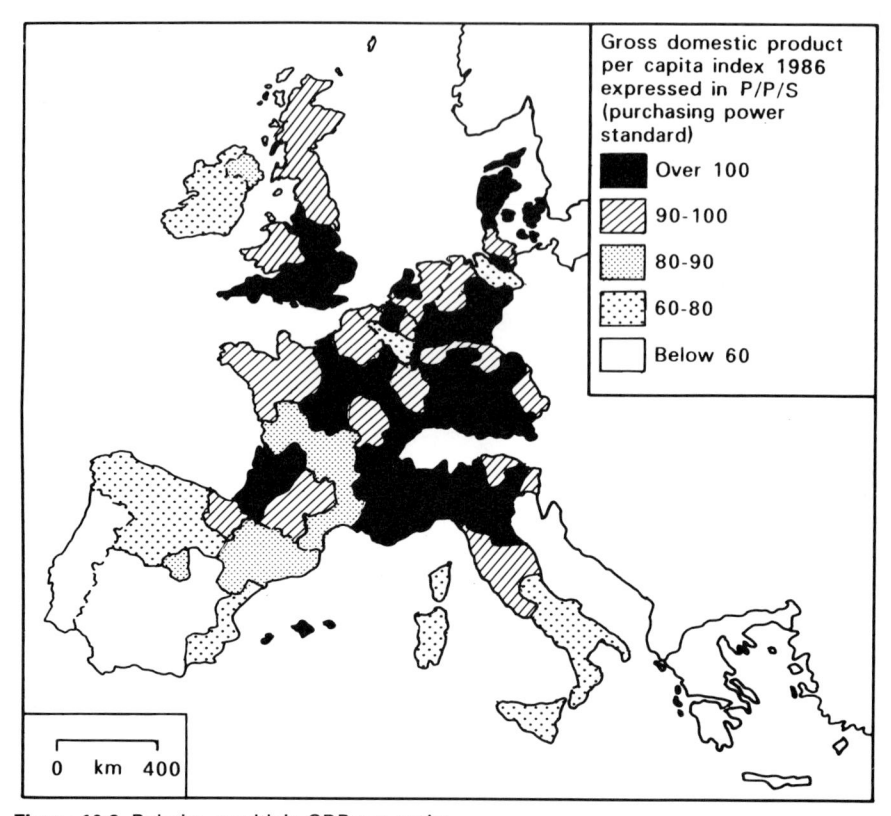

Figure 10.3 Relative wealth in GDP per capita

The Rhineland

Geographically perhaps the most significant part of this heartland is the Rhineland, partly because of its centrality, but also because it illustrates the profound consequences of political unity upon economic strength. The Rhineland was for long denied the unifying forces from which most of the European nation-states emerged. The states were either grouped around a cultural heartland as in France, or sheltered behind natural physical frontiers as in England. Even Germany and Italy, which emerged late as nations, had a cohesive culture and were brought to nationhood by the political cores of Prussia and Piedmont respectively. Moreover, Italy had the advantage of natural frontiers, and Germany a political and military force of great strength in the Hohenzollern kings of Prussia.

By contrast the Rhineland is a zone of physical convergence without any natural frontiers and, denied any political cohesion, became a buffer state between two strong cultural heartlands. By the Treaty of Verdun, AD 843, Charlemagne's Empire was dismembered into three parts; the Western Kingdom evolved into modern France; the Eastern Kingdom of the Saxons became the Holy Roman Empire and later Germany; and the Middle Kingdom or Lotharingia (fig. 10.2) was a narrow corridor in between. It covered the present area of Belgium, the Netherlands, Luxembourg, the German Rhinelands and the Saarland, Alsace-Lorraine, Switzerland, Burgundy, and Northern Italy. These emerged at various periods as smaller and less powerful states. Belgium was under Spanish and later Austrian rule and was a late developer into statehood in 1832. The Netherlands became independent in the sixteenth century. Switzerland, a confederation with four official languages, but with a majority of German-speaking people, debated whether to join the German Zollverein during the nineteenth century, as did the Grand Duchy of Luxembourg. Alsace-Lorraine and the Saarland, and specifically their mineral resources, have been a major source of friction between Germany and France over the last 150 years, and have changed hands on a number of occasions. The Rhinelands have been a contentious zone for centuries; a zone of convergence, political change and instability; the cockpit of Europe.

The locational advantages of the Rhineland are now realised. Its mineral resources, industrial zones, great urban centres, waterways and network of other communications have been allowed to realise their potential because of the development of the Community. From being an unstable frontier zone, its economic unity 'always prescribed by geography, always prevented by history' (the Schumann Declaration) has become real. The frontier image and the restrictions of politically separate economies have been removed by the EC. The cohesion provided by a well integrated system of transport links, the proximity of the majority of home energy and heavy industry production and the possession of three national core areas (fig. 10.2), has made the Rhinelands the keystone of the Community's economy. In place of the unstable Rhinelands of nineteenth-century Europe, a 'super-core' region on a continental scale has emerged.

The economic regions of the EC

The consideration of the economic heartland leads into a more detailed analysis of four types of regions: dynamic growth centres; stable diversified areas; older industrial regions with varying degrees of maladjustment; remote peripheral areas with harsh environments.

Regions of dynamic growth

These are regions with a high level of economic activity, having experienced sustained population growth, a rapid industrialisation rate, and a high degree of urbanisation. The Rhinelands and three other national economic cores - Greater Paris, the upper plain of Lombardy and the Manchester-London axis - together constitute this economic heartland (fig. 10.2). There is a case for adding others of more recent rapid economic growth, such as Hanover-Brunswick, Marseilles and Lyons, which form extensions not too far removed from the central axis. Economic life revolves around the 'city region' and, as illustrated in the previous chapter (fig. 9.8), the principal great cities of the community lie within a broad axis stretching between Manchester, Paris, the

Rhineland and Milan. However, the very nature of rapid growth is associated with the questions of pollution and congestion. Rotterdam and the Lower Rhine have considerable pollution problems, whilst the growth of metropolitan areas such as the Randstad and Paris poses enormous planning problems.

Stable diversified regions

Large tracts of accessible land close to the core areas have few economic problems. These are balanced communities in which a prosperous agricultural base is combined with moderate urban growth, light industry and a wide-range of service employment. Market towns and cathedral cities act as the focal points of these regions, in which economic growth is less rapid and can be absorbed more easily. Stability and diversity are the keynotes here. Most new development comes from the new mobility which motorways and other forms of fast transit have brought. The diffusion of industry and the development of new residential areas and commuting have brought these regions into easy contact with the core areas and have given them a new prosperity. Much of East Anglia, and the lowland countries of England fall into this category. Amiens, Reims, Troyes and Orléans act as centres for rich agricultural hinterlands in France, as do Wurzburg and Bamberg in West Germany and Piacenza and Verona in the plain of Lombardy.

Old depressed industrial regions

Many industrial areas can be found close to the 'axial growth belt'. These areas such as the Nord, Saar and Sambre-Meuse coalfields, and even the Ruhr, have structural problems related to the decline of industries of nineteenth-century origin like coal, steel and textiles. There is dependence upon imported raw materials and a narrow industrial base. There is also often an unattractive environment of man-made dereliction and obsolescent buildings. Although these are problem regions within their own national context, on a European scale they are relatively much less significant and restructuring is aided considerably by their position close to the growth axis. Central Scotland, South Wales and north-east England are three of the earliest locations of the industrial revolution. They are more remote from the main areas of growth in the UK, and the contraction of heavy industry has been exacerbated by the distance from the major European markets.

The maladjustment stems from a loss of economic vitality, but with investment, new industries can be brought into these areas. These are usually labour-intensive, light and consumer-goods industries which use the existing abundant labour supply. Retraining of the labour force and the provision of new houses, roads and services, is necessary to replace the obsolete and unattractive environment. There is great potential but industrial restructuring and adjustment is needed.

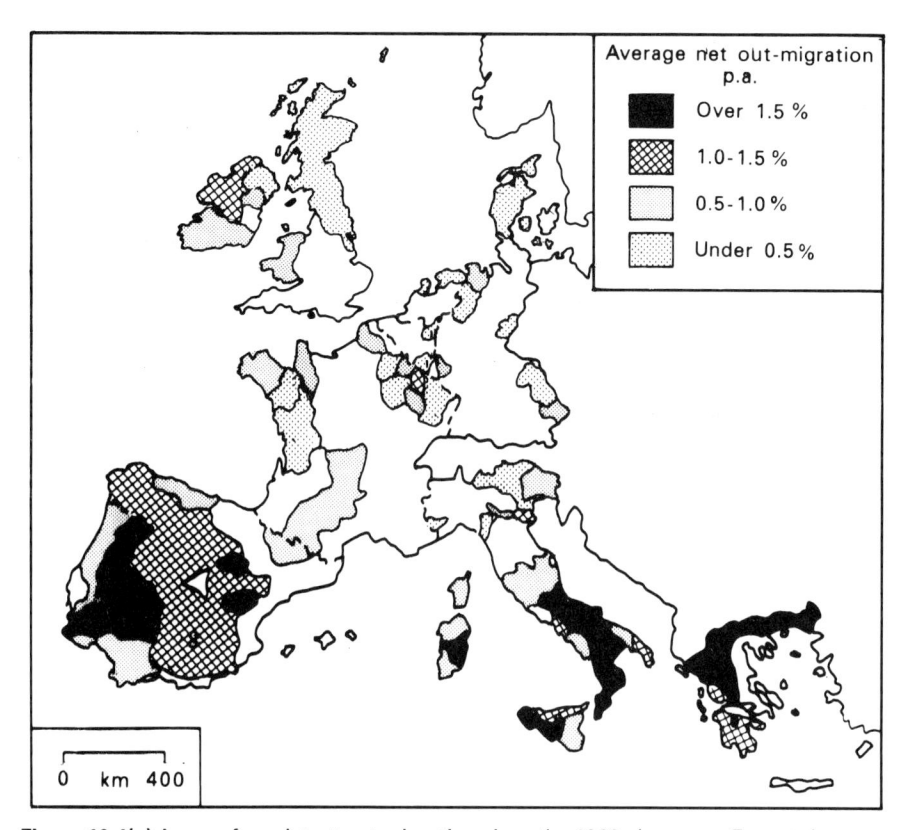

Figure 10.4(a) Areas of persistent out-migration since the 1960s (source – *Eurostat*)

The under-developed periphery: remote regions and harsh environments

The most serious problem areas are those relatively under-developed regions which are physically harsh, are remote, have a marginal agricultural base and a lack of industrial employment. Occasionally there are small pockets of mineral resources which have stimulated development such as the small coalfields of the Massif Central and tin-mining in Cornwall, but after a short period contraction and the resulting unemployment have only aggravated the local situation. Persistent out-migration from these regions is a common symptom of their problems (fig. 10.4(a)).
(a) Southern Italy (the Mezzogiorno) is a typical example. It is remote from the main-stream of economic activity, and has a desiccated environment, inefficient agriculture, over-population, inadequate transport and a limited local market owing to the low living standards. The rugged and semi-arid nature of much of Spain, Portugal and Greece, and difficulties of communication, give rise to uneven development between the different regions. Historical and economic problems show a striking resemblance to Italy's Mezzogiorno.

(b) The Massif Central, the Highlands of Scotland, Ireland and much of northern Britain generally are remote upland areas with harsh environments. They are plateaux and moorlands, with marginal farming and scattered settlements, and are geographically isolated. The Central Hercynian uplands, the Ardennes, Eifel and Vosges, are of the same general type, though more heavily forested.
(c) The Dutch and North German heathlands of Groningen, Oldenburg and Luneberg have infertile acid soils and a poor agricultural environment.
(d) In Lower Bavaria and Thuringia the upland environment has combined with the political isolation of being close to the East German and Czech frontier zone.
(e) Western France, west of a line from Normandy to Marseilles, is characterised by an over-dependence upon agricultural employment. As structural reform and improved techniques of farming are introduced, agricultural employment is now declining quite steeply, but there is little compensating growth in industrial and tertiary employment. Income levels are lower than the national average, and industrial activity is limited to a few centres, including Toulouse, Bordeaux, Limoges and Nantes. Net out-migration remains a severe problem over large areas including the Massif Central, western Brittany and the Loire-Poitou region.

National Regional Policies

The development or revitalisation of these peripheral or depressed regions has been accepted as a necessary policy in Western Europe since before the Second World War. However, the intensity of the problem varies greatly as does the approach. In the UK schemes began as early as 1934 to aid mining and industrial areas with high unemployment and developed into a wide-ranging system by the 1950s. In France the highly centralised government system set up a coordinated national economic planning system from 1947 onwards. In the Netherlands, planning has been dominated by the imbalances caused by urban density in the Randstad and has been devoted to deconcentration policies. In Italy, the intractable historical, political and socio-economic problems of the south stimulated a massive programme of regional assistance in the shape of the Cassa per il Mezzogiorno in 1950. By contrast, in Spain, Portugal and Greece, there has been a minimal attempt at regional planning.

Economic Potential in the European Community

The spatial structure of the European Community economy may be examined in the light of a number of factors which include: the loosening of industry from traditional and energy-based locations; the enhanced significance of transport for reduction of distribution costs and lower unit costs of production, and the progressive removal of physical and fiscal barriers to the free

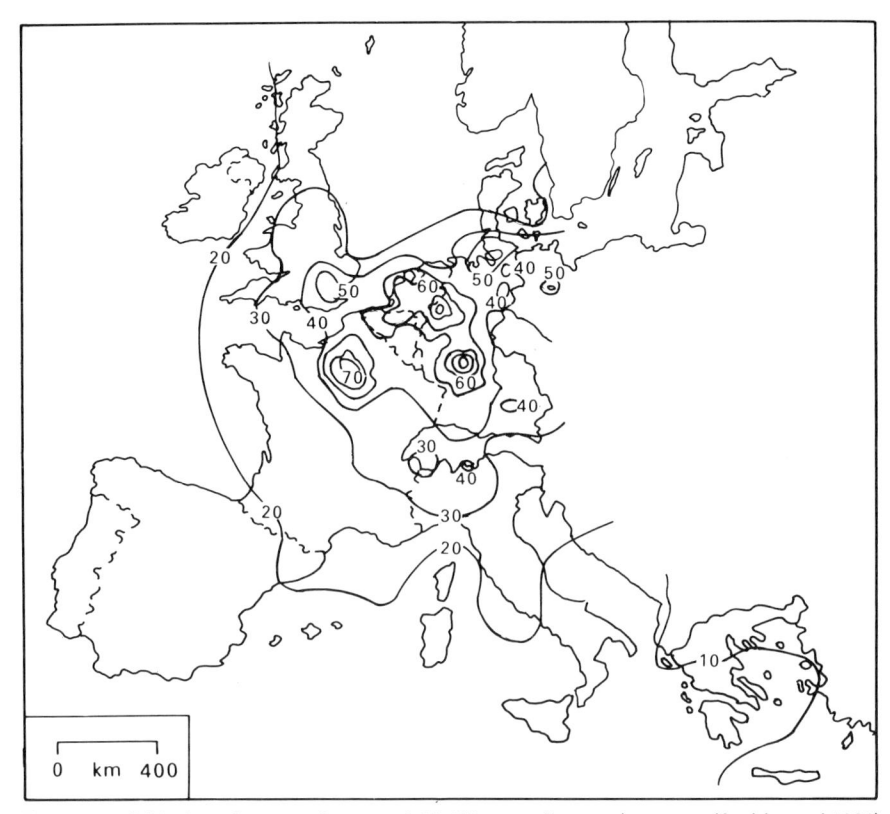

Figure 10.4(b) Index of economic potential in Western Europe (source – *Keeble et al 1982*)

flow of goods created by the movement towards economic integration. This has led increasingly, over the past forty years, to the concentration of productive resources into a limited number of areas in the geographical centre of the European Community, and particularly in the land area of the original six member states.

The core-periphery concept has been applied to the European Community by Clark 1969, and Keeble 1982, by means of accessibility coefficients based upon regional income data and distance costs, giving an index measure of a region's relative potential for economic activity (fig. 10.4(b)). The European Community of twelve in the late 1980s shows an extensive plateau of high economic potential in north-west Europe, with the Rhinelands as the most favoured region, but significant outlying peaks around the original national core area of the Paris basin and southeast England, and the plain of Lombardy. By contrast, the southern, western and northern peripheries are marked by extensive areas of low potential, particularly in Ireland, Iberia and Greece. The Aegean islands and Portugal have the lowest absolute values. This observed pattern has been analysed by Friedmann and Myrdal who suggest that processes of diffusion of wealth to the periphery may be hindered by a whole range of political and economic processes. The concentration of political and financial power in core areas means that they retain a disproportionate amount of new investment, that economic activity is encouraged to locate near its market in the heavily populated core regions with their existing wealth surplus, that selective migration from peripheral areas loses them their skilled labour and young enterprising people, and that the benefits of improvements to the transport infrastructure means an 'increasing peripheralisation of the periphery' as improved access benefits the core region first and foremost. These four processes amongst others, are behind Myrdal's concept of 'backwash', that is the tendency for a flow of factors back to the accessible core where the process of cumulative causation works positively to enhance the existing surplus.

An effective regional policy is therefore considered as economically and politically necessary in order to redistribute resources from the core so that the benefits of the EC economy may be transmitted to those less prosperous regions of the periphery.

Expenditure on regional policy has increased immensely, particularly during the 1960s and 1970s. All countries rely heavily on positive inducements such as capital grants and loans, subsidies on capital and current expenditure, tax concessions, training grants and the provision of infrastructure. This important principle is intended to move 'work to the workers' in the designated assisted regions. The United Kingdom also applied a policy of negative controls in the south east region for industrial and office building until the 1970s in an attempt to move development away from the congested south east into northern development areas. This policy has also been followed by France and the Netherlands in an attempt to limit growth in cities such as Paris and the Randstad. A fashionable strategy in the 1960s was the growth-pole aimed at encouraging the development of backward areas, and relieving congestion in major metropolitan areas. Based partially upon the ideas of Francois Perroux, this made use of the idea of investment in urban areas of 'propellant' industries. Significant use was made of the concept in Italy and France. Two major problems have persisted. Regional development is a slow complex process, and many policies such as that in Italy, the Cassa per il Mezzogiorno, were created with ideas that the problem of the south could be solved within ten years. Very large disparities between north and south still exist, and the Cassa has now been in existence for over thirty years. The time-scale of regional development is extremely long. Secondly, within the growth pole, the 'propellant' industries such as steel, and car assembly were intended to stimulate the growth of ancillary specialist component and service industries which in themselves would stimulate employment. This again would stimulate local demand setting up the multiplier effect. There is evidence that this effect has been much lower than expected, particularly since the recessions of 1973 and 1980. Hence the term 'cathedrals in the desert' has been applied to some of these large projects in southern Italy and France.

UK Development areas

The British classification was a three-fold division into special development

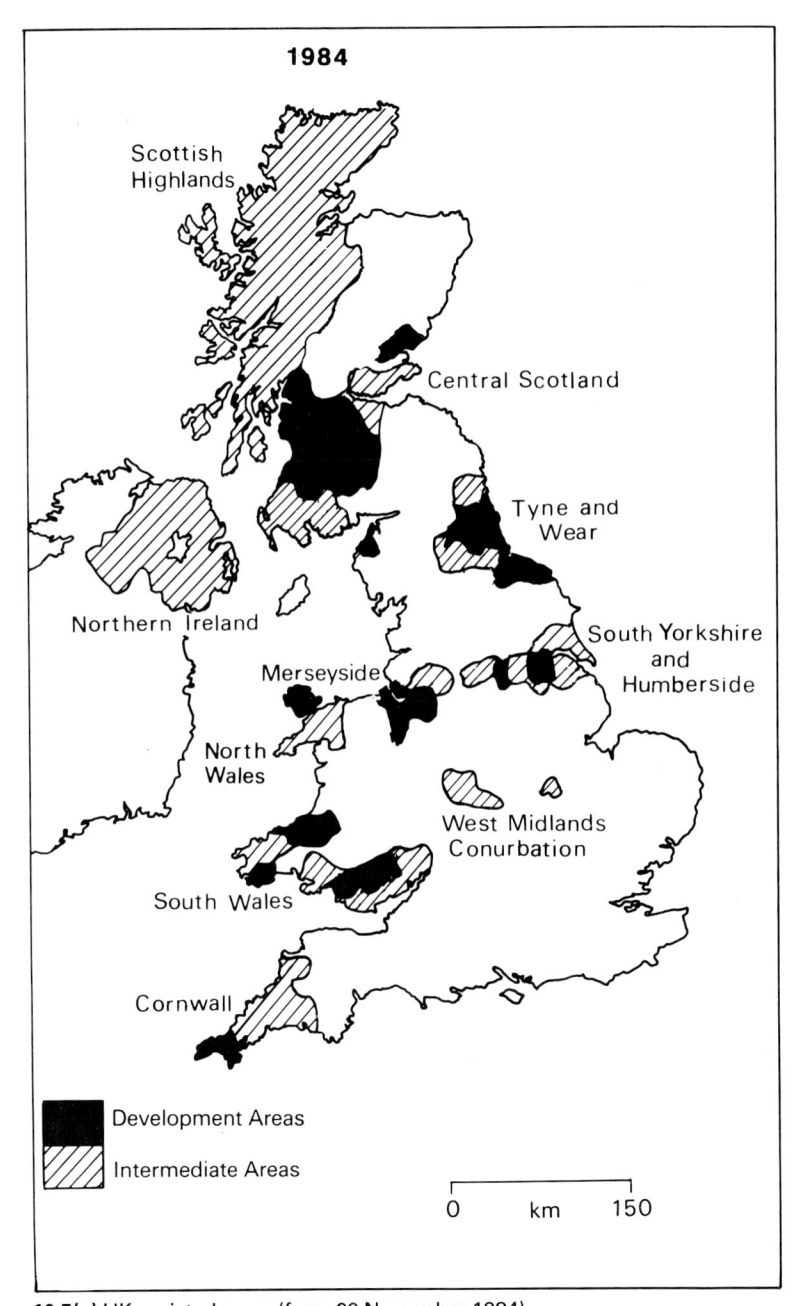

Figure 10.5(a) UK assisted areas (from 29 November 1984)

areas, development areas and intermediate areas. Originating in the 1930s, government policy was coordinated in 1966 by an Act which placed forty-four per cent of the population of the United Kingdom under development schedules. It included the whole of Britain north of a line drawn from Liverpool to the Humber and west of the Welsh border. These development

198   THE NEW EUROPE

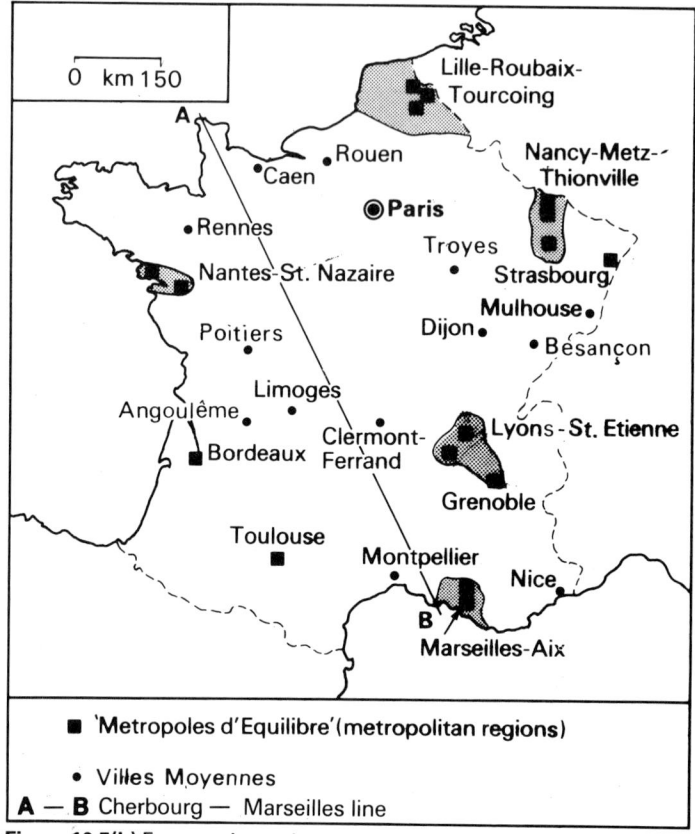

Figure 10.5(b) France: city regions and development poles

areas are designated usually on the basis of a rate of unemployment above the national average. Industrial companies are encouraged to move into the area with grants for factory construction, working capital, and machinery installation. The special development areas include severely declining coalfields which receive more substantial assistance. It has been questioned as to whether regional aid in the UK has been spread too thinly over large areas to be effective. Since 1982 there has been a change of emphasis, and the large-scale extent of the development areas has been dramatically reduced and reclassified, now covering only twenty-six per cent of the population (fig. 10.5a). The concept of the growth-pole has been adopted, where assistance is channelled into areas with a high growth potential. These include western Scotland, Tyne and Wearside, Merseyside, South Wales and Humberside. In addition enterprise zones in densely populated urban areas with high unemployment, have been designated. The enterprise zones include Clydeside, the lower Swansea valley, the Isle of Dogs, Dudley and Corby.

France

French planning policies are extremely well developed with a series of five year plans, under the direction of DATAR, in answer to three major problems. These are:
(a) the need to develop the under-industrialised sections of the country west of the Cherbourg-Marseilles line, (fig. 10.5b) to make its agriculture more efficient, and to develop the industrial and service infrastructure;
(b) to renovate older industrial areas such as the Nord and Lorraine;
(c) to correct the imbalance caused by the dominance of Paris. The excessive concentration of administration, commerce, industry and wealth in the capital has created a serious gap between Paris and the other French regions, referred to by J F Gravier as 'Paris et le desért francais'.

Planning takes five main forms:
(a) **Twenty-two planning regions** were created by combining the small départements into larger regions so that planning could be coordinated on a larger scale. These planning regions include Aquitaine, Burgundy, Languedoc and Brittany and are reminiscent of the ancient French provinces.
(b) **Industrial conversion grants** provide finance for the conversion and renovation of industry in the older coal-mining, textile and steel areas and to assist relocation away from Paris to centres like Rennes, Toulouse, Nantes and Clermont Ferrand. Since 1984, fifteen special development poles have been set up to attract investment into the worst affected industrial areas such as Lorraine.
(c) **Rural planning agencies** coordinate planning in rural areas. The French call this 'Amenagement du territoire' (management of territory). Finance is given for 'Remembrement', the reorganisation of farms into larger more efficient units, for improvement of farming techniques, more efficient marketing and the development of cooperatives. Forest management is encouraged. Planning corporations exist for overall management of development in tourism, agriculture and services such as CNABRL in Languedoc. National parks such as the Cevennes, and regional parks like the Camargue, have been created. Brittany and the Massif Central are areas of 'rural renovation'.
(d) **The Paris region** development plan called PADOG (1960) and the Schema Directeur (1965) revised in 1969 and 1975 (Chapter 19).
(e) **Metropoles d'equilibre** are cities which have been selected for expansion as administrative, commercial, industrial and cultural centres to counterbalance Paris (fig. 10.5b). There are eight of them such as Marseilles and Toulouse designed to act as 'Growth Poles'. 'Grand Projects' such as the industrial pole of Fos-Marseilles are designed as showpieces for French planning (figs. 3.4 and 3.5). A further number of intermediate regional centres have also been selected for development, particularly to attract tertiary employment.

Since 1975 there has been drastic modification of the territorial planning programme as France has been affected by economic recession. Since 1981 in

The Camargue Regional Nature Park in the Rhône Delta

particular there has been much more attention paid to smaller-scale, local regional projects and to conservation of town centres.

European Community Regional Policies

1951–1973

Throughout the buoyant period of economic growth, most regional development was left in the hands of national governments with their own specific regional problems. One of the articles of the Treaty of Rome was concerned with 'aid to promote the economic development of regions where the standard of living is abnormally low and where there is serious unemployment'. However, most financial assistance was given through structural funds such as the EAGGF or the ECSC. The Common Agricultural Policy, through the guidance section of the EAGGF, gives specific aid to farming in less-favoured areas. This includes farm modernisation, infrastructure, irrigation and intensification schemes, and training for farm workers, which has had a significant impact in peripheral areas such as the Mezzogiorno. The ECSC, initially set up to integrate and increase coal and steel output, has become increasingly concerned with modernisation and rationalisation of production, retraining of redundant workers, construction of housing, and schemes to attract new industries into the depressed areas.

Since 1973, with the accession of the UK, Ireland and Denmark, and the onset of economic recession, unemployment and the new international division of labour, a major change in attitudes took place. The three new members, and particularly the UK, insisted that a regional aid system should

	Total (million ECU)	Percentage of total
West Germany	152	3.6
France	708	16.6
Italy	1946	45.7
Netherlands	–	–
Belgium/Luxembourg	–	–
United Kingdom	591	13.8
Ireland	235	5.5
Denmark	260	6.1
Greece	364	8.5
Total	4256	100

Figure 10.6 EIB loans in 1985-86

be established by the Community. In May 1973 the Thomson Report (produced by the Commissioner for regional development, George Thomson) identified types of problem areas and announced the establishment of a Regional Development Fund (ERDF). It marked a key change in direction by the Community. Several important funds and agencies have been established to aid regional development.

(a) **The European Investment Bank (EIB)** has a separate identity, but works closely with the Commission and member governments. Its aims, as laid down in the Treaty of Rome (Articles 129 and 130) are:
 (i) to grant loans for projects in under-developed regions of the community such as the Mezzogiorno. The Taranto steelworks, and Naples to Reggio railway are examples.
 (ii) to grant loans and assistance for the modernisation, extension and reorganisation of particular industries and for the development of energy supplies.
 (iii) to finance projects of joint interest to member states, which may be difficult to finance by one member country alone. The cross-frontier motorways (Nice to Genoa or Paris to Brussels) are an example of this kind of project, or the trans-Alpine Mont Blanc tunnel between Italy and France. The improvement of communications between member states is of key interest to the Community.

During the 1960s EIB loans were used primarily for roads, motorways and other infrastructure developments, but since 1973 75 per cent of its efforts have gone to designated assisted areas in the European Community. Italy, France, the United Kingdom, Ireland, and latterly Greece and Spain, have become the major beneficiaries (fig. 10.6). The fund has been augmented by the NCI (New Community Instrument) which gives loans for priority energy and industrial projects. In addition, substantial loans are now given to smaller firms, in line with encouragement for entrepreneurial skill and new competition in the market.

(b) **The European Social Fund (ESF)** gives aid to specific sections of the population, in particular stimulating training and employment for young people, women, the handicapped, and migrant workers. The declining steel, textile, shipbuilding and clothing industries are assisted to restructure and rationalise by retraining of workers for other employment. Small business and new technology enterprises receive assistance where they provide much-needed new employment in the problem regions. It has a greatly enhanced role in the high unemployment situation of the 1980s. Two-thirds of the Fund's resources go to correct structural weaknesses and to improve employment in the less developed parts of the Community.

(c) **The Regional Development Fund (ERDF)**. Since the 1973 enlargement of the Community to nine, pressure for regional aid increased for a series of reasons such as serious regional imbalances in countries like the United Kingdom, the peripheral nature of member states like Ireland, the energy crisis and the onset of economic recession, the development of high unemployment in formerly prosperous regions, and the recognition of the inner city problem. The introduction of the Thomson report in May 1973 was followed by a protracted argument over the funding of the proposed regional policy. In a period of economic recession it was difficult to persuade weaker states to set up another costly fund. Eventually the Regional Development Fund became operative on 1 January 1975, with an initial grant of 600 million pounds over three years. The criteria upon which regions were to receive aid were as follows:

(i) a lower gross domestic product than the Community average, plus at least one of (ii), (iii) and (iv).

(ii) heavy dependence on agricultural employment in declining industries with at least twenty per cent of local employment in such a category.

(iii) a persistently high rate of unemployment over a number of years

(iv) a high and sustained rate of emigration (10 per 1000 each year) averaged over a long period.

All those areas aided by the member states themselves now automatically qualify for the Regional Development Fund which is therefore complementary to national regional policies. It can be seen from these criteria that those areas qualifying for regional aid (fig. 10.7) are essentially the Community's upland and remote rural regions. Core-periphery contrasts are immediately striking. But the complexity of the situation is shown by the inclusion of problem areas which exist closer to the Community's growth axis, amongst them the old industrial regions of Belgium, West Germany and northern France.

By 1975 the Regional Development Fund was allocated on the basis of agreed quotas to the member states with the greatest regional imbalances.

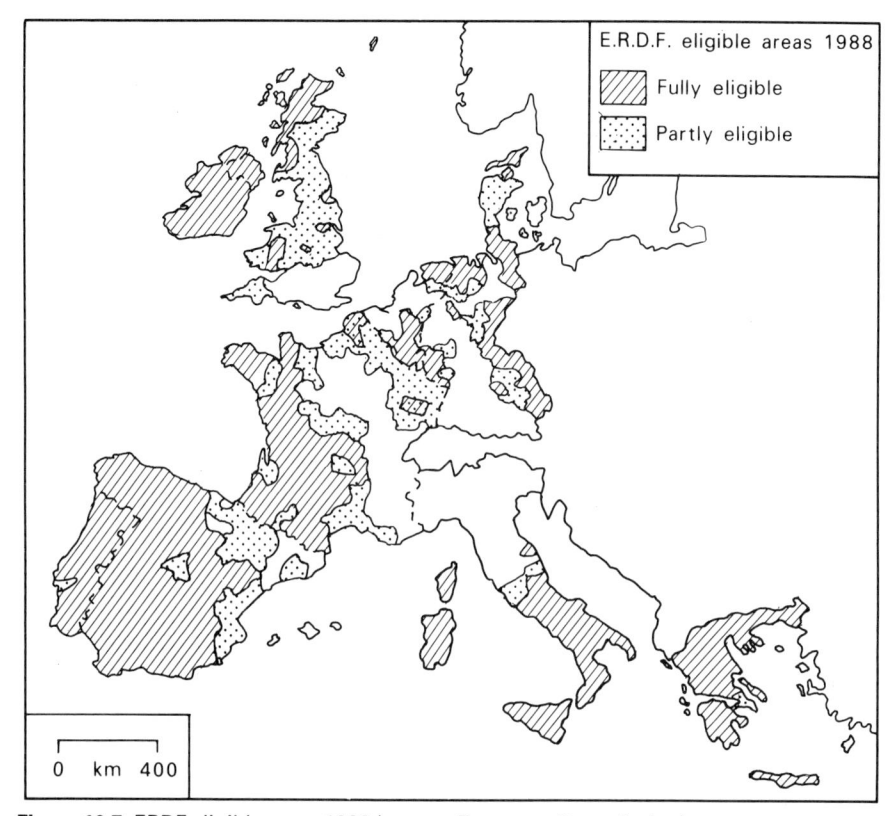

Figure 10.7 ERDF eligible areas 1988 (source - European Commission)

Thus Italy received 39 per cent, the UK 27 per cent, France 17 per cent and Ireland 6 per cent. The individual regions receiving the aid may be seen clearly in figure 10.8.

From 1978 onwards there were improvements in the fund and greater flexibility. The fund was doubled to £1.2 billion and a small non-quota section was added which gave assistance to specific regions outside the quota allocation. The European Commission also put forward proposals for assistance to shipbuilding towns in the UK, for steel centres in the UK, Belgium and France, and for the Mezzogiorno. Greek entry in 1987 necessitated a reallocation of the quotas as follows: Mezzogiorno 44 per cent; the UK northern and western regions 29 per cent; Greece 16 per cent; Ireland 7 per cent. Since 1985 there has been a greater concentration of regional assistance, and the quota system has given way to a priority funding of the most backward regions which contain some 50 million people. These include Ireland, Northern Ireland, the Mezzogiorno, Greece, Portugal, Corsica, and much of Spain. One significant innovation is the Integrated Mediterranean Programme (IMP) which is operating from 1985 - 1990. This coordinates aid from the EC to the new member countries, Spain, Portugal and Greece, but also to the Mediterranean regions of southern central Italy, and parts of

	1980	1985		1980	1985
North (UK)	4.2	2.7	Lazio (Italy)	2.3	1.5
Scotland (UK)	6.0	6.9	North-West (UK)	2.4	3.5
Apulia (Italy)	3.9	3.8	Brittany	2.6	1.4
Campania (Italy)	6.4	13.7	Sardinia	1.6	1.3
Ireland	6.5	5.0	Calabria (Italy)	2.2	5.0
Sicily	6.1	3.9	Nord/Pas de Calais	4.7	0.2
Abruzzi (Italy)	3.3	2.8	Midi-Pyrenees	–	2.6
Northern Ireland	4.4	1.9	French Overseas		
Venezia Giulia	1.7	–	Departements	–	3.4
Wales	2.5	3.9	Basilicata	–	4.2
			Greece	–	16.8
			Percentages of total fund	60.8	84.5

Figure 10.8 Percentages of Regional Development Fund to specific regions

southern France. It pays particular attention to agricultural modernisation, forestry, rural diversification, and tourism.

Since the fund began in 1975 there have been considerable criticisms levelled at it. The amount of money available is small. In 1984-85 the fund used only eight per cent of European Community expenditure, which is tiny by comparison with the large amounts of money spent on agriculture. Inflation and recession rendered the amount of aid available relatively even less effective than before. The rapidly increasing problem of urban deprivation is as yet ignored. The identification of problem areas and the indices used to measure these differ considerably between the member countries, and it is therefore difficult to be entirely objective in the amount and direction of regional assistance. The ERDF recognises the basic need for the transfer of resources in order to compensate for regional disparities. However, it is at present inadequately funded and unable to carry out the task in a coordinated and effective manner.

11
North-Rhine Westphalia: crossroads of Europe

North-Rhine-Westphalia is the administrative region (fig. 11.1) which contains both major growth elements and problems of industrial restructuring, and is situated along the principal growth axis of the European Community.

The Ruhr coalfield: development

The Ruhr coalfield is one of the world's best examples of a heavy industrial region based upon coal. It produces thirty per cent of the coal and nearly one-fifth of the steel of the enlarged European Community, and it makes a significant contribution to the oil-refining, petrochemicals and heavy engineering industries. It lies close to the centre of the so-called 'growth zone' of the European Community. Its development began from the eleventh century onwards when the medieval cities of Cologne, Duisburg, Essen and Dortmund originated as trade centres along the 'Hellweg', the ancient east-west line of migration through Europe. With the increasing importance of the Rhine routeway, the Ruhr has been a major route focus for hundreds of years. In addition, iron-making was practised immediately to the south in the forested Sauerland. Although coal was mined from very early times, it was in the 1830s that the first shafts were bored in the concealed coalfield areas to the north of the river Ruhr, and since that time mining activity has been moving steadily northwards into the Emscher and Lippe valleys.

The 'take-off' period was the 1870s when the 'Fettkohle' or coking coal in the Ruhr was found to be of extremely good quality and suitable for the improved iron-furnaces. The unification of Germany in 1871 was responsible for a great burst of confidence, railway construction and industrial expansion, with coal-mining increasing from 5 million tonnes in 1870 to 60 million tonnes in 1900. With the construction of the Dortmund-Ems and Rhine-Herne canals, the advantages of the Rhine were improved considerably for heavy industry based on local coal and water-borne raw materials. By 1913 coal output had reached 115 million tonnes and iron and steel production 7 million tonnes. The Ruhr's great industrial significance is shown by its resilience in twice recovering after a great deal of its capacity has been either destroyed or dismantled. After the First World War coal output reached 127 million tonnes and iron and steel 13 million tonnes in 1938, and after the Second World War

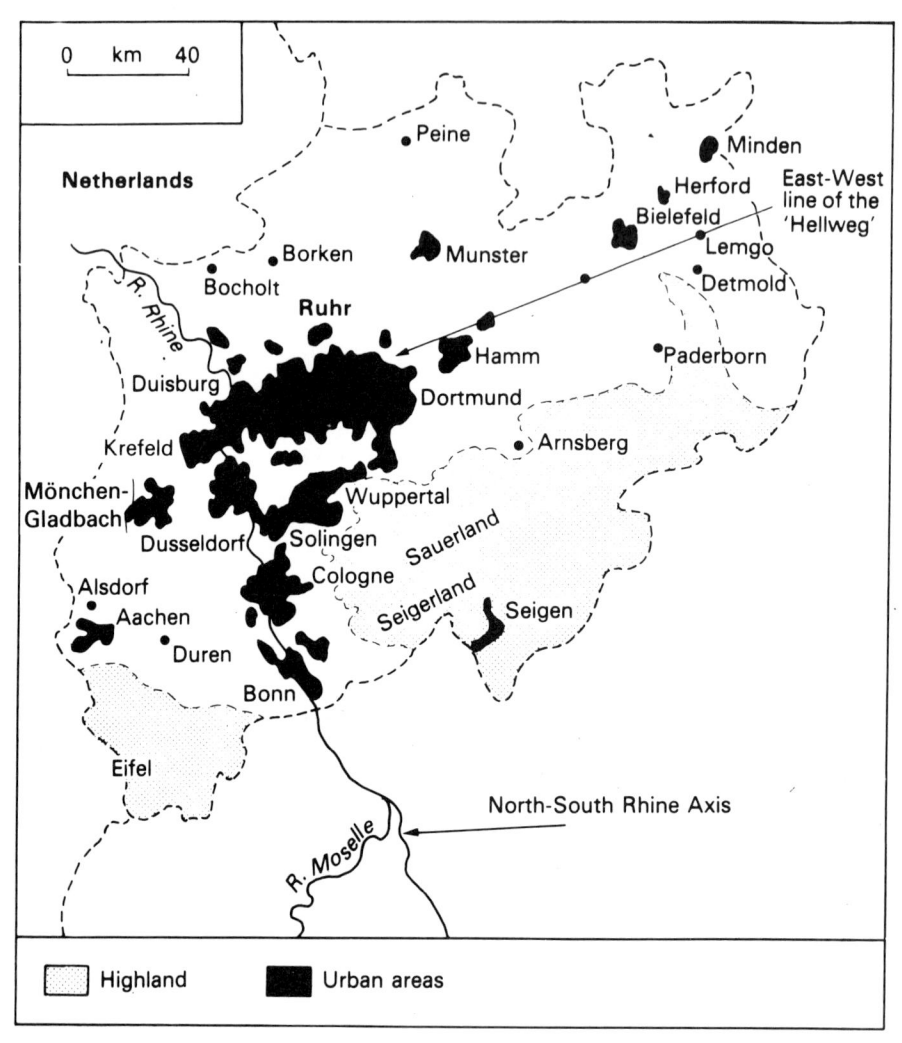

Figure 11.1 North-Rhine Westphalia

coal output reached the peak post-war production level of 125 million tonnes in 1956 (fig. 11.3).

A distinctive feature of the Ruhr is its great density of settlement. It is an amalgamation of over twenty towns and cities, a complex conurbation or polycentric city stretching east-west from Hamm to Geldern and north-south from the River Ruhr to the Lippe (fig. 11.2). The Ruhr Planning Region (SVR) contains over six million people, great cities like Dortmund, Essen and Duisburg, and eighteen 'free cities' in all. In addition, Cologne, Düsseldorf, Bonn and many other associated cities make up the Rhine/Ruhr Conurbation of some ten million people.

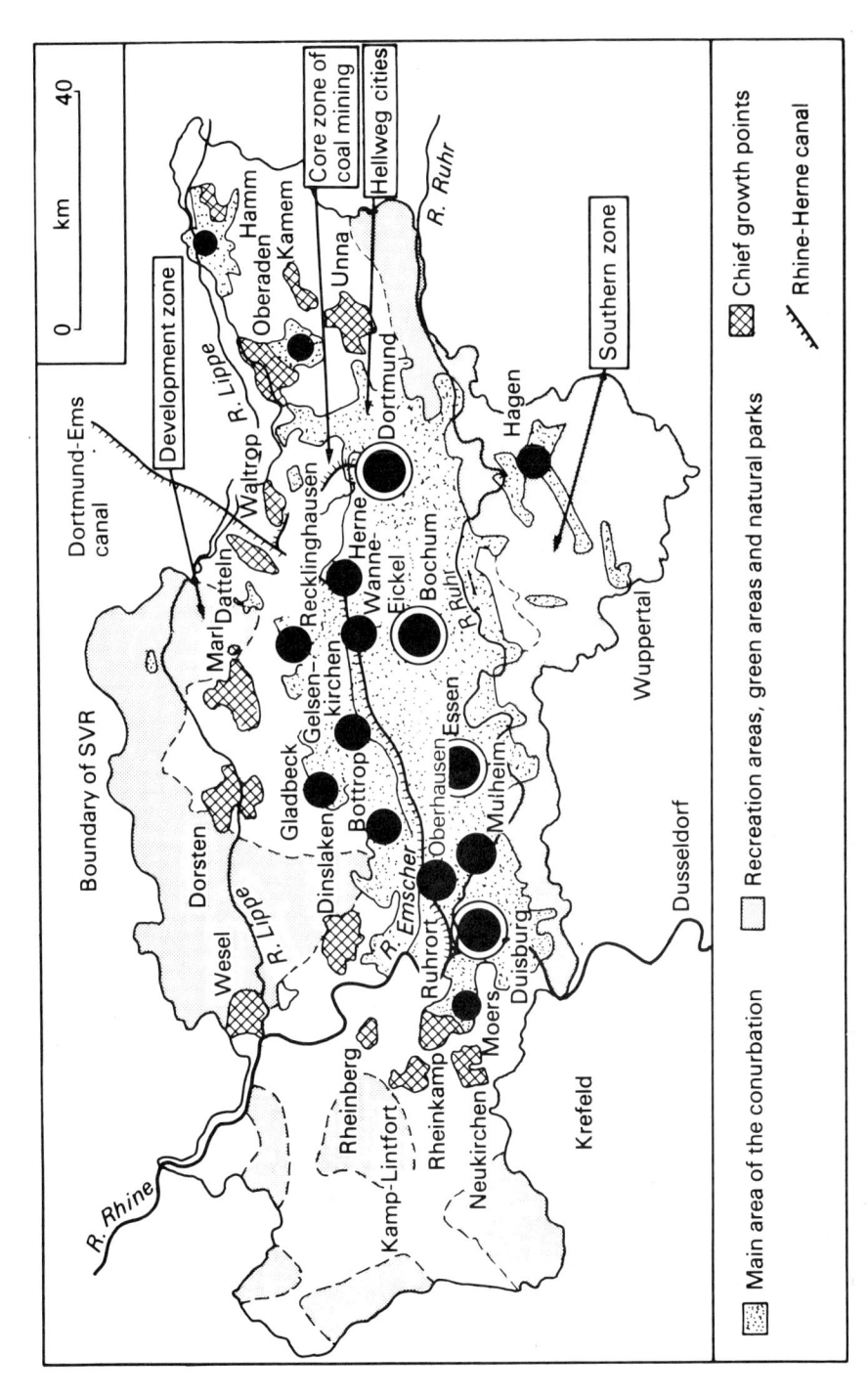

Figure 11.2 The Ruhr Planning Region (SVR) with four major planning zones

'Prosper 10' at Bottrop in the Ruhr: a new colliery which began production in 1981

The decline of coal

The extent of the coalfield can be seen in fig. 11.2. The real problem has become the increasing depth at which mining has to be carried out. The coal seams dip to the north and the nineteenth-century 'adit' mines are exhausted, whilst the average depth of mines today is nearly 1000 metres in the northern concealed part of the coalfield. The progressive migration northwards into the deeper mines has resulted in increasing costs at a time of competition from other cheap energy sources. The Ruhr is in a much better position than most established coalfields in Western Europe, because its coal is by no means exhausted. There are proven reserves of 65000 million tonnes. There are two other circumstances in which it is much superior to other coalfields. The first is in the great variety of types of coal, over fifty varieties, and the second is the extremely high level of productivity achieved by large-scale mechanisation. However, the Ruhr has been affected by the same factors as the other major coalfields in the trend during the 1960s to diversify sources of energy, and especially by the cheap competition from oil. Coal production had fallen to below 110 million tonnes by 1967 and 68 million tonnes by 1986. The mining force has fallen from 500 000 in 1956 to under 100 000 in 1986; the number of pits has fallen from ninety-nine in 1964 to twenty-six in 1986 (fig. 11.3). Associated with this has been a remarkable rise in productivity as inefficient mines were closed or amalgamated and the larger mines remained, with productivity rising by over fifty per cent, thus consolidating the Ruhr's productivity lead in the European Community.

There has been an accompanying structural reorganisation. Twenty-six of the coal companies amalgamated in 1969 to form Ruhrkole AG, which now controls 94 per cent of total hard coal production in the Ruhr basin. A real social problem was that the mine closures affected employment patterns tremendously, with over 40000 coal miners being made redundant since 1950. More significantly, the Emscher valley (fig. 11.2) has suffered more than the other areas. The impact has been concentrated by the nature of the coalfield. The area south of the River Ruhr has long lost its mining capability as the shallow 'adit' mines of the nineteenth century ceased to be used; the area north of Dinslaken and Recklinghausen contains the newer, deeper, and more modern mines. The central Ruhr towns of the Emscher valley such as Gelsenkirchen and Bottrop have been affected most, as their employment and economy was based almost exclusively upon coal-mining, to a far greater extent than in the other parts of the conurbation.

A reduced labour-force, fewer collieries, the operation of the coalfield as a single unit, and greatly increased mechanisation have made the Ruhr a highly productive coalfield. The rapid decline has halted and annual production has stabilised at about seventy million tonnes (fig. 11.3). The energy crisis has led to a reappraisal of the position of coal. West German law requires that the electricity industry uses a fixed percentage of coal and steel per annum. The coal industry in the Ruhr is therefore protected to a certain degree and an increase in coal demand is widely predicted for the 1990s. The existing mining infrastructure is now robotised and automated with very high technology. Hence the Ruhr is poised to take advantage of continuing changes in the energy market.

Industrial change

Structural problems facing the Ruhr are not confined to the coal industry. The steel industry has been adversely affected by the higher cost of Ruhr coal as compared to cheaper American coal and by competition from the coastal steel-works, which rely upon imported iron-ore. Since 1974 production has fallen (fig. 11.3), in common with other areas in the European Community, largely because of the recession in world demand for steel and the competition from low-cost producers outside Western Europe. 1981 saw the low point in the industry. However, the steel industry has shown remarkable resilience and has one major advantage, the importance of the local steel-using market. Steel-making has migrated heavily to the points of greatest cost-effectiveness, that is, the Rhine water frontage on the West, and to the Dortmund-Ems and Rhine-Herne canals in the east. Steel-making has ceased in the central parts of the Ruhr, at Bochum and Essen. The Rhine frontage at Duisburg-Ruhrort and Rheinhausen now produces three-quarters or Ruhr streel and Dortmund the rest. In addition, there is the manufacture of special electric steels at Remscheid, Krefeld and Bochum.

The comparative advantages of water-front locations, owing to the very substantial imports of Swedish iron-ore, coal and other raw materials via the Rhine, has affected the present-day location of the Ruhr steel industry. In addition, there has been considerable reinvestment in new steel plant and

Year	Coal Production Million tonnes produced	Year	Steel Production Million tonnes produced
1850	1	1860	0.1
1870	5	1913	7
1900	60	1938	13
1913	115	1957	18
1938	127	1965	23
1956	125	1970	28
1965	115	1974	34
1970	96	1977	25
1974	84	1981	26
1977	74	1986	24
1981	77		
1986	68		

Year	Operative coal mines	Year	No. of coal miners
1850	200	1950	500 000
1890	175	1956	494 000
1950	150	1964	325 000
1961	120	1969	183 000
1964	99	1974	150 000
1969	57	1977	128 000
1974	45	1981	111 000
1977	32	1986	85 000
1981	28		
1986	26		

Figure 11.3 Ruhr coal and steel production

reorganisation into larger units, Thyssen, Mannesmann, Hoesch and Krupp being the largest firms. Companies like Hoesch are now associated with the Ijmuiden works of Hoogovens on the Dutch coast, thereby gaining the advantages of scale. The central area, at Bochum and Essen, has become a steel-using area with vehicles, mechanical engineering, metal fabrication and electrical engineering; growth industries such as electro-technical, chemicals, plastics; and consumer industries such as clothing. Essen has a number of substitute metal industries such as aluminium and zinc-smelting.

The other major case of industrial decline has been textiles and clothing. Its workforce has dropped from over 350 000 in 1965 to less than 150 000 in 1985. However, the heavy chemical industry has had remarkable success in adapting to changed circumstances. The loss of the chemical industries on the Middle Elbe after the partition of Germany gave considerable benefit to the Ruhr. The industry has abandoned its traditional dependence upon coal and coke by-products, and since 1950, oil has been the major raw material transported by pipeline and by the Rhine routeway from Rotterdam. The growth of oil refineries and petro-chemicals at Marl-Huls and Gelsenkirchen are associated with this. Light, high value chemical industries such as pharmaceuticals have rapidly developed.

Settlement and planning

The other main problem is the unplanned, high density and obsolescent environment. The development stages of the coalfield are largely responsible for the nature of urban development. The Ruhr may be divided into four main zones: the southern zone; the Hellweg; the core coal-mining zone; and northern development zone (fig. 11.2). The regional planning authority for the Ruhr (SVR) has been active in promoting schemes for balanced industrial development, new housing schemes, new town re-development, green areas, leisure parks and reclamation of tips and spoil heaps. Its activities vary, however, according to the needs of the four zones.

The southern zone coincides essentially with much of the valley of the river Ruhr and the low hills to the south. Here very few coal mines and little heavy industry remain, and population is much less dense than in other parts of the Ruhr. The area consists of the southern residential suburbs of Essen, Bochum and Dortmund, reservoirs on the river Ruhr, and four large leisure parks which have been laid out to serve the cities to the north.

The Hellweg cities of Duisburg, Essen, Bochum and Dortmund are large, well-developed urban areas with good shopping facilities, cultural and historic cores, and with important commercial interests, company head offices, tertiary activity, and a high proportion of professional people. They are therefore wealthy urban communities, although surrounded by extensive housing areas of monotonous design or obsolescence and by heavy industrial areas. Their problems, therefore, involve planning for high density living, but they present much less of a problem than the area immediately to the north.

The core coal-mining zone along the Emscher valley has the greatest problems. Rapid nineteenth-century growth meant that villages such as Gelsenkirchen grew into monofunctional mining towns with heavy industry. Others such as Oberhausen, Recklinghausen, Herne and Bottrop had proportions as high as fifty and sixty per cent in mining employment. Both air and water pollution are problems and the Emscher is one of the worst polluted rivers in Western Europe. As much as forty per cent of the land is old mining land and is often derelict. Obsolescent housing, unsightly heavy industry and the absence of green space are the principal features of the area, quite apart from the economic problem of redeployment of its labour and considerable migration into the outer areas of the Ruhr.

To the north lies the development zone. This is the most recent coal-mining area stretching northwards towards Munster (fig. 11.1) and although the mines are deep, the pits are large, modern and efficient. This is the newer concealed coalfield, lying at a depth of up to 1300 metres, and the coal is often converted at the pithead into electricity for the industries further south. Heavy industry has not developed here to any large extent but instead there

The Western Hellweg shopping precinct in Dortmund. It carries the name of the ancient east-west routeway across Germany, and symbolises the nodality of North-Rhine Westphalia.

are many light and consumer industry factories. The zone's share of the Ruhr total population is only about fifteen per cent, and so there is considerable space for development. Major growth points for both new towns and industrial complexes are Wesel, Dinslaken, Dorsten, Marl-Huls and Datteln (fig. 11.2).

Other manufacturing towns

There are a number of important sub-zones and groups of towns which are outside the Ruhr Planning Authority area, yet which have close associations with it (fig. 11.1).
1. The first group comprises the textile towns on the west bank of the Rhine. Krefeld is a traditional centre for silk and velvets, and Mönchen-Gladbach and Rheydt have textile machinery works, and clothing factories.

2. To the south-west lies the city of Aachen with its small coalfield and the huge brown coal deposit on the Ville ridge west of Bonn and Cologne, which is open-cast mined.
3. To the south lies the Sauerland and Siegerland and the deep valleys of the Sieg and Wupper cut out of the Rhine plateaux. Here are towns related to the early importance of local raw materials and power, and the later specialisation and geographical inertia which has led to two highly complex manufacturing concentrations. Both the manufacture of high quality cutlery at Remscheid and Solingen, and locks and keys at Velbert can be traced to the smelting of local iron-ore deposits worked with charcoal from the forests, and these industries survive now because of accumulated skills and specialisation. The old-established textile manufacturing towns of Elberfeld and Barmen were based upon the abundance of water power in the Wupper valley, urban growth having now merged them into the conurbation of Wuppertal.

The cities along the Rhine valley

It is along the Rhine valley, however, that the greatest growth area is found. That part of the Rhineland which lies immediately to the south of the Ruhr is

The Königsallee: an elegant retail street in Düsseldorf

dominated by three cities - Düsseldorf, Cologne and Bonn (fig. 11.1). Düsseldorf is referred to in West Germany as the 'Rhinegold', a term which indicated its wealth and reflects its image as an example of the German economic miracle. It was not a medieval city and developed only during the seventeenth century as a minor principality, from which it has grown into the modern banking, financial and commercial centre of North-Rhine Westphalia. As a financial centre it has a remarkable expertise amongst its 'prominenz' or establishment, and it is also the administrative capital of North-Rhine Westphalia. An enormous amount of rebuilding has taken place since 1945 to create one of the most modern cities in Europe. The 'Königsallee', an expensive shopping street set in park-like boulevard surroundings, exemplifies the city's wealth. There are also impressive high-rise offices and headquarter buildings of the major Ruhr companies, Krupp and Thyssen, as well as those of international companies such as IBM and the Chase Manhattan Bank. Not only does the city act as the financial capital of the Ruhr but it also has its own industries, principally engineering, and the huge chemical complex at Leverkusen is just to the south.

Cologne is an ancient Roman settlement which flourished as a medieval trade centre of the Hanseatic League, based upon its position where the river route from Flanders into the Rhinelands crossed the fertile loess embayment and the Hellweg. The railway development of the nineteenth century, however, gave the city its modern importance. It is still both an important railway junction and an inland port. As a commercial and business centre it is a rival to Düsseldorf, but in addition it has a very varied industrial base, ranging from iron and steel, oil and petrochemical production, to engineering and car manufacture. It also has a wide range of consumer goods including leather, cosmetics, clothing and chocolate, which are an unusual feature in the region and which originated in the traditions of medieval crafts of the old city.

Farther south again is the city of Bonn, capital of West Germany. It is situated at the point where the Rhine emerges from its gorge, and is a relatively small city of medieval origins. Its bishopric, university and medieval prosperity were based largely upon its function as the historic seat of the Principality of Cologne. Although the Federal capital since 1949, it has remained relatively small with light industries, and its population is now 300 000.

North-Rhine Westphalia: the advantages of centrality

The 'Land' of North-Rhine Westphalia is the richest and most populous part of West Germany with a total population of seventeen million within which is the Rhine-Ruhr Conurbation with over ten million people. The Ruhr Planning Region has over six million people, whilst the Düsseldorf and Cologne city regions have one and a half million people each. There is a distinction to be made between the growth of the long-established cities and outer suburban and commuter areas, and the structural problems which have caused a slackening of growth and even decline of population in the central coalfield areas. The central core of the Ruhr coalfield, the Emscher Valley, has in fact lost some 300 000 people from towns like Gelsenkirchen, Bottrop and Bochum, but these have been involved in a sub-regional movement to the cities of Düsseldorf and Cologne, and to the new towns and suburbs on the edge of the coalfield. Any loss of population in the centre has therefore been compensated within the region.

The Ruhr coalfield has had several major advantages in its structural and economic reconstruction.

(a) It is part of the Land of North-Rhine Westphalia, the single most populous, wealthy and industrious region in West Germany, and it has benefited by its association with the state government in Düsseldorf. For instance, since 1945, four new universities have been established in the Ruhr where none existed before, all part of the changing image of this heavy industry region.

(b) The influence of the Planning Authority (SVR) can be seen in the changing quality of the environment. Although much remains to be done, the green areas, nature parks, reclaimed spoil-heaps and new urban motorways are a witness to its effect since it was set up in 1920. Of the total area of the Ruhr Planning Region, only 24 per cent is urbanised; 53 per cent is open land and farmland; over 20 per cent is woodland and leisure areas. There is plenty of room available for gradual, planned and comprehensive redevelopment within the whole area (fig. 11.2).

(c) The nature of the coalfield is of great significance. It has few of the problems of thin seams and low productivity associated with many of the other European Community coalfields, and has traditionally been the most productive in Western Europe. There has been a steady reduction in the number of coal-mines, rather than the massive reductions which have been seen elsewhere. Modern investment places the Ruhr coalfield in a strong position for any up-turn in demand.

(d) The regional distinctiveness of the Ruhr owes much to the intensive inter-dependence of its industry and transport system. The steel and chemicals complex has been strengthened by new steel-using industries, by petro-chemicals and oil-refining, and by light engineering and consumer-products industry. In particular, the car industry is now represented by over 20 000 workers in three factories of General Motors. Television and electronics factories have also migrated into the old central steel-making area of Bochum and Essen. The 'rust belt' image that has been associated with the Ruhr is very much a partial one and as economic adjustment has proceeded rapidly, may well be transitory.

(e) The Rhine waterway has remained central to West Germany's economy and has also become the growth-axis of the European Community (chapter 10), underlining the position of centrality which the Ruhr possesses. The West German autobahn network makes possible rapid transport to other parts of the Euro-core.

(f) Finally, the Ruhr lies on the historic lateral axis of commerce and city development, the Hellweg (fig. 11.1), and is associated with great cities

like Cologne, Dortmund and Düsseldorf. The long established structures of these cities, with their commercial wealth, financial services and shopping facilities, enables them to adapt to change and their growing industries are able to absorb the excess population from the Ruhr coal-mining areas. In addition, the large consumer market of seventeen million people is a major factor in the ability of the region to adapt its employment structure. The tertiary sector has grown rapidly (51 per cent) and has now overtaken the secondary sector (46 per cent) for the first time. This is a remarkable commentary upon the changing economic patterns in this erstwhile dominantly heavy industrial area.
(g) North-Rhine Westphalia is a core region on a European scale, referred to by Parker as the 'super-core', and effectively at the crossroads of Europe.

12

The Middle Rhinelands: three city regions

Varied resources

The Middle Rhine valley of West Germany stretches from Wiesbaden and Frankfurt-on-Main in the north to Karlsruhe in the south, and includes the right-bank tributary valley of the Neckar, in which is the city of Stuttgart (fig. 12.1). The French city of Strasbourg also lies within the physical region. Thus defined, it is a most important economic area lying within the European Community core. It has been a traditional zone of urbanisation from Roman times, and throughout the Middle Ages towns like Worms, Mainz, Speyer and Heidelberg were associated with bishoprics or universities, and later there were royal residences and planned cities such as Karlsruhe. There is an abundance of resources, and four principal factors may be identified as contributing at various stages to its development.

A rich and diverse farming region

This is a most productive agricultural area, a fact reflected in the densely populated countryside, with large and numerous villages. Particularly along the foothills, on the loess lands and on the sheltered terraces there is a wide variety of agriculture. The Rhine-Main plain to the north of Frankfurt, the Wetterau, is extremely fertile and is seventy per cent arable, whilst to the west of Frankfurt is the Rheingau with its orchards and vineyards. The Kraichgau between the Rhine and Neckar is a loess-loam undulating lowland with arable lands, orchards and vineyards. By contrast the alluvial plains along the rivers are heavy clays liable to flooding, with damp water meadows. The advantages of the region are not only the wide variety of landscape and terrain, but also the occurrence of loess, which is easily cultivated and has a relatively high fertility, together with a greater amount of sunshine and hotter summers, than in the more northerly parts of Germany. In particular, the cultivation of the vine depends to a great extent upon south-facing slopes, and these are found at the point where the Rhine makes its westward turn to enter the gorge. As a result, the south-facing slopes of the Taunus ridge overlooking the Rhine have an almost continuous covering of vineyards for some twenty miles from Wiesbaden to the gorge, giving one of the most famous wine-producing areas in West Germany - the Rheingau. Here originate the

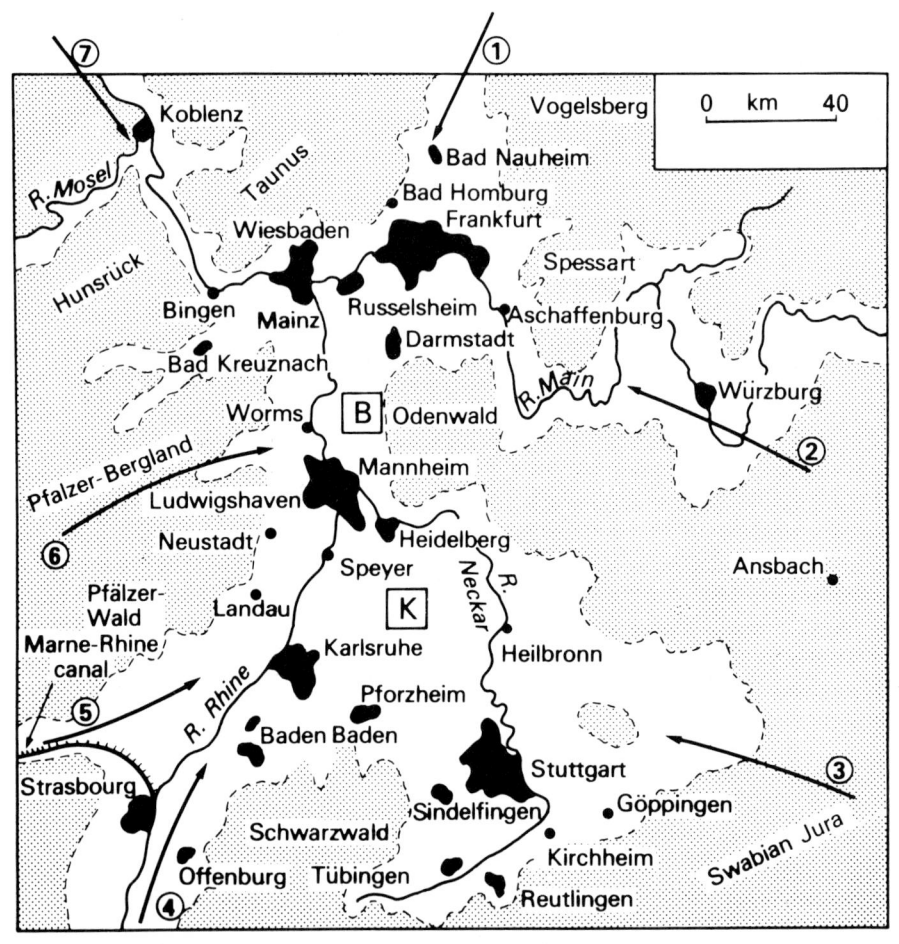

Figure 12.1 The Middle Rhinelands: zone of convergence. Major routeways into the middle Rhinelands.
1. The Hessian corridor   2. Main valley from Nuremberg   3. Foreland route from Augsburg and Munich   4. Upper Rhine route from Basle   5. Saverne gap from Lorraine   6. Route from the Saarland   7. Rhine route from Rotterdam and the Ruhr.   K: Kraichgau   B: Bergstrasse

most famous names in German wines, such as Johannisberg and Rudesheim.

Farther south, another intensive area of cultivation is the Bergstrasse, the loess-loam foothills of the Odenwald between Darmstadt and Heidelberg. This area on the eastern side of the Rift valley cultivates vines, tobacco and fruit, and almost opposite on the western side are the Worms - Nierstein - Oppenheim vineyards on the Haupterrasse, backed by the Pfalzer Bergland.

The Kraichgau between the Rhine and Neckar is limestone covered with loess and is one of the most intensively farmed areas in south Germany. Cereals are more in evidence because of the gently undulating or even nature of the land above level-bedded limestones. In addition there are fodder crops, sugar-beet, fruit, hops and the vine. The landscape is completely cleared and farming is very intensive, with large prosperous villages and a high rural density of population. It is upon this prosperous countryside that the initial wealth of the Middle Rhinelands is based.

Convergence of routes

The convergence of routes is a second advantage. Since medieval times the Rhine valley corridor has become an area of convergence as a link between the North Sea and Alpine passes, and between France and Austria. The position of the Rhine-Main valley, surrounded by the Hercynian mountain blocks yet with major gaps from north, south, east and west, has created the greatest junction in middle Europe. The traditional trade routes of the Middle Ages have been superseded by the railways and autobahns. Figure 12.1 illustrates the passageways, the Hessian Corridor, the Main valley from Würzburg, routes from Munich via Stuttgart, the Upper Rhine route from Basle and the Alps, the Belfort gap from Lyons and the Rhône, the Saverne gap from Paris, Lorraine and the Saar, and the Rhine waterway route from the Ruhr and Benelux countries. The importance of the corridor has been increased immeasurably by the development of the European Community, as the whole effect of the convergence of routes has to be looked upon on a European scale rather than as hitherto on a German scale. The Rhine valley is now the central axis of the European Community rather than being a frontier zone, a change which has been stressed in chapter 10.

Flourishing cities

The concentration of human resources is the most significant reason for the importance of the Middle Rhinelands. Rather than depending upon mineral resources, it has developed around the numerous medieval cities and bishoprics, and has flourished because of the cultural and economic activity associated with a continual inflow of trade, expertise and new ideas. The earliest development of towns was during the Roman period, when the Rhine was the frontier of civilisation, and later the medieval bishoprics, such as Worms, Speyer and Mainz, developed on the west bank of the river. Heidelberg is the chief east-bank medieval and university town which has developed at the junction of plain and foothills along the Bergstrasse. Baroque towns are usually associated with the numerous principalities and minor states which existed up to the nineteenth century. Karlsruhe is the best example of a planned city, founded in 1715 by the Margrave of Baden, and has a well-preserved radial pattern of roads leading to the Royal palace. Darmstadt was the seat of the Elector of Hesse, whilst Mannheim was originally founded as a capital city by the Elector of the Palatinate, and its rectilinear pattern of streets reflects the original planned town. The trade, culture and economic activity of this river region have been developed over a thousand years, and they have depended to a great extent upon the traffic carried by the river Rhine, although the effects of this have been seen to a much greater extent during the present century.

The Rhine

The river Rhine dominates the inland waterway system of Western Europe (fig. 8.2). Long-distance commerce between Basle and the Netherlands has been considerable since the Middle Ages. It declined later largely because of the considerable sums of money required by the strategically positioned toll-enforcing castles along the line of the river, but there was a revival in trade during the nineteenth century. The river was first freed from tolls during the French Revolutionary wars, and in 1868 was made an international navigation channel. Navigation was vastly improved during the nineteenth century by deepening, bedrock blasting and straightening of the channel. The next step was to improve the tributaries as the converging arteries of the river. The Rhine - Marne canal was built from Paris via Strasbourg, the Main itself was improved up to Frankfurt, and with the advent of Ruhr coal and industrial products the trade south into the Middle Rhine region rapidly increased in volume. The growth of industry and cities has been directly affected by the Rhine acting as a factor of convergence. This is an industrial region far from the coast and the river has played a particularly relevant part in bringing the advantages of low-cost water transport. The advent of the European Community has done much to emphasise the centrality of the region. Since 1960 enormous strides have been made. The Main is now navigable for 1500-tonne barges as far as Bamberg, and the new Rhine - Main - Danube canal goes as far as Regensburg. The Neckar is navigable (1350

River	Major inland ports	Traffic volume (million tonnes)
Rhine	Mannheim	9.3
	Ludswigshaven	9.4
	Karlsruhe	6.7
	Mainz	4.1
	Wiesbaden	1.1
Main	Frankfurt	7.8
	Offenbach	1.2
	Aschaffenburg	1.0
	Würzburg	1.6
	Bamberg	1.2
Neckar	Heilbronn	5.7
	Stuttgart	3.2
For Comparison (*Lower Rhine*)	Duisburg Ruhrort	20.5
	Duisburg Works	20.7

Figure 12.2 Middle Rhine inland ports - traffic volume 1980s average

tonnes) as far as Stuttgart. On the west bank the Rhine - Marne canal is supplemented by a link southwards to the Rhône via Belfort. The Rhine itself is navigable up to Basle for 2500-tonne barges. The major river ports are Mannheim, Ludwigshaven, Karlsruhe and Mainz (fig. 12.2). Other smaller ports are Wiesbaden and Speyer (fig. 12.1).

On the river Main the major port is Frankfurt, but several other ports such as Würzburg and Bamberg are expected to increase their level of traffic as the Rhine - Main - Danube canal through to Regensburg becomes integrated with the Rhine waterway system. Most inland ports in the region have a four to five days' transit time to Rotterdam and the coast. Thus, some comparison may be made from the following progression: the limit for ocean-going craft is Cologne; barges of 7000 tonnes reach Duisburg, 5000 tonnes Mannheim, and 2500 tonnes Basle. If the total river traffic flows in the Middle Rhine region (fig. 12.2) are added together, they are roughly comparable to the trade of Duisburg. Although such figures highlight the importance of Duisburg to the Ruhr, the Middle Rhine region nevertheless is a very substantial focus of water transport.

The development of the Rhine - Rhône canal is one of the more significant aspects of Europe integration. It will benefit the Rhine corridor, giving new impetus to cities like Lyons, but most significantly will enhance the nodality of the Rhine routeway by linking the North Sea and the Rhineland urban-industrial axis, with the Mediterranean Sea.

Three city regions

There are three city regions of special note: Greater Frankfurt; Ludwigshaven - Mannheim; and Stuttgart. Each of these illustrates different characteristics and deserves separate consideration.

The Rhine - Main Complex: Frankfurt-on-Main

The area lies in a triangular zone bounded by Frankfurt-on-Main, Wiesbaden - Mainz and Darmstadt (fig. 12.3). Throughout history there have been continuing factors which have created a focal area of transport routes and economic activity. Mainz was a Roman town at the confluence of the Rhine and Main, but was overtaken by Frankfurt (Franks' ford) during the Frankish colonisation. Frankfurt developed faster due to its establishment as a 'Konigshof' and it its more central position on the plain for the routeways which developed during the medieval commercial period. From the tenth century onwards the Rhine - Main valley was centrally placed in the Holy Roman Empire and became a focus for exchange between the Netherlands and Italy. Frankfurt was also a 'Reichstad', a free city in the Empire with considerable industry and commerce. Mainz became an important archbishopric and was fortified, and both towns benefited from the great medieval fairs. Wiesbaden lies on the north bank of the Rhine at the point where the Taunus mountains come close to the Rhine. It was a fortified town and seat of the Counts of Nassau, but became famous during the eighteenth century when its hot salt springs were popularised and it developed into a health resort and spa. The third apex of the triangle, Darmstadt, originated as the seat of the Elector of Hesse, and its planned origins are reflected in its rectilinear street plan.

The nineteenth and twentieth centuries have underlined the importance of the region in terms of access. Frankfurt is a focal point in the German railway network with the largest railway station in the Federal Republic, and it has developed industry on a large scale with engineering, electrical, chemical and consumer goods. The increase in Rhine traffic began during the nineteenth century, with the development of port facilities at Osthafen above the old city on the north bank of the Main. Today the West German Autobahn system focusses on the Frankfurt Cross, and the outlines of the city region can be discerned in the triangular network of motorways which covers the whole area (fig. 12.3). In addition, the airport is one of the busiest in Europe, with

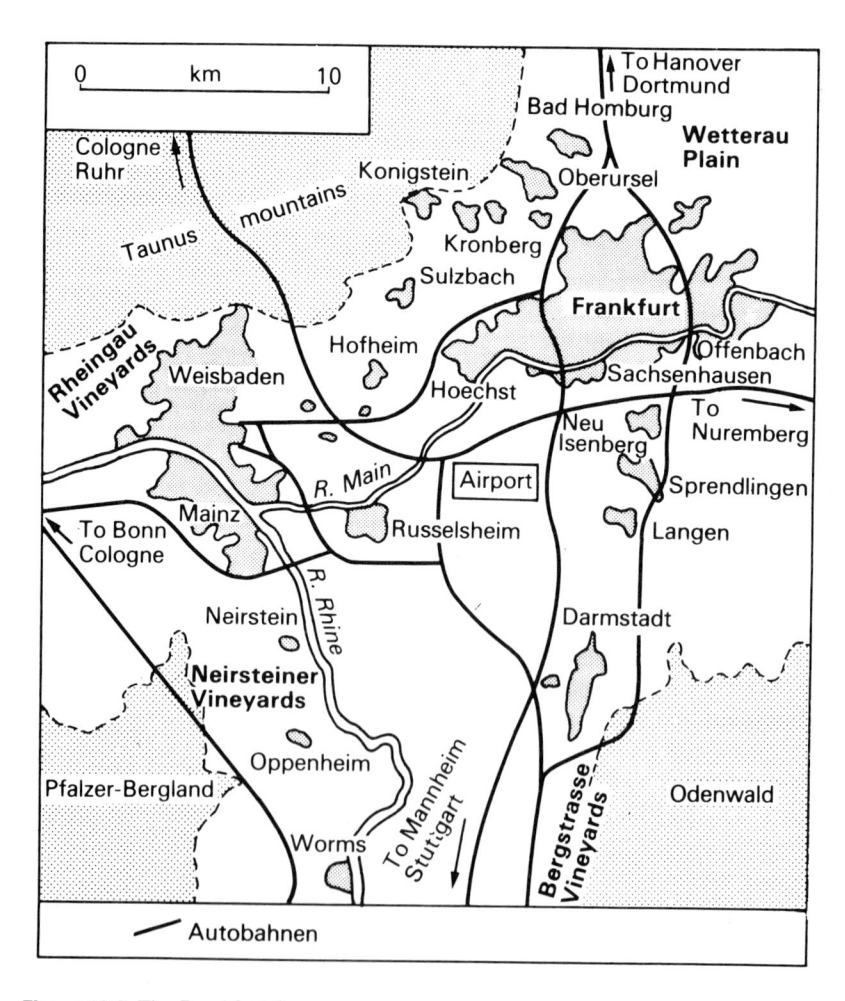

Figure 12.3 The Frankfurt Cross

twenty million passengers per year, emphasising the city's role as an international communications focus.

The most important present-day function of Frankfurt, however, is finance and commerce. It is the banking centre of West Germany, with the head offices of the German Federal Bank and no less than 148 German and 114 foreign banks. Its stock exchange is the largest on the continent, and eleven major fairs are held there each year.

Today the city has spread outwards and suburban growth has reached such proportions that to the north the formerly small spa towns which grew up at the foot of the Taunus massif - Bad Homburg, Konigstein and Oberursel - have become dormitory suburbs. New townships such as Nordweststadt lie four miles to the north. Along the River Main, west towards Wiesbaden, are the chemical complexes of Hoechst, the Opel car plant at Russelsheim and a number of other towns like Hofheim, so that the seventeen kilometres between Frankfurt and Wiesbaden are suburban in nature. To the south, Offenbach and Sachsenhausen now form part of the city. The residential town of Neu-Isenberg farther south is associated with the airport, motorway junction area and state forest, all of which occupy a broad belt of land just south of the Main. To the east along the Main is the jewellery-making town of Hanau. The zone thus described is closely linked in an economic sense, and rapid transport has ensured that all parts of the area are inside one hour's journey from Frankfurt. Commuting has developed on a large scale with complex movements of workers into the factories and offices of the city, which is ranked with Düsseldorf as a business centre. The whole city region has over 1.5 million people and is one of the most important growth areas in the European Community.

Mannheim - Ludwigshaven

The twin cities of Mannheim and Ludwigshaven are the most important inland ports in the area, dependent upon the Rhine for imports of raw material and major heavy chemical centres. They are perhaps more akin to the heavy industrial centres of the Ruhr than the other cities in the Middle Rhine. This is certainly true, but there are also very ancient settlements in the area. Landau and Neustadt are examples of settlement along the Weinstrasse, the foothill zone on the west bank corresponding to the Bergstrasse on the east. Worms, with its famous Liebfraumilch wines, Speyer and its textile industry, and Heidelberg, all lie within a dozen kilometres of the two main cities. Mannheim itself was founded as a princely residence at the confluence of the Rhine and Neckar rivers, by the Elector of the Palatinate in 1720, and became a great theatrical and musical centre.

Mannheim's modern growth dates from the increase in Rhine traffic when it became the head of navigation for 5000-tonne barges in 1885. The advantages of river ports as input and processing points dependent upon low-cost waterborne raw materials are shown clearly here. Mannheim is a major distribution and transhipment point, a large railway junction, and one

The German autobahn system between Frankfurt and Kassel

of the largest inland ports in Europe (fig. 12.2). Heavy chemicals, petrochemicals, dyestuffs, synthetic fibres, pharmaceuticals and grain milling are important. The twin inland ports of Ludwigshaven developed on the west bank, the combined urban complex having over 1.5 million people.

Stuttgart and the Neckar valley

The city of Stuttgart shows many of the features already mentioned. It is a city region based largely upon human resources and the convergence of routeways (fig. 12.4), and at least part of its prosperity derives from the rivers Rhine and Neckar. Its site is an accident of history, and was chosen for a palace and fortified town from which the Dukes of Wurttemburg were to rule their principality from 960 AD onwards. The city was built in the baroque style and has many elegant buildings. Its general situation must be one of the most beautiful in Europe for it stands in a saucer-shaped valley on the south side of the river. Neckar with wooded and vine-clad hills around acting as natural boundaries. To the south-west lies the Black Forest, and to the south-east, the Swabian Jura.

This would appear to be hardly the situation for a major industrial city, but even in the medieval period it became a focal point for Alpine routes from the south and routes from the Rhine via Heidelberg and the Neckar valley. In addition, the east-west routes from the Danube valley and Munich passed across the Neckar valley and Stuttgart. This pattern, like that of Frankfurt, was re-emphasised during the nineteenth century when the railway system

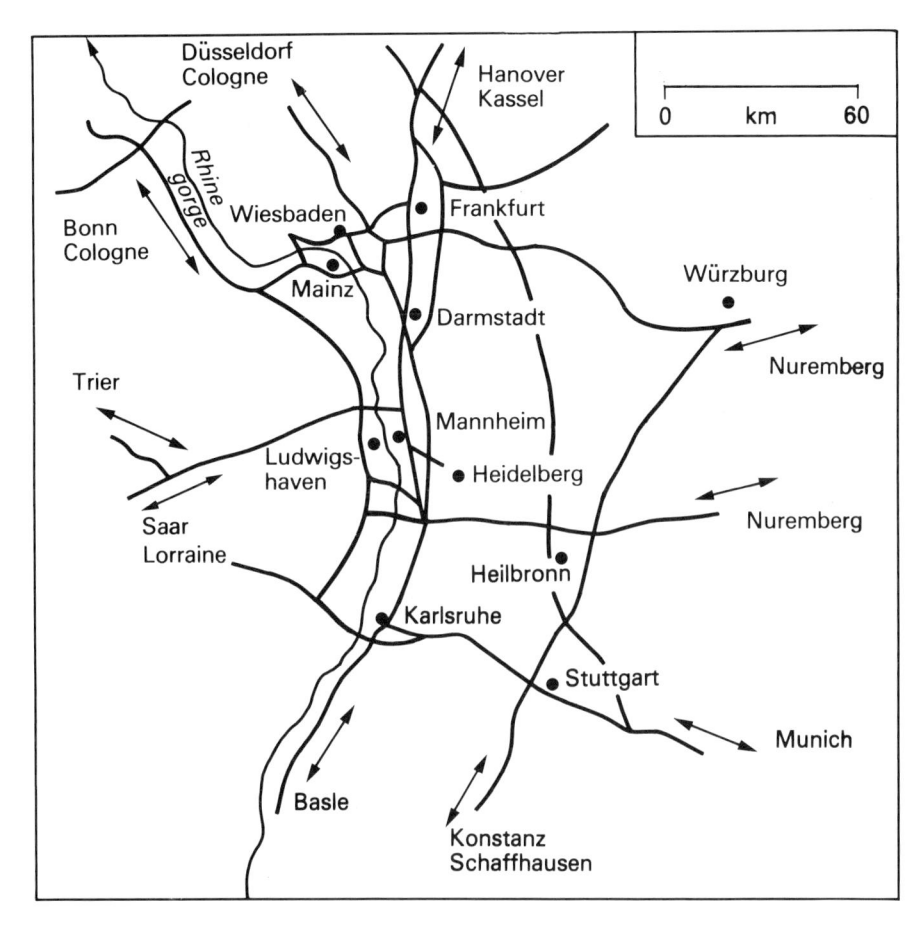

Figure 12.4 The autobahn network in the Middle Rhinelands

made the city into a focal point of South Germany, and the Neckar became navigable for barges of up to 1350 tonnes.

The industrial wealth of the city traditionally was based on textiles from local wool, but the present-day industrial pattern shows dependence upon the production of high value goods, the assembly components using skilled labour and the minimum of raw materials. Stuttgart is so far from raw materials that even with Rhine - Neckar transport, costs are relatively high, and skilled labour provides the one real resource. The city is the centre of a densely populated rural hinterland which provides a pool of labour, but in addition, there are many foreign immigrant workers. The machine-tool, automobile components, electrical engineering, electronics and precision and optical instruments are supplemented by chemicals and pharmaceuticals, electrical domestic equipment, textiles, footwear and food-processing. Daimler-Benz, Mercedes, Audi and Porsche cars are built here, and Stuttgart is a major centre of the West German car industry. The IBM computer company reflects the science-based and sophisticated nature of the manufacturing. Many industrial townships lie to the north-east of the city in the Neckar valley itself; Unterturkheim, Cannstadt, Feuerbach and Zuffenhausen. Some idea of the great significance of industry can be gauged from the fact that Stuttgart's industrial output is fourth among West German cities, and it produces one-third of the gross domestic product of the Land of Baden-Wurttemburg.

Stuttgart has considerable industrial, regional, political and cultural significance as the capital of Baden-Wurttemburg, and within the city region are some two million people. Industrial satellite towns, based upon local supplies of labour, extend to a large distance up the valleys surrounding the city, north to Ludwigsburg and Heilbronn, south to Tubingen and Reutlingen, east to Goppingen and west to Pforzheim. Within this Stuttgart region there is a very dense movement of traffic, and the city is therefore a good example in southern Germany of a 'landeshaupstadt', or regional service centre. It is a highly specialised city with an extensive tributary rural area in the Neckar valley. Finally, it has a natural physical zone of influence, lying between the Odenwald to the north, the Black Forest to the south-west, and the Swabian Jura to the south-east. Like Frankfurt, Stuttgart has developed and prospered because of its position at the hub of a communications network and its nodality in human terms. It is the perfect example of a city which has prospered because it was there.

Summary

There is a north-south division along the Rhinelands, although perhaps not as dramatic as in other parts of the Community. There is a contrast between the Ruhr, with its heavy industry re-adjusting to twentieth-century conditions, and these city regions of the Middle Rhinelands. Centres like Frankfurt and Stuttgart are much favoured because communications, markets and labour supply have replaced raw materials as the dominant factors in industrial location. The Frankfurt - Mannheim - Stuttgart corridor has both pleasant environmental conditions and a high level of accessibility and is quoted by Hugh Clout as being of one of the new 'industrial boulevards' of Western Europe.

13

Belgium: a study in regional contrasts

Local variations

Discussion of the European growth axis tends to obscure examination of local variations within it. Belgium is such an example. Although Belgium lies almost totally upon the central belt of economic activity, there are, nevertheless, four geographical themes within the country (fig. 13.1) which illustrate major contrasts.
1. The Brussels-Antwerp growth axis.
2. The coalfield belt from Mons to Liège which has similar problems to its western extension, the Nord coalfield of France.
3. The highland area of the Ardennes which is an area of out-migration, although there is an interesting variation from the norm.
4. The language and cultural division between Fleming and Walloon complicates the economic differences within the country. Changes in recent years have swung the balance of ascendancy from Wallonia to Flanders, creating considerable intergroup tensions.

The Brussels - Antwerp growth axis

This is the most densely populated part of Belgium, and although the port and capital city are some forty kilometres apart, they are beginning to show all the signs of conurban linkage as Antwerp rapidly becomes similar to Rotterdam as a major input point. Brussels has grown enormously since becoming the effective administrative centre of the European Community. They are connected by the Willebroek canal, railways and the E10 motorway, and show signs of creating a future conurbation on the southern side of the Rhine delta, similar to the Randstad on the north. Indeed, if Ghent is included, a traingular growth area can be seen to be emerging (fig. 13.2).

Antwerp

The port and city is some eighty kilometres from the sea, but has a modern deep-water channel through the Western Scheldt estuary, which is the southern arm of the Rhine delta. The Dutch delta plan will benefit Antwerp in terms of further waterway improvements and will also give better transport links with Rotterdam.

228   THE NEW EUROPE

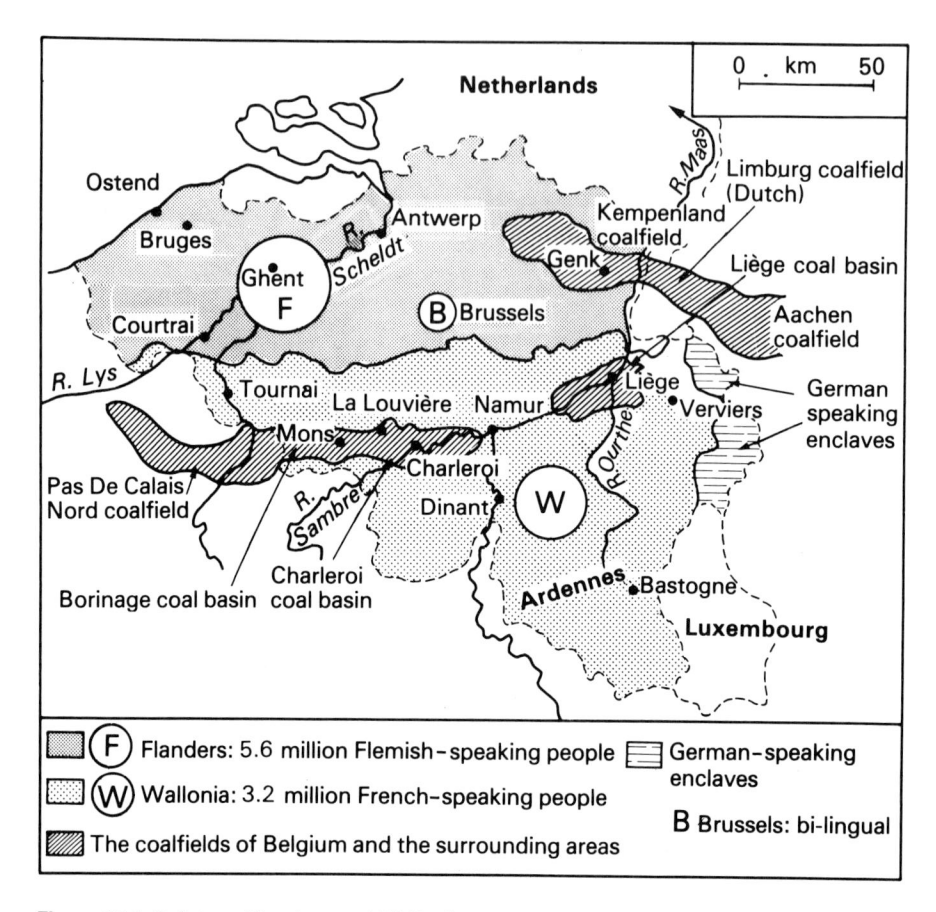

Figure 13.1 Belgium: Flanders and Wallonia

Antwerp is a very good example of a medieval city with the original sixteenth-century fortifications marked by boulevards which now enclose the present city-centre. The old city was the principal port in the whole region up to the sixteenth century, but by the Treaty of Westphalia, 1648, the Scheldt was completely closed to sea-traffic, thus ensuring the rise and dominance of Amsterdam to the north. As a result, Antwerp and many other Belgian ports declined, and it was not until the early nineteenth century, with Belgian independence (1830) and increasing industrial traffic into the Rhine-Scheldt delta, that Antwerp began to prosper again. Growth was then rapid and in the late nineteenth century the earlier walls were replaced by the 'enceinte', an elaborate defensive complex with forts. This larger ring now contains the city centre, inner areas and railway termini.

Since the Second World War, development has been particularly rapid. Residential districts on the east bank in particular fan out along the main roads east towards Turnhout and south towards Mechelin. Road tunnels under the Scheldt have led to modern developments on the west bank.

Industrial growth has followed the development of the port, and large areas of land along the river are available for industrial development. To the north is a large and comprehensive dockland with oil refineries, car assembly plants, and the processing of imported foodstuffs - mainly tropical products. To the south of the city along the Scheldt are shipyards at Hoboken and a variety of heavy industry along the river Rupel and Willebroek canal towards Boom. These include heavy ceramics, cement, chemicals, textiles and brickmaking. There are also precision industries, such as photographic processing, diamond cutting, radio and electronics, which have grown with the increasing sophistication of the city's industrial capacity.

Antwerp has emerged as an international port, third in European rank, with a hinterland largely complementing that of Rotterdam. This has a radius of up to 400 kilometres extending through Belgium into north-east France and to Aachen, and including a small part of the southern Netherlands. The modern importance of the city is based upon communications. There are no raw materials, but there are canal links to Brussels, Charleroi, Liège and the Meuse valley, and the Kempenland. Motorway links are becoming increasingly important, in particular the Antwerp-Brussels route, and the Antwerp - Liège - Aachen route, which follows the line of the Albert canal. The whole agglomeration has a population of three-quarters of a million people and is one of the fastest-growing areas in Belgium.

Brussels

The early extent of the city is indicated by the almost complete polygon of boulevards which marks the old walled city, and which now contains the administrative, commercial and retail sectors of the central area as well as the oldest parts of the city, which are a considerable attraction for tourists.

Industrial locations are highly zoned, lying along the Senne valley to the north-east and the south-west. Along the Willebroek canal to the north, reaching towards Mechelin, are heavier processing industries, timber, chemicals, heavy metals and food processing, whilst to the south along the Charleroi canal lie textile works, engineering and cable works and the Clabecq steelworks. In addition, however, the city's light and specialised industries are immensely varied and widespread, including clothing, jewellery and cosmetics (Brussels is a fashion centre), and pharmaceuticals, electrical goods, printing and miscellaneous consumer industries.

The agglomeration now stretches along the Senne valley and towards the south has extended around and beyond the Forest of Soignies, so this now forms an enclave of green belt surrounded by suburbs and commuter villages. Expansion eastwards to Louvain and south to Wavre has created an intensely suburbanised zone. Brussels is a bi-lingual island (French and Flemish) just within the Flemish-speaking part of Belgium. Near Wavre, the city's expansion has crossed into the French-speaking (Walloon) section of the country. Brussels has gained its dynamism from being a regional and political centre for much of East Flanders, although latterly there has been an increasing

230   THE NEW EUROPE

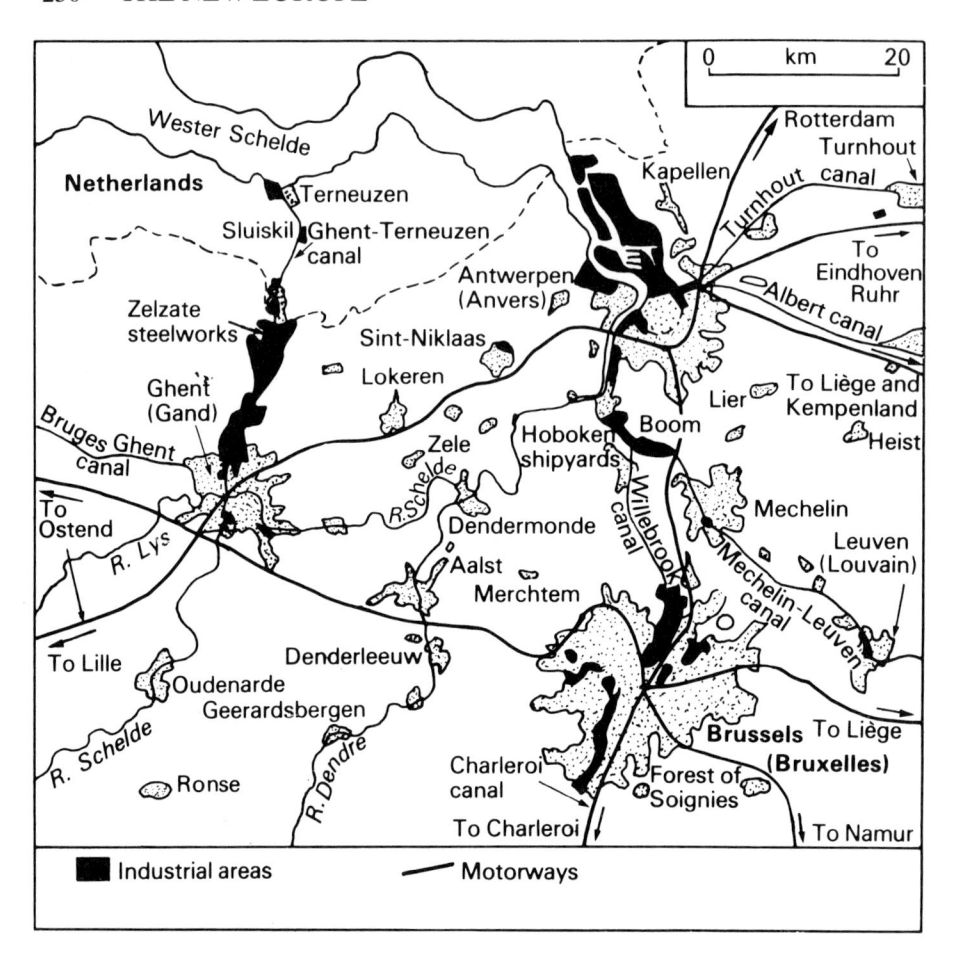

Figure 13.2 The Brussels-Antwerp-Ghent growth area

movement from all parts of Belgium, the attraction of the city contrasting with the decline areas of the south. Then there is its function as national capital of Belgium, now supplemented by its growth as administrative centre of the European Community. There is major expansion of the tertiary sector, and most of Belgium's insurance, commerce and finance companies are in Brussels. The Berlaymont Building, which houses the European Community Commission, is the nucleus around which much activity occurs. The factor of cumulative causality is reflected in the need for multi-national industrial, professional and commercial companies to have their head offices at this central point. Since 1983 a building boom, often stimulated by British capital, has exerted tremendous pressure upon land in the centre of Brussels as well as in the suburbs. Brussels has a population of over one million people and with Antwerp to the north constitutes a very definite growth axis.

Brussels, with the Grand Place in the foreground

Ghent

If Ghent is considered, with its 50 000 tonne capacity ship canal to Terneuzen on the Scheldt estuary, and other canal links to Bruges, plus the Ostend - Ghent - Brussels motorway, then there is a third urban - industrial nucleus. Ghent has 250 000 people, and has cotton and synthetic fibres industries. Along the Terneuzen canal, in an industrial area reaching almost to the Scheldt estuary, are shipbuilding yards, chemical plants, oil refineries, paper works and the integrated Zelzate iron and steel complex.

The Belgian coalfields with particular reference to the Borinage and Liège

Figure 13.1 illustrates the relationships of the smaller coalfields of north-west Europe, an almost continuous band stretching from Douai to Aachen. Whilst these are much less important than the major British coalfields and the German Ruhr, they were nevertheless a major factor in the nineteenth century as national sources of energy during the industrial revolution. The problems of each area today vary in intensity according to the extent of exploitation, the nature of the coal seams and the degree of exhaustion, but all have come under considerable pressure because of the alternative sources

	1961 Production (million tonnes)	Number of pits	1970 Production (million tonnes)	Number of pits	1981 Production (million tonnes)	Number of pits	1986 Production (million tonnes)	Number of pits
Kempenland	9.6	7	7.1	5	5.8	5	5.6	5
Sambre-Meuse	11.9	47	4.3	15	0.3	1	nil	nil

Figure 13.3 Coal production by the Kempenland and Sambre-Meuse coalfields

of energy and the need for low-cost fuel. Of all the Belgian coalfields the Kempenland, around Genk and Hasselt, has the longest term future (figs. 13.1 and 13.3). It produces six million tonnes of coal per year from only five large collieries. In the Sambre-Meuse valley coal-mining has now disappeared (fig. 13.3). Because of early exploitation the best seams are now exhausted and with difficult mining conditions and relatively small-scale mines, these are high-cost coalfields. Decline has been so rapid that there have been massive problems of readjustment in the industrial towns which stretch for nearly 140 kilometres across Belgium. There are four regions: the Borinage, centred upon Mons; the Central basin around La Louvière; the Charleroi basin; and the eastern or Liège coalfield. Some measure of the decline may be seen in the fact that in 1953 there were still 136 pits and 120 000 men employed in coal-mining in Wallonia (compare fig. 13.3 and fig. 2.3).

Liège and the eastern basin

Although the basis of expansion in Liège and the eastern basin was coal, the district has always been based upon more than extractive industry and this is the reason for its greater economic resilience during the last twenty years. The city was an ancient bishopric and city-state and as such has traditionally been a major service centre for the eastern regions of Belgium. It covers the Meuse valley and its confluence zone with the Ourthe, Amblève and Vesdre, the Verviers textile area and Ardennes foothills to the south, the fertile Hesbaye to the north-west and the Pays D'Herve to the east of the Meuse. It is the cultural capital of Wallonia and third city of Belgium, with a conurbation population of 500 000.

In the Liège basin, the same problems of productivity arise as in the rest of the Sambre-Meuse valley, and coal mining has now ceased. The worst problem is the legacy of the 'old industrial landscape' with masses of spoilheaps on the Hesbaye plateau above the deeply trenched Meuse valley, where most of the settlement lies. However, the industrial revolution has left Liège with much more than a mining economy. John Cockerill, an Englishman, was responsible for the first blast furnace in 1832, locomotive manufacture in 1835 and a large part of the metal-working tradition of the city. He was the first in Belgium to use the Bessemer process in 1863 (*The Times* – Wednesday 31 May 1972). His company has now become Belgium's major steel-manufacturing concern. Cockerill has the bulk of its steel-making capacity at Seraing and Jemeppe upstream from Liège, and at Chertal downstream of the city. It also has factories at Charleroi and is now known as Cockerill-Sambre, with over fifty per cent of Belgian steel-making capacity. Other important metal manufacturing industries are zinc smelting, tubes, cables and small-arms, aircraft engineering and heavy electrical machinery. 'Geographical inertia' is characteristic of the whole area, with the metal-based industries originally dependent upon local charcoal and water power from the streams running down to the incised Meuse valley, and subsequently, upon coking coal and local iron ores. Other important industries are chemicals, glassware and tyre manufacture.

Verviers and Eupen, along the Vesdre valley to the east, form an associated industrial area manufacturing woollen textiles, originating on local wool from the Ardennes, and water power from reservoirs along the valley.

Despite the obsolescence of the industrial environment, unplanned piecemeal development crowded in the river valley, and relentless decline in industrial employment, Liège has already largely adapted to the decline in coalmining and steelmaking, and has bright prospects for the future. The reason is its location on the major link routes betweeen the Rhine delta, West Germany, and France. Raw materials are now imported via the Albert Canal, particularly coking coal from the Kempenland and Ruhr. The river Meuse provides a very useful means of transport for heavy goods. More important are the motorway links (fig. 8.6). The motorways from Ostend, Brussels and Antwerp lead to Aachen, Cologne and Frankfurt. The 'Autoroute De Wallonie' (E.41), links Liège with the Sambre-Meuse towns and the Lille-Paris motorway. There are major development plans along two axes. Heavy industry is zoned along the river, with a nuclear power station at Tihange, up-river from Liège, an oil refinery on the Albert canal, and petrochemical works and fertiliser plants, which will join the predominantly steel and heavy engineering works of the river valley. New light industry estates are zoned near the motorways on the plateau on the outskirts of the city. Computers, electronic components, clothing, fibreglass and ceramics are produced at green field sites such as Hauts Sarts, the industrial estate north-east of Liège. There are three other smaller industrial estates lying adjacent to the Brussels motorway (E5).

Charleroi

With its surrounding satellite towns, Charleroi is an agglomeration of some 400 000 people and its industrial structure has much in common with Liège. Although coal production has disappeared there is a steel industry, chemical works and, more recently, a plastics industry. This was also the centre of the Belgian glass industry, based upon local sands. Like Liège, Charleroi has a sufficiently broad and diversified manufacturing base to avoid the worst effects of the decline of coal. Engineering, vehicle and aircraft components, electronic calculators, printing and food processing are represented on new industrial estates near the city.

	Miners	Number of Pits
1948–50	30 000	28
1956	24 000	9
1965	8 000	5
1974	1 500	1
1977	nil	nil

Figure 13.4 Changes in the Borinage coalfield

Mons and the Borinage: La Louvière and the Central Basin

The Borinage stretches from the Belgian border to the city of Mons, provincial capital of Hainault and the area to the east including La Louvière is characterised by the same problems. These areas have experienced considerable economic decline and harsh adjustments have been necessary because of their over-dependence upon a single activity, coal-mining. Here there was little industrial development of any sort, and in 1953 coal-mining accounted for fifty six per cent of employment. The seams of coal were almost exhausted, mines were very deep and in one particular colliery, Rieu De Coeur, galleries were specially refrigerated over one and a half kilometres below ground. As a small-scale producer (the Borinage produced 5.9 million tonnes in 1927, the peak year) it was a very high-cost coalfield. Not only was there heavy competition from oil and natural gas, but increased cross-frontier competition and tariff-free conditions which arrived with the European Community meant that by 1958 Ruhr coal, even after being transported to Charleroi, was considerably cheaper. With the reduction in Atlantic freight rates, American coal could be sold at Charleroi for substantially less than Borinage coal. Productivity was low, production was small-scale and inefficient from many small pits, and Borinage coal could not be made competitive. The Conseil National Des Charbonnages worked out a reorganisation and contraction plan (fig. 13.4). Coal-mining has seen a dramatic decline and is now completely extinguished (fig. 13.5). The ECSC aided the region in two ways. Loans were given, often in association with the Belgian Government, to lay out new industrial estates and encourage investment in new plant. The ECSC also shared with the Belgian authorities the cost of retraining workers for new skilled jobs. Belgian Government legislation (1959, 1966 and 1976) has given the area development status, and modern growth industries have been attracted to the area.

The changes in the region are threefold:
(a) The Borinage became a region of out-migration, mainly to Antwerp and Brussels and the employment force actually declined by nearly fifty per cent. There has also been marked demographic stagnation and an ageing population.

(b) Secondly, however, considerable industrial diversification schemes began to take effect (fig. 13.5). There are six industrial estates with consumer and light industries - pharmaceuticals, electronics and telecommunications factories. Since 1960 over fifty factories with 8000 new jobs have been created. There is a vehicle assembly plant at La Louvière and an oil refinery and petrochemical plant at Feluy to the north. With increased employment in services, the industrial structure now approaches the Western European norm.
(c) With the introduction of fast electrified rail services between Mons and other parts of Belgium, many of the area's inhabitants now commute to work in Charleroi, Liège, Brussels, and to Valenciennes and other French towns.

The Borinage has adapted well, although the hardship was considerable during the 1960s. From being a mono-functional coal-mining area, it now has a solid industrial base. Many of the environmental problems still remain, as the landscape of the industrial revolution cannot be obliterated overnight. Its position and population are its main future resources: five million people, a large consumer market, live within fifty kilometres of the provincial capital, Mons. The area lies at the junction of two motorways - Brussels - Paris and

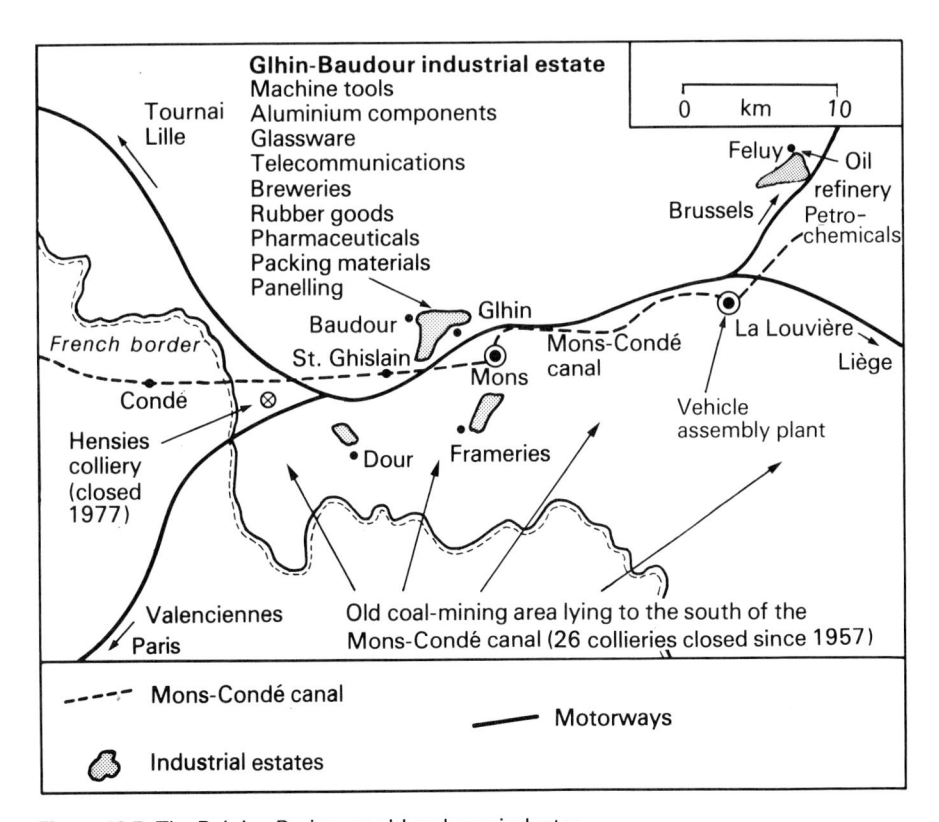

Figure 13.5 The Belgian Borinage: old and new industry

La Louviere, Belgium. The interchange between E 41 Wallonia motorway and the E 10 towards Paris

the Wallonia motorway (E41) (Liège - Mons - Paris) (fig. 13.5). The raw materials may have vanished, but the infrastructure is vastly improved and Borinage lies along a major growth axis.

The Ardennes Massif

The Ardennes is a heavily forested upland (fig. 13.6), lying at about 350 metres OD but reaching 600 metres in the Haut Fagnes near the West German border. With rainfall reaching a maximum of fifty-five inches, it is a zone of marginal agriculture and experiences persistent out-migration. Most settlements are small and confined to the valleys of the Semois, Ourthe, Amblève and Vesdre, tributaries of the Meuse. Farms are small and mainly in pasture, with some cereals (oats, rye or barley), and potatoes and fodder crops. In the sheltered areas such as the Semois valley, tobacco is grown and dried on local farms. Farms are abandoned yearly and the whole area is the least densely populated in Belgium with a population of under 200 000. Forestry is an important occupation and timber is a major resource of the Ardennes, in areas such as Beauraing and Gedinne south of Dinant. There are isolated areas of economic activity such as limestone quarrying near Marche.

The drift of population is a continuing feature of the life of the area, but it is

Figure 13.6 The Belgian Ardennes

now alleviated by two factors. One is the outstanding scenic beauty of the Ardennes which has led to a considerable tourist industry, and the second, inter-related, is the proximity of the region to densely populated lowlands of Belgium, Holland, and West Germany. Its wealth potential increasingly lies in its landscape. With the attractions of woods and forests, numerous chateaux, and beautiful valleys like the Semois and Viroin, many towns have developed a tourist function. Dinant, La Roche, Bouillon, Houffalize and Spa (the original mineral springs have given their name to all towns of this type) all have a tourist function, and are supplemented by others with a market role such as Bastogne, Marche-en-Famenne and St. Vith.

Depopulation is also being partially reversed by the 'weekend cottage'. The second house is a popular idea in Europe, and the Ardennes are ringed with the dense urban populations of the Meuse valley, Brussels, northern France, and the Rhinelands. Barvaux in the Ourthe valley particularly, and the Dinant and Marche areas of the Condroz have large areas of sub-rural development which, if allowed to spread unchecked will rapidly spoil the landscape it is designed to enjoy. In addition, the whole northern section of the Ardennes has fallen within commuting range of Liège and even Brussels, and towns such as Spa, Theux, Aywaille and Remouchamps, in the Amblève and Lower Ourthe valleys, are becoming dormitories.

Accessibility was always limited, with one major north-south routeway (N4) and railway line which ran from Namur through Marche, Bastogne and Arlon into Luxembourg. However, the isolation is now being removed with two motorways largely completed from Brussels (E40) and Liège (E9) which run south towards Luxembourg. These are of considerable importance for the economic revitalisation of the Ardennes.

The regional and cultural dichotomy of Belgium: Flanders and Wallonia

Considerable economic differences between northern Belgium (the Brussels-Antwerp-Ghent region) and southern areas like the coalfields of the Meuse valley and the Ardennes uplands have been outlined above. Belgium is a bi-lingual and bi-cultural state, and the boundaries between the Flemings and Walloons approximately correspond to the economic lines of demarcation between growth areas in the north and areas of decline in the south (fig. 13.1). The history of the two cultural groups adds another dimension to the dichotomy. Flanders was, in medieval times, an economic core area region. The great ports of Bruges and Antwerp were part of the Hanseatic League. Trading links with Italy, and the Flemish cloth trade were all part of one of the most successful commercial areas in Europe at that time.

Since the nineteenth century, however, economic power, based upon the Sambre-Meuse coalfield and its steel, engineering and chemical industries was concentrated in the south. The French speaking south, covering the provinces of Luxembourg, Liège, Namur, Hainault and south Brabant, is known as Wallonia. French language and culture was dominant during the nineteenth century and Belgian administration, culture and teaching was in French. The French-speaking Walloons were an élite both culturally and in terms of prosperity. Liège in particular was the nerve-centre of Walloon economic power.

The Flemish provinces lie to the north, covering West and East Flanders, Antwerp, Limbourg and North Brabant. During the nineteenth century the Flemings were effectively second-class citizens. They lived in an agricultural area with a stagnant peasant economy. Their language was a dialect of Dutch, and spoken by only about fifteen million people in the world. Thus it was at a disadvantage in relation to French as a major language. After Belgium became one country in 1830, the 'Flemish Movement' set out to achieve parity for their language in their own country, and this, although a slow process, was achieved by the 1960s, when for all legal, educational and administrative purposes, Flemish and French became equal in national status. Belgium was divided in two parts by a language line, with Flemish spoken north of it and French to the south. Brussels is a bi-lingual island just north of the dividing line.

During the last forty years major demographic changes have added to this complete reversal of ascendancy of the two groups. The population of Flanders has been rising by over seven per cent each decade, whilst that of Wallonia increased by only 2.6 per cent. The density of population in Flanders is now double that of Wallonia. This has reinforced the Flemish majority position; they are now about sixty per cent of the total population, and the Walloons are very conscious of their diminishing relative importance within the state. Secondly, the reversal of economic fortune from Wallonia to Flanders has reinforced the picture of 'two nations'. Sluggish growth, unemployment, declining industries, obsolescent factories, declining infrastructure, an ageing population, and a 'black country' image, is typical of the depressed Sambre-Meuse valley. By contrast, Flanders now has a larger share of the country's wealth. The integrated steel mill at Zelzate, the development of Antwerp as a port and large-scale industrial area, and the linking of Ghent to the sea are examples of the dynamism of port locations, and the Rhine delta in particular. The newer productive coalfield of the Kempenland lies in the provinces of Antwerp and Limburg. The position of Brussels with its service infrastructure in the European Community is an additional factor in the growth of the Brussels - Antwerp axis. Some 81 per cent of American investment has gone into Flanders and the Greater Brussels area, only 19 per cent into Wallonia.

Summary

This picture of regional contrasts within Belgium attempts to show that not all regions within the European growth axis share equally in prosperity. Southern Belgium is one such region, but its process of adjustment is assisted greatly by its accessibility and its proximity to the European core.

14

Randstad Holland: the Ring City

The concentration of population

One of the most urbanised and wealthy regions in the European Community is the western part of the Netherlands, covering the provinces of North and South Holland, Zeeland and Utrecht. The concentration of population may be seen by comparing the total population of the Netherlands, 14.5 million people in 1986, with the population of 6.7 million in the four provinces (fig. 14.2). Nearly half the country's total population is concentrated into twenty per cent of the country's area.

Specifically, the ring city is formed by two major urban regions, the cities of Amsterdam, Utrecht and Haarlem in the northern arc, and Rotterdam and The Hague in the south (fig. 14.1). Together with a number of smaller units interposed between, such as the historic cities of Leiden and Delft, the new industrial centres of Zaandam and Ijmuiden, seaside resorts such as Scheveningen, Katwijk and Zandvoort, and other towns, the whole begins to take on the shape of a broken ring, or horseshoe.

The reasons for this major concentration of population lie mainly in the central position of these cities in relation to north-west Europe. The delta region of the Rhine and Maas, opening out to the North Sea, has traditionally been important for trade, commercial and industrial activity since the medieval period when Amsterdam was one of the foremost ports of the Hanseatic League. This was underlined during the colonial period when Amsterdam became the world centre of the diamond trade. The cities now have a role as the outlet for the Rhine basin, the major industrial, transport, and population axis of the community. The Randstad cities are literally at the centre of European integration (figs. 10.1, 10.2).

This concentration of population, even in an extremely prosperous city region, causes problems and creates the need for very careful planning. There are four convenient themes by which to look at the geography of this dynamic region.

1. The unusual pattern of urbanisation in this polycentric city region, based upon distinct settlements originally sited at dry points above the unreclaimed marshlands of the Rhine delta.
2. The Port of Rotterdam and its industrial development with the attendant pollution problems.

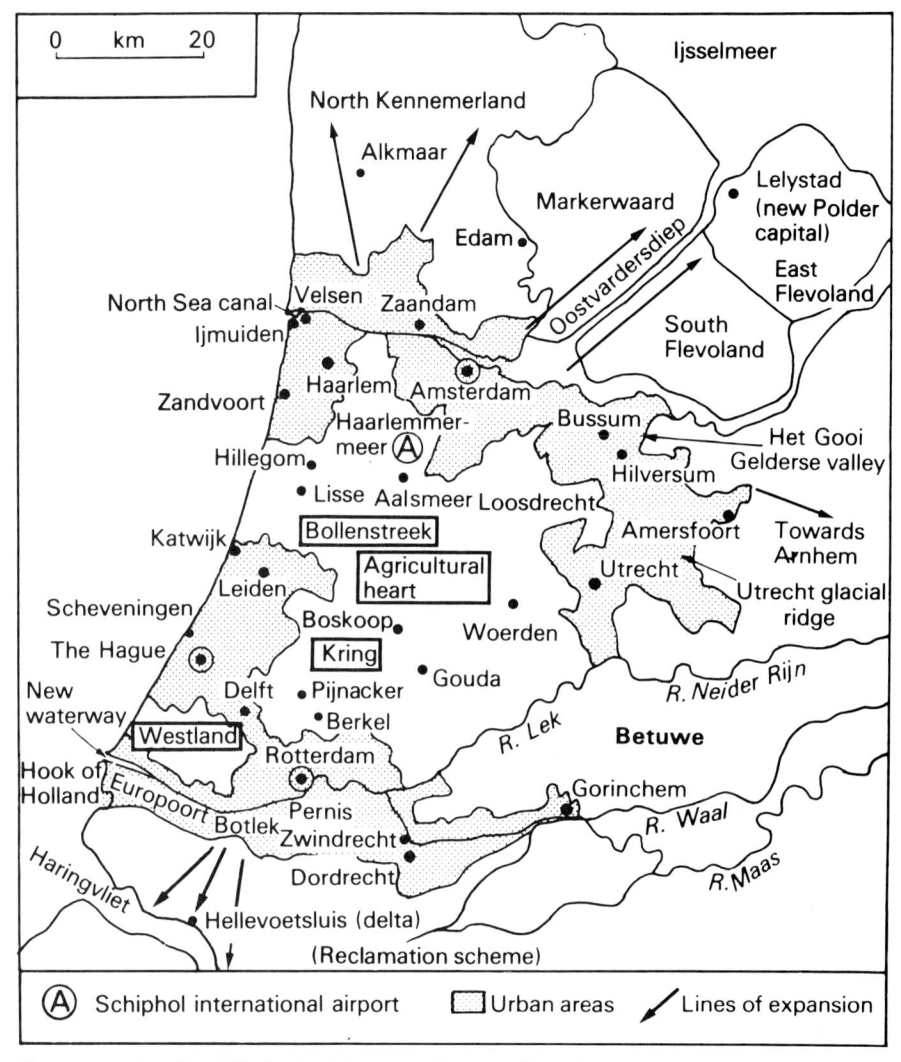

Figure 14.1 Randstad Holland: cities, greenheart and development plan
(A = Schiphol International Airport)

3. The needs of agriculture in this very fertile part of the Netherlands which produces forty per cent of the country's total food.
4. Severe competition for land and Regional planning.

The urban structure of the Randstad

The Randstad cities form a distinctive horseshoe shape which are in danger of coalescing except perhaps on the south-east, where an open section exists between Dordrecht and Utrecht. Its unique character derives from its 'rim' structure around the green centre, but also from its functional and hierarchical development. Unlike other European cities such as Paris or London, the

City	Population total of agglomeration (thousands)	Functions of principal city
Amsterdam	1036	Financial and commercial centre. The main service and retail centre, entirely metropolitan in character.
Rotterdam	1031	Industrial, transhipment, storage terminals, import and export.
The Hague, including Delft	681	Administrative capital. Centre of Government and international agencies.
Utrecht	463	Historic university town, ecclesiastical centre, now provincial capital and regional service centre and communications centre.
Haarlem	233	Residential and regional centre. Engineering industry.
Leyden	166	Historic university town and regional centre for mid-western Randstad.
Dordrecht	185	Industrial town and a satellite for Rotterdam.
Hilversum	111	Residential and commuter town for Amsterdam.
Ijmuiden, including Velsen and Beverwijk	135	Outlet for North Sea canal with fishing port and steel works.
Zaandam	137	Industrial outlier to Amsterdam at the inner end of the North Sea canal.
Total large municipalities	4177	
Randstad (Total)	5093	(includes 70 municipalities)
Total for North and South Holland, Zeeland and Utrecht	6658	
Total population for the Netherlands	14150	

Figure 14.2 The Randstad hierarchy: a summary, 1988

Herengracht, in the historic centre of Amsterdam, showing seventeenth century merchant houses

multitude of functions normally carried out within the 'central business district' of a capital city is distributed here between several cities. There is a hierarchy of centres of different sizes, some seventy municipalities in all (fig. 14.2), which can be conveniently grouped as follows:

Amsterdam

This historic city is the cultural, financial and commercial capital of the Netherlands and is the most highly metropolitan in character of all the Randstad cities, with a population of over one million. The city has a wide range of banking, finance and commerce, and luxury shopping facilities. As a tourist centre it has museums, art galleries and luxury hotels. The industries within the city are those of printing, fashion clothing and diamond-cutting. The distinctive city centre, with its semi-circular structure bounded by quiet tree-lined canals, is a major tourist attraction.

The expansion of the city has created several sub-zones on the periphery. To the south is the international airport at Schiphol and the new residential areas of Amstelveen and Sloetermeer, whilst its commuter zones lie farther east around Bussum and Hilversum, in the undulating wooded hills of the Het Gooi.

On the north side of Amsterdam is the important industrial region around the North Sea canal. The canal was opened in 1876 to provide better communications from Amsterdam to the sea. There are two industrial complexes associated with it, Ijmuiden-Velsen and Zaandam. At Velsen is

Amstelveen, new suburban residential complex on the southern side of Amsterdam

the integrated Iron and Steelworks of Hoogovens, with blast furnaces, rolling mills and tin-plating mills. The complex is based upon cheap imports of raw materials, coke from the USA and iron ore from Sweden and North Africa. Nearer Amsterdam is Zaandam, where the North Sea canal reaches the city. Once important for shipping, it is now concerned mainly with processing of imported raw materials and foodstuffs of colonial origin and also local dairy and vegetable products.

Haarlem

An historic city which was associated with the Dutch war of independence against Spain in the sixteenth century. Haarlem is now mainly a residential and commuter area for Amsterdam and is the regional centre for South Kennemerland.

Utrecht

Utrecht is the one city of the Randstad which is a considerable distance from the sea, and has become an inland communications centre. It is a railway focus at the eastern end of the conurbation and has now become a motorway junction. It is, however, one of the oldest cities of Holland and also an ecclesiastical and university town, and a provincial capital. Suburban development is taking place eastwards towards Amersfoort.

The Hague agglomeration

The two historic cities of Delft, famous for pottery, and Leiden, with its university, are very nearly joined by ribbon development to The Hague. Zoetermeer to the east, and Wassenaar and Scheveningen to the north are residential outliers for the Hague agglomeration. The Hague itself is the seat of government of the Netherlands and has most of the administrative and public bodies, and many international agencies such as the International Court of Justice. Otherwise it is very much an attractive residential city, and has been called the largest village in Europe.

Rotterdam

Of all the centres so far described, Rotterdam has experienced the most tremendous growth in size and international importance. As the raw materials input point and transhipment centre, it is a most dynamic industrial growth area, and the largest port in the world. It will be dealt with in detail in the next section.

Rotterdam: Europe's leading port

Although the picture already given indicates the considerable number of industries associated with the Randstad, there are two marked concentrations. One, already mentioned, is the belt along the North Sea canal from Ijmuiden and the Velsen Steelworks to Zaandam. By far the more important is the thirty kilometre stretch of water from the Hook of Holland to the city of Rotterdam itself along the New Waterway and thence along the distributaries of the Maas and Waal as far as Dordrecht (fig. 14.3). In 1962 Rotterdam moved ahead of New York in terms of cargo tonnage handled and in the decades since it has become the pre-eminent world port. Figure 14.4 illustrates the comparisons with other leading European ports. As late as the Second World War it was a port of only moderate importance and in 1945 was in ruins as a result of bombing. It has a long history but its growth to prominence can be seen in three stages, in each case initiated by a major change in transport factors.

The early port, a dam on the River Rotte, itself a right-bank tributary of the tidal Maas, was a medieval fishing village with a very long and tortuous channel to the sea. Growth was slow and unspectacular, although by the end of the eighteenth century Rotterdam had grown to 50 000, the second largest city of Holland. It was a prosperous pre-industrial port with the 'Oudehaven' tucked easily into the centre of the city.

The ultimate dominance of Rotterdam over both Amsterdam and Antwerp was ensured by the cutting of the New Waterway. This is wide, lock-free and cuts straight through the sandspit of the Hook of Holland to a point on the

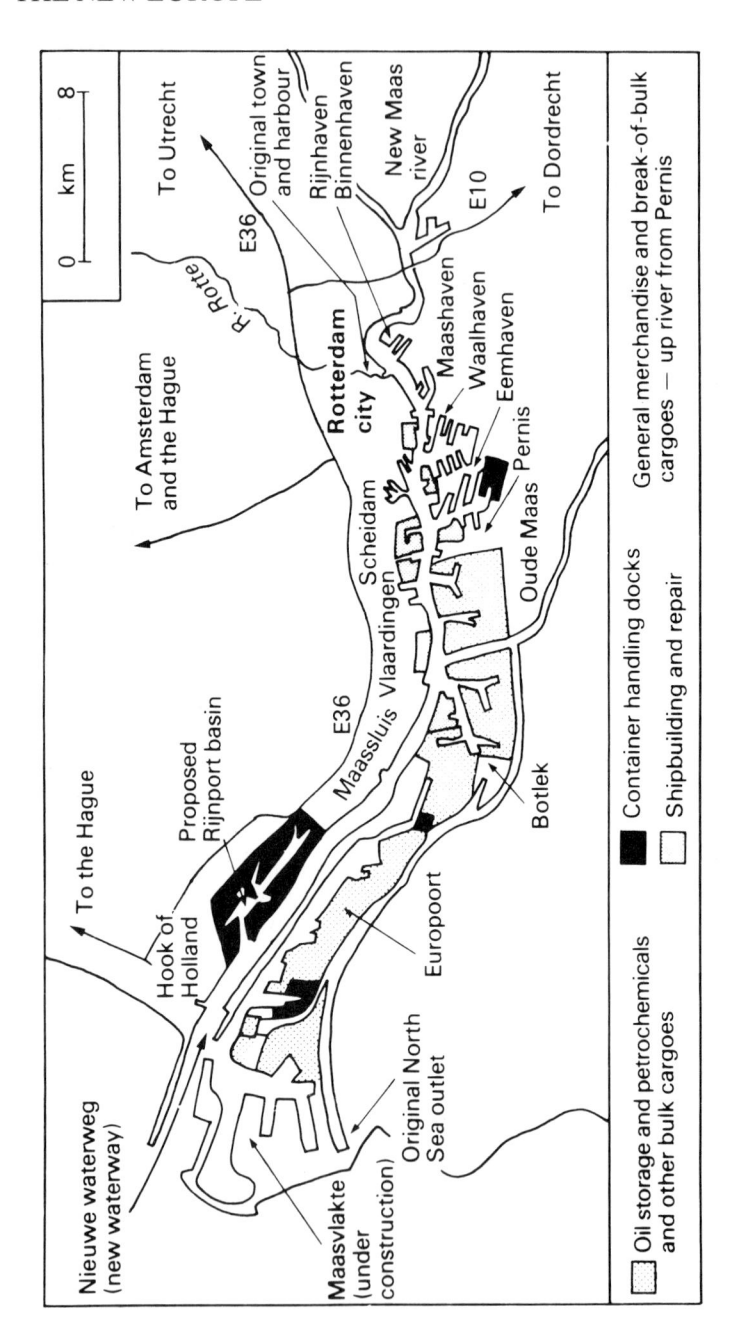

Figure 14.3 Rotterdam: Europoort

coastline where deeper water allows larger ships to enter the harbour. It became operational in 1872 and gave a new lease of life to the port at a critical time. The arrival of the steamship needing deeper channels coincided with the development of the vast hinterland of the port and the growing lines of waterway communication along the Rhine axis. At the same time as the Rhine and its tributaries became Europe's commercial artery, navigable as far as Switzerland, there was also rapid development of the Ruhr coalfield, Lorraine, the Saar, Limburg and the Sambre-Meuse valley. Rotterdam became the input and output point for industrial Europe, and to this major locational advantage was added the technical superiority of a modern deep-water channel. The docks of this period lie on the south bank opposite the old medieval harbour Rijnhaven, Binnenhaven, Spoorweghaven (Railway harbour), Waalhaven and Maashaven. These are small by modern standards, and date from the nineteenth century. They indicate the approximate extent of the port before 1945.

The really significant change came after 1946 when new opportunities presented themselves. Rotterdam's access to the Ruhr, and its position as the outlet for the Rhinelands, the most populous part of Europe, matched its adjacency to the English Channel, the busiest stretch of water in the world. Even more important was the development of the European Community and its geographical axis along the Rhinelands. The whole Rhine delta assumed a new economic importance, and Rotterdam was one of the most central points on this axis. However, the single most significant change was the shipping revolution which developed from the great increase in size of ships during the 1960s and the increasing specialisation of cargo transport, with a distinction being made between bulk cargoes and break-bulk or transhipment cargoes.

The rapid decline of coal during he post-war period and its replacement primarily by oil was the great opportunity for ports like Rotterdam to become the reception, storage, and refining points for West European oil. Imports of crude oil rocketed from 2.3 million tonnes in 1938 to 61 million tonnes in 1967 and 130 million tonnes in 1981, indicating that Rotterdam's growth has been due in large part to her oil imports. The pipeline system to the Rhinelands (fig. 2.7) is an important supplement to the port facility which has helped to create a tremendous number of processing industries, such as oil refineries, petrochemical works, chemical plants and plasttics fabricators. The port was well placed for the further boost to the bulk trade when the Suez Canal closed in 1967 and the era of the super-tanker began. Rotterdam's deep-dredged channels enabled it to become pre-eminent as a terminal port for raw materials, including crude oil, mineral ores, scrap-iron, timber and fertilisers, grain and coal, often requiring further processing. To accommodate the huge bulk-carriers, the deeper water areas from Pernis downstream have been developed since the 1950s, and by 1966 bulk carriers of over 200 000 tonnes d.w.t. could enter the port. The Pernis area is a complex of storage facilities, oil refineries and petrochemical works, followed by Botlek (fig. 14.3), and Europoort which came into use during the 1960s with a capacity for 300 000-tonne oil tankers and 125 000-tonne grain carriers. The final phase is the reclamation of 2500 hectares of land at Maasvlakte, an ambitious scheme to provide even more land for industry and port facilities. The post-war development of bulk shipping can be seen as a succession of newer, larger dock basins and industrial complexes, each one downriver from the original

Europoort, Rotterdam, looking seawards towards the new development on flat estuarine land at Maasvlaakte

Rotterdam city docks and nearer to the sea, with over thirty kilometres of waterfront in all.

However, the upper harbour is important and adds the other element - the break of bulk, transhipment, or 'gateway' function (fig. 14.3). The development of containerised cargo-handling has revolutionised the loading and unloading of the higher-value processed or manufactured goods which are packaged in smaller units. The 'gateway' port function is necessary principally to handle, transfer, and despatch to a variety of destinations. Traditionally this was a cumbersome inefficient operation, consuming time and labour. With the container, the high-value component or complete product is moved through the port as quickly as possible by efficient cargo-handling equipment. Larger ships have meant greater economies, provided that 'turn-round time' has been cut. For example, on the Atlantic run 'dead time' in port has been cut from 70 to 20 per cent, with consequent saving on costs. On the dockside large areas of flat stacking space are needed for stacking and marshalling and also an efficient system of inland transportation. During the 1960s Rotterdam re-equipped itself for containerisation with the Europoort container dock so that it has become the foremost container port in Europe and even takes goods destined for the UK, formerly handled by London. The upper harbour

A

Port	Cargo tonnage (million tonnes) 1986
Rotterdam	399
Marseilles	72
Antwerp	49
Le Havre	39
Genoa	39
Hamburg	32
Amsterdam	21

B

Year	Cargo tonnage (million tonnes)
1938	42
1946	8
1955	66
1966	130
1973	233
1976	279
1986	399

C

Commodity	Per cent trade (by volume)	
Oil	58	
Mineral ores	12	
Coal	4	bulk cargoes 84
Cereals	4	
Other bulk cargoes	6	
General cargo (break of bulk)	16	

Figure 14.4 Rotterdam: **(A)** comparative trade figures **(B)** growth of trade **(C)** principal trading commodities, 1986

has largely been adapted for containers. This is the small group of basins on the south bank of Rotterdam city, Rijnhaven, Binnenhaven, Spoorweghaven and Maashaven (fig. 14.3) where the clutter of merchandise is replaced by huge cranes with 50 tonnes capacity, standardised boxes, and large storage zones. Waalhaven and Eemhaven are larger docks, the first one modernised, and the second built during the 1960s purely for container traffic. An extension downstream on the north bank is Rijnpoort. All the docks above Pernis are concerned with break-bulk cargoes. The successful operation of the 'gateway' function requires excellent communications to all parts of Europe. The river Rhine and its tributary systems of canals and navigable rivers are Rotterdam's principal advantage. In addition, the system of railways and autoroutes allows fast transit to all parts of the European Community. The upper harbours are not only cargo-handling. Waalhaven and Schiedam are also the major shipbuilding sections of the port and its main liner terminals. There are also many other industries: glass-making, car assembly, brewing and distilling, and food processing (chocolate and margarine) in the dock areas.

Without any doubt, Rotterdam's growth and prosperity is based firmly upon its bulk handling facilities and in particular, oil (fig. 14.4). One of the main problems is pollution. Water pollution is a menace and the Rhine is heavily polluted here by sewage, industrial and power station effluents from up-river and from West Germany (60 000 tonnes of chemicals per day are estimated). A more serious matter is air pollution at Botlek, Pernis and Europoort, where the oil refineries, chemical and petrochemical plants foul the air with hydrocarbons and sulphur dioxide. It is the downstream residential areas which suffer most - Vlaardingen, Maasluis, Rozenburg and the Hook of Holland. The original plans for heavy industry at Maasvlakte have had to be modified because of the success of local action groups, and the land reclamation has been much slower than originally anticipated. Finally, there are constant worries about housing and recreation, which relates to the ever-present problem in the whole Randstad: competition for land. In Rotterdam, industrial considerations have often completely outweighed any others. The symptom of this is the outward movement of Dutch people from the port and industrial area of the Rhinemouth since 1960.

In conclusion, it is relevant to consider the future of Rotterdam. Will improvements in pipelines allow Marseilles and Genoa to reduce Rotterdam's oil hinterland? Rotterdam's bulk oil imports may not be a means of continual growth, as was assumed during the 1960s. The port and city will probably remain the major throughput point of the European Community because of its superb junction position between the North Sea and the Rhineland axis. Principal future growth will probably be based upon its high value break of bulk container trade.

The Agricultural Heart

Western Holland is also a key agricultural area and formed the original core of the extensive schemes of land reclamation which began in the thirteenth century. The culmination of Dutch enterprise in this direction came in 1852 with the draining of the Haarlemmermeer of 18 000 hectares. The future of the open centre of the Randstad (the Greenheart), traditionally one of the most productive Dutch farming areas, causes considerable concern. The essence of the continuing planning problem is how to ensure continued urban development without giving up the agricultural land inside the Randstad.

Most of the Greenheart (fig. 14.1) is polder land below sea level. The principal characteristics of these polders are rich silts and clays, a reclaimed landscape of rectangular fields, and an intensive and varied agriculture. There are many lakes which are used for recreation, sailing and fishing, and have areas for picnics and camping. Water drainage is essential through a system of ditches and dykes and the former importance of windmills is shown by their continued existence alongside the more modern pumps.

The intensive horticulture with which the Dutch have traditionally been

Waalhaven, Rotterdam. These are the adapted upper docks and enclosed basins near the city centre which handle general cargo and container ships, the entrepôt function.

concerned stems from the need to produce high value foods on a relatively restricted land area, and also from the demand which built up during the nineteenth century for large quantities of food by both the Randstad population, and the nearby industrial countries of Belgium, the United Kingdom and Germany. Much of the Netherlands total crop of market garden products is grown here. There are considerable areas of specialisation, perhaps the best known of which is the small area between Leiden and Haarlem, the Bollenstreek (Lisse, Hillegom and Keukenhof), famous for flowers and bulb production. Aalsmeer near Schiphol airport is concerned with cut flowers (roses and indoor plants) and Boskoop to the south of Leiden is a specialist area of ornamental conifers and shrubs. Two major zones of vegetables occur: Westland lies between the triangle of the cities of The Hague, Rotterdam and The Hook, and contains the major concentration of glasshouses (1400 hectares in all). The predominance of glasshouses gives the area an urbanised appearance but it grows a large proportion of Dutch tomatoes, cucumber, lettuce, leeks, carrots and spinach, etc. The Kring district north-west of Rotterdam (Berkel and Pijnacker) specialises in salad crops. In addition to vegetables, there are extensive areas under fruit, both soft and orchard, largely around Utrecht and to the south of the Neider Rijn. Melons and grapes are cultivated under glass.

Whilst horticulture has been traditionally the most significant form of farming, arable production, dairy farming and recreational use also make major demands upon the limited space of the Randstad. High yields are characteristic, potatoes, sugar beet and wheat being the principal crops.

252   THE NEW EUROPE

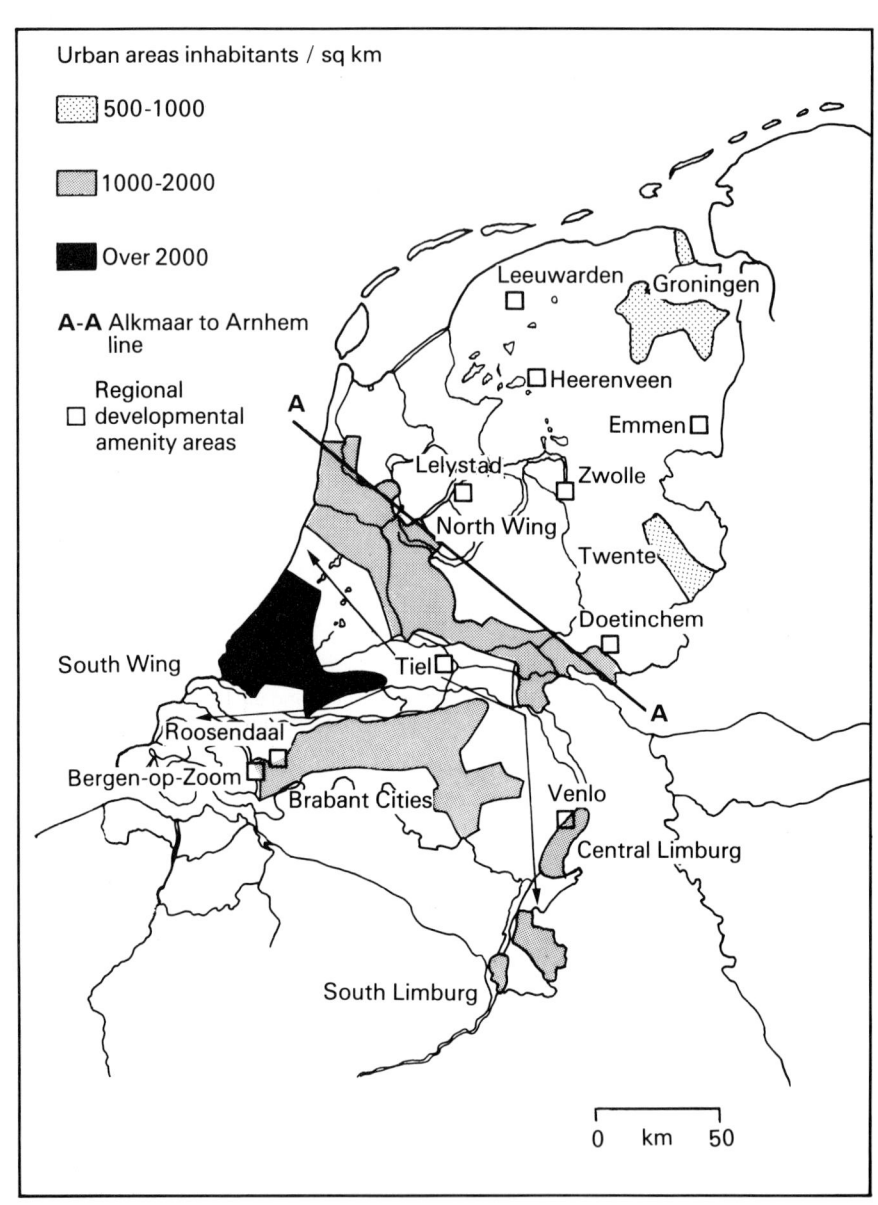

Figure 14.5 Projected population model for the Netherlands AD 2000
(Source - *Report on Physical Planning for the Netherlands, 1966*)

Much arable production is geared to the supply of food-grains for cattle, particularly around the Haarlemmermeer near Amsterdam. This area has light sea-clay which is ideal for retention of nutrients, there is less wind-erosion, and with effective drainage it can be ploughed easily and the water table can be controlled. Dairy farming is also important throughout most of the inner Randstad. The southern region of the Betuwe between the Lek and Waal, the Gouda cheese region, and the Loosdrecht between Amsterdam and Utrecht are primarily grassland areas containing one-fifth of Holland's total cattle, and large numbers of pigs and poultry.

Competition for land

It must not be supposed that the Randstad open heart will inevitably remain agricultural, as there are tremendous pressures upon land. Recreation is one of these. The relatively short dune coastline extending through Scheveningen to Zandvoort has a great width, often extending inland for five kilometres, but is in great demand from the Randstad cities for recreational purposes, and is also accessible from German cities in the Rhinelands. The Utrecht ridge of glacial sands, stretching from Hilversum between Utrecht and Amersfoort, are low hills and wooded heaths and provide parkland and scenic amenity areas. Some measure of land pressure can be gauged, however, from the fact that residential development from the city of Utrecht has spread on to the open land on the eastern side of the ridge.

Most critical of the pressures is the demand for land for housing and industry. The heavy industry and port installations of Rotterdam have been expanding at such a rate that, if continued, it is estimated that 6000 hectares of extra land will be needed by the year 2000. The ancient cities and their municipal areas are rigidly defined and the outer suburban areas and the agricultural heart are experiencing strong growth pressures. The population of the Randstad cities had risen to 6.7 million by 1986. They are already crowded and short of land for all purposes.

Regional planning

Some form of regional planning was a necessity from 1945 onwards. The historic cities are preserved as nucleii with buffer zones between them, and the agricultural heart in the centre is preserved. The only growth allowed in this central area is limited within existing historic towns such as Gouda or Woerden. The present morphology of the Randstad therefore is dominated by the horseshoe shape of urbanisation with an effective 'green belt' inside. The 1948 report *The Development of the Western Netherlands* contained the then revolutionary proposal that in order to maintain the 'Greenheart' future growth be guided radically outwards along the main transport routes. There are four main avenues of development (fig. 14.1). To the north of Ijmuiden and Haarlem is the peninsula of North Kennemerland, where growth could continue as far as Alkmaar, the old cheese-marketing town. On the east, the Utrecht ridge must be preserved as open land but beyond it are a number of very suitable nucleii from Amersfoort to Arnhem, Nijmegen and along the Rhine and Waal rivers almost to the German frontier. There is considerable potential for light industry here and this is a very important future area for development. To the north-east of Amsterdam lies the polder of South Flevoland and the city of Lelystad. This is a new area which, although retained mainly for agriculture, will nevertheless provide open space directly adjacent to Amsterdam, the present plan being that the Amsterdam -

Lelystad strip will be urbanised. Along this strip runs the main highway from Amsterdam to Groningen, via Lelystad, and also the Oostvaardersdiep, the extension of the North Sea canal from Ijmuiden and Amsterdam. The final avenue lies to the south of Rotterdam and The Hague into the delta of the Rhine and Maas. The whole area is dyked against the sea (the delta scheme) and the formerly disastrous floods, and the Haringvliet has great potential as an alternative waterway. New towns at Hellevoetsluis on the Haringvliet along with the extension of new roads sooth towards Middelburg have ended the isolation of the delta. These various radial schemes are positioned to relieve the points of maximum congestion: Amsterdam - Haarlem north and north-east; Utrecht eastwards; Rotterdam and The Hague southwards.

The 'spreading policy' or deconcentration of 1951/52 designed to disperse population to the underdeveloped eastern provinces of Groningen, Friesland, Drenthe, Overijssel and Gelderland (AA line fig. 14.5) and to the old industrial area of Limburg in the south, was a natural corollary of the guided radial development of the Randstad cities.

The First and Second Reports on Physical Planning (1960 and 1966) were designed to cope with large-scale growth pressures within the polycentric Randstad. The four avenues of radial development, combined with buffer zones between the major cities and restriction of development in the Greenheart, have already been mentioned. The 1966 Second Report also contained a remarkable population projection, Netherlands 2000 (fig. 14.5), which assumed continuous economic and population growth and the need to control vigorously the growth of the Randstad cities. To that end, the Randstad came under an investment levy with forty per cent tax on all development projects, clearly designed to slow down its growth.

However, from 1973-76, in the Third Report on Physical Planning, attitudes changed. There was a major reappraisal of policies, which had been overtaken by events. The economic recession, a rapid decline in the birth-rate, and a reduction in immigration, suggested that there was no longer such a pressing need for population dispersal from the Randstad cities. Indeed, with environmental pressure for conservation of historic buildings, and the need for renovation of inner city areas such as the Jordaan in Amsterdam, it was realised that the physical, social and economic fabric of the major cities was now under pressure. From 1976 to the 1984 Structure Plan, the following policies are now followed:

(a) the preservation of the Greenheart and the maintenance of buffer zones between the cities are continued, albeit with less rigidity.
(b) continuing radial decentralisation plans in the city region.
(c) the removal in 1983 of the forty per cent investment levy in the Randstad.
(d) revitalisation and urban renewal of twelve inner city areas, including the Jordaan in Amsterdam; continued suburban development on peripheral housing estates such as Biljemeer to the south of Amsterdam.

The Randstad cities form a unique physical, morphological, functional and economic unit in the Rhine delta. Their significance is reflected in the very detailed urban and regional structure plans which have been adopted and modified by the Dutch government since 1945.

15

Denmark: dairy specialist and gateway to northern Europe

Rural-urban change

Denmark is renowned for its dairy production based essentially upon the successful export of butter, bacon and eggs, and fresh, frozen and canned foods associated with livestock – cattle, pigs and poultry. The total population of 5.1 million people has traditionally been rural, living in small market centres, coastal ports and agricultural villages. One of the remarkable characteristics of modern Denmark, however, is the enormous growth, industrialisation and increasing importance of Copenhagen, now a city of 1.4 million people, twenty-eight per cent of the country's total population. The purpose of this chapter will be to examine Denmark with this contrast in

Landscapes and land-use

The extremely complicated glacial history of Denmark has produced two areas of landscape, one in West Jutland and the other in East Jutland and the islands (fig. 15.1).

West Jutland

The outwash plain and older morainic landscape of the Saale/Riss glaciation which covers the western part of Jutland form low sandy hills rising from the North Sea sand-dunes and coastal marshes. They were for long covered in heath interspersed with birch and oak - picturesque but wasteland and agriculturally unproductive. Soils are podsolised, known locally as 'blegsand' (bleached sterile ash-coloured sand) often with hard pan and accumulations of sour humus or peat. The reclamation of this area from the late nineteenth century onwards, is one of the success stories of Danish energy and of comparable importance to similar Dutch areas. The growth in population in Europe, coupled with rising standards of nutrition, caused a rise in the demand for dairy produce: this led to an increasing need for more agricultural land which, in turn, began the systematic reclamation of the western heathlands. The clearance of woodland and burning of the heath was followed by deep ploughing to break up the hard pan, regular draining and heavy application of fertilisers. The landscape of heather moor is now

256 THE NEW EUROPE

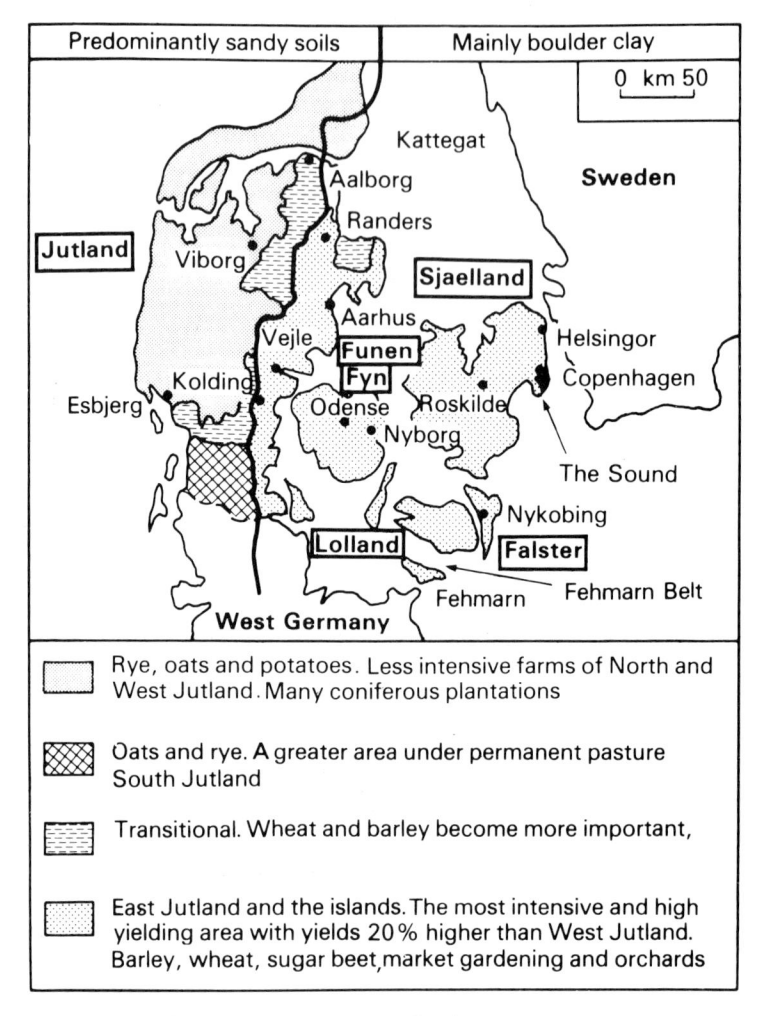

Figure 15.1 Denmark: farming economy generalised

replaced to a very large extent by coniferous plantations, fields of pasture and crops, and some of the most modern farms in Denmark. These modern farms and the straight roads which lead to them are often dispersed throughout the countryside and are outside the old villages, a pattern indicative of late reclamation and secondary settlement. So successful has been the reclamation of the heathlands that today improved farmland covers over seventy per cent of the country. This 'inner colonisation' of West Jutland has made it possible for industrialisation and towns to spread with small loss to farmland.

East Jutland and the islands

A younger moraine landscape exists in East Jutland and the islands of the Danish Archipelago, that is the area east of the main stationary line marking the limit of the later Weichsel/Wurm glaciation. Here there is much more

Type of land use		Percentage of total
Arable	62	Total land in
Permanent pasture	7	agricultural
Fruit/other crops	1	use 70 per cent
Woodland	10	
Heathland	11	
Urban	9	

Figure 15.2 Denmark: land use 1986

boulder clay, giving rise to a brown forest soil, which even here is a highly improved soil due to the rational application of manure over the centuries, continued soil ventilation by cultivation, and the practice of crop rotation. Thus there is now a man-made farmscape which bears little relationship to the original temperate forest, marsh and heath.

The development of intensive arable-livestock farming

Denmark's role as a high quality dairy-food producer is illustrated by its man-made farmscape, principally created by the organisation and the vigour of the farmer, government, and the whole population. The creation of this system is a result of the economic changes which were associated with the period post-1880. Danish farmers could not compete with large imports of low-cost wheat from the new world grasslands. The system of large manorial and crown farms and peasant subsistence farms existing side by side began to change dramatically. After a deliberate land policy to encourage the break-up of the large estates, and the establishment of family smallholdings, the typical Danish farm now has an area of between ten and thirty hectares. There is, however, considerable variation across the country and in the sandy poorer soils of West Jutland, the individual farm tends to be much larger.

Alongside these changes in tenure to a tenant-farmer system, Denmark also turned to intensive dairy farming as the best way of utilising her small area, and utilising the real advantage in her position, that of proximity to the two greatest industrial nations in Europe, Germany and Great Britain, who required large quantities of food. It is upon the cultivation of fodder crops that Denmark relies for her efficiency, since it must be stressed that nature has not really endowed her with excellent conditions for dairy farming. The climate is not particularly suitable for grass growing, with a rather inadequate precipitation of below 25 inches (63.5 cm), and spring comes later than in most dairy countries, giving a shorter growing season. Throughout the winter live-stock must be stall-fed on fodder crops, some of which must be imported. The intensive use of land through arable farming and fodder crops is therefore essential in order to gain the maximum foodstuffs from a limited area (fig. 15.2). Grain crops, wheat, rye, barley and oats, usually occupy

56 per cent of total agricultural land; 14 per cent is under rotation grass; 22 per cent is under pulses and root crops including beet, mangolds, kohlrabi and potatoes; and only eight per cent is under permanent grass. This is most unusual amongst dairy farming economies, which are usually predominantly pastureland. Denmark is probably the most intensive dairy farming country in the world. There are variations: West Jutland still has a greater acreage under rye; East Jutland and the islands have higher yields (fig. 15.1); around Copenhagen peas, beans, carrots and market garden and salad crops are important; the islands contain 75 per cent of the area of fruit and market garden crops. The main theme is animal husbandry, with 85 per cent of crops used as stock feed. The highest yields come from root crops, which are at home in the Danish soils and climate with its long autumns, and roots and grass provide rough coarse feed. Cereals are invaluable because they supply the carbohydrate for concentrated feedstuffs. Proteins are supplied mainly by oil seeds which have to be imported.

Danish cattle are principally of two national breeds. Black and white milch cows are found mostly in West Jutland because of their hardiness and ability to thrive in areas of poor grass. The Danish Red breed is widespread throughout the country. Jerseys and Shorthorns are increasing in popularity. Yields of milk are very high and the vast bulk (61 per cent) goes for butter-making. Cheese-making consumes thirteen per cent, and the rest goes for cream, condensed milk and milk-powder, with the intensive nature of the operation apparent as the skimmed milk is returned to the farms for pig-feed. The number of pigs has increased considerably to 9.4 million, the main variety being the Landrace. Skimmed milk from the dairies is added to grain, potatoes and sugar beet toppings to feed these animals which produce the

The Danish dairy-farming landscape with modern farm buildings, Friesian cattle and rotation grassland

high quality ham and bacon for which Denmark is famous. The third element in the economy is a battery-rearing operation associated with smaller farms of less than ten hectares: eggs and poultry meat account for twelve per cent of the total value of farm produce.

The most important single characteristic of Danish farming is its dynamic efficiency. The cooperative movement is carried to its ultimate limits in the loan of machinery, the dissemination of research, and the organised processing and marketing of produce in creameries and factories. Government aid, the systematic construction of Esbjerg as a port for trade with the UK, agricultural schools, and price guarantees are matched by very high inspection standards. The significance of dairy products is shown by the fact that farm produce provides about twenty-eight per cent of Denmark's exports by value. Agriculture is very dependent upon export markets for its continued prosperity, and Denmark's dilemma in the 1960s was that her two greatest markets, the UK and West Germany, were within different tariff groupings, EFTA and EC. The accession of the UK with Denmark to the EC meant that these two major markets were assured. Denmark's principal contribution to the Community is as a very significant supplier of dairy foods. However, this picture is changing as the balance between agriculture and industry/services is altering quite dramatically (fig. 15.3).

Industrial and urban changes

The traditional picture of rural Denmark has now changed considerably, with rapid industrialisation and urbanisation. There has been a marked change in the employment structure of the country. The number of workers in industry and the services has risen whilst those in agriculture have fallen dramatically. Industrial production is now much more significant than agriculture in terms of exports (figs. 15.3 and 15.4).

Denmark has a major proportion of its total population concentrated within the capital city. In 1981 Greater Copenhagen had a population of 1.4 million out of 5 million, approximately 28 per cent. The development of the city may be considered under four headings: site and origins; the establish-

Export	*Percentage of total exports*	
	1981	1986
Food and beverages	30.9	27.9
Minerals and fuel	3.2	3.0
Crude materials	7.3	6.4
Machinery and transport	24.9	24.6
Other manufacturers	33.7	38.1

Figure 15.3 Denmark: principal exports

| | Percentage of employed population | | | | |
Employment sector	1950	1970	1977	1981	1986
Agriculture	25	11.5	9.8	8.5	6.2
Manufacture	28	32.5	31.5	27.1	27.0
Services	47	56.0	58.7	64.4	66.8

Figure 15.4 Denmark: employment

ment of the 'Freeport' in 1894; modern residential expansion and the city as a tourist attraction; development as a communications centre linking northern and western Europe.

Copenhagen and the Baltic Sea

The old city of Copenhagen (fig. 15.5) was built as a safe anchorage and deep water port on the narrow strait between Zeeland (Sjaelland) and the small island of Amager, and it developed an early commercial importance as one of the major trading ports of the Hanseatic League in the thirteenth century. Its traditional role was as entry port into the Baltic Sea via the Ore Sund, the sound between Denmark and Sweden. The part of Sweden directly opposite Copenhagen was Danish territory (the province of Scania) until 1660, and thus the city was originally in a much more central position to Denmark as a whole.

The Kiel canal and the 'Freeport'

A change of critical importance came with the opening of the Kiel canal in 1894. This provided the city with serious competition by giving a more direct entry into the Baltic, and therefore threatened its role and prosperity as entry point to the Baltic. One answer to this problem was the development of the 'Frijhaven', Copenhagen Freeport, which was built at the northern end of the existing city harbour where it joins the Ore Sund. This has now become one of the largest entrepôt ports of Europe, with over fifty per cent of total Danish imports including petroleum, coal, metal ores, timber, fertilisers, oil seeds (agricultural and industrial raw materials). Perhaps more significant was the resulting industrial expansion - over forty five per cent of Denmark's industrial workers are now employed in Copenhagen. Shipbuilding and repairing, marine engineering, oil storage for bunkering, and the generation of electricity are large-scale activities. There are also grain mills, milk and meat processing factories, and cattle-cake as a by-product, vegetable canning, tobacco processing, and brewing (particularly Tuborg and Carlsberg lagers) and fertiliser (phosphates), all of which have a clear association with agriculture. The industrial base, however, is very varied, with engineering, electricals and electronics, automobile assembly, printing, textiles, shoes,

Glassware exhibition, illustrating the modern designs of Danish consumer industry

Bridge over the Little Belt linking the Danish island of Funen with Jutland

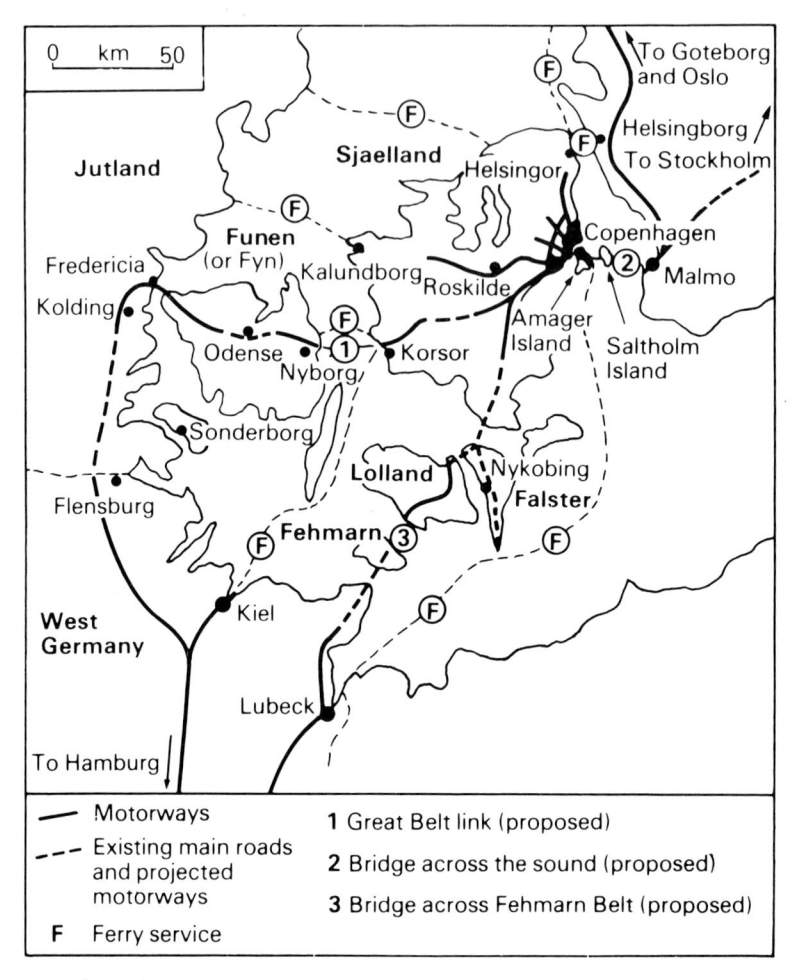

Figure 15.5 Copenhagen and the Danish islands: developing transport links

rubber, soap, paint and pharmaceuticals. Special mention must be made of furniture, stainless steel and glassware, high quality Danish consumer goods specialities.

The expansion of the city

The expansion of the city since the eighteenth century has dramatically changed the balance of its structure and morphology. Amager Island was the initial area of expansion with Christianshavn and Sundbyern to the south of the old city. In the present century, however, the west and north have figured largely in an exceptional rate of growth, with the main area of suburban expansion in the districts of Fredericksberg and Gentofte. These are added to Copenhagen itself to give a total population of 1.4 millions. The inner city is surrounded by the old canal and a ring of gardens of which the Tivoli is the most famous. A thriving tourist industry is based upon the attraction of the old city, the Christiansborg and Rosenberg palaces and the specialised traditional industries such as the Royal Danish procelain factory, and silverware and other fine craftwork.

Copenhagen as a communications centre

A large part of the increased importance of the city since 1960 is its developing function as a communications centre. Ferry services across the Sound via Helsingor to Halsingborg, Malmo and Halmstad in Sweden, and to Travemunde (Lubeck) illustrate the position of the city as a bridge between continental Europe and Scandinavia. The international airport at Kastrup on Amager island is just six miles south of the city centre and in addition to European flights, there is the Polar route to Canada and Tokyo. Three major projects will increase the nodality of the city as a European - Scandinavian land-bridge (fig. 15.5): (1) a fixed link over the Great belt (railway or six-lane motorway) to link Zeeland with Funen and Jutland and thence to West Germany; (2) a bridge across the Sound to Malmo via the Danish island of Saltholm (where the new international airport is planned; (3) a third bridge over the Fehmarn belt via Falster and Lolland across to Puttgarde on the German island of Fehmarn and thence to Kiel and Lubeck. These fixed links would be a critical step forward for the city, and would help continue its present growth. The single most significant of the three is the link across the Sound which could eventually result in the fusion of Malmo and Copenhagen to create a conurbation, the first in Scandinavia.

16

Piedmont, Lombardy and Liguria: the economic core of Italy

The wealthy north

The administrative regions of Piedmont, Lombardy and Liguria form the extreme north-western corner of Italy, and represent a small proportion (some twenty per cent) of the total area of the country, yet they occupy a dominant place in the Italian economy. Piedmont originated as the Duchy of

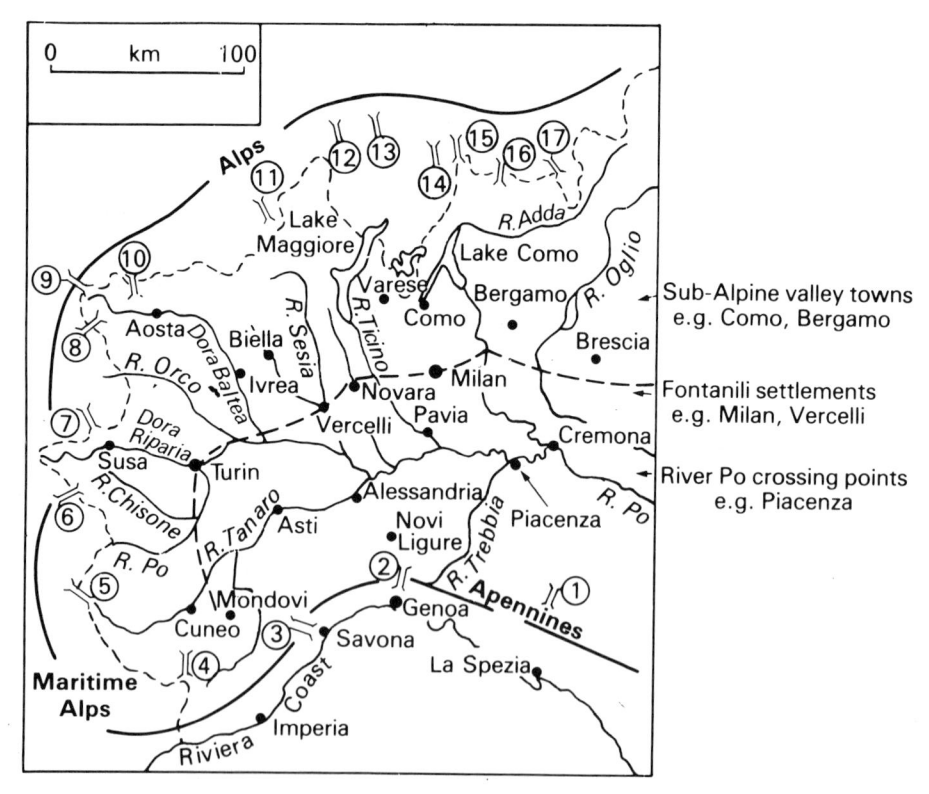

Figure 16.1 Lombardy, Piedmont and Liguria   1. Cisa pass 2. Giovi pass 3. Altare pass 4. Tenda pass 5. Maddalena pass 6. Mt. Genevre pass 7. Mt. Cenis pass 8. Little St. Bernard pass 9. Mt. Blanc tunnel 10. Grand St. Bernard pass 11. Simplon pass 12. St. Gotthard pass 13. Luckmanier pass 14. San Bernardino pass 15. Splugen pass 16. Maloja pass 17. Bernina pass

Savoy and from the sixteenth century its main city, Turin, has exercised an important influence over the formation and subsequent industrialisation of Italy. Lombardy is centred around the ancient Duchy and city of Milan, although it covers much of the central part of the Po basin. Liguria is a narrow coastal strip with the port of Genoa as its major focal point. Structurally the three regions are dissimilar. The Ligurian coast forms part of the Maritime Alps and Apennines and is a narrow coastal strip with fast flowing streams and steep valleys making communications very difficult. The apparent uniformity of the Po basin masks a landscape which is rich in contrasts (fig. 16.1). It varies from the sub-alpine hilly margins, with long tributary valleys such as the Dora Baltea and Dora Riparia, to the upper and lower plains divided by the line of fontanili springs and broken by the occasional low hills of moraine, such as the Serra D'Ivrea north of Turin, and the deeply incised hill country of Monferrato and Le Langhe, south of Turin. Climatically there are also considerable contrasts. Liguria is sheltered from the north by the mountains and enjoys a dry summer and a mildness of climate more truly mediterranean and reminiscent of southern Italy. On the other hand, the plain of Lombardy has cold winters of continental origin, hot summers (24°C) with a tendency to thunderstorms and a well distributed rainfall. This stimulating climate can be related to the traditional energy of the people of northern Italy.

Overriding all these differences, however, is a common feature. This is the domination of the Italian economy by the three cities of Genoa, Milan and Turin. During the Middle Ages the significant factors were the rich endowment of a well-watered plain and climate beneficial to large-scale agriculture. The control of water supplies both for irrigation and energy has been an essential feature throughout. Equally as important has been proximity to the passes into France, Switzerland and northern Europe, and maritime access via Genoa. This has encouraged enterprise, new ideas, commerce and industry, best illustrated by the extraordinary vitality of city life throughout the region's history. This proximity to north-western Europe is now physically reinforced by road tunnels, such as the Mont Blanc, and by the growing network of autostrada (fig. 16.2). The European Community has brought a fundamental change in attitudes and in economic integration, which has benefited Italy as a whole enormously. This populous triangle, rich in farmland, industry, and city life, is closest to the West European growth axis, and has developed a dominating position as the economic core of Italy.

Agriculture

The extensive plain drained by the Upper Po and its tributary network, the encircling mountains and their foothills, and south-facing Ligurian coast, exhibit a wide variety of farm landscapes (fig. 16.3). The whole area is one of the European Community's major warm-temperate food-producing zones. The principal divisions are outlined in the following pages.

The Piedmont plain

The Piedmontese plain of fine fluvio-glacial deposits and alluvium is centred on Turin, but describes a wide arc from the Cuneo basin in the south, and is continued eastwards along the valley of the Po to the basin of Alessandria. Non-irrigated cereals are the dominant type, with wheat, maize, potatoes, beans and fodder as the main field crops; the fields are often lined with tree crops of apples, pears, plums or vines. The numerous tributaries of the Po give the area a well-watered look, and the river terraces are broken by steep bluffs commonly in woodland, whilst the flood plains are often irrigated. The plain of Alessandria, in particular, specialises in market gardening, with sugar beet at Marengo, and is the chief granary of Piedmont. The farms are substantial holdings, often over fifty hectares, with large, prosperous looking farmhouses, grain silos and outbuildings.

Monferrato hill country

To the south of Turin lie the foothills of the Apennines, the Monferrato and Le Langhe hill country. The northward-flowing river Tanaro follows an attractively varied landscape of cuestas, vales and rolling hills, with many hill-top villages and market towns such as Mondovi and Ceva, which lie on the route to the coast at Savona. The speciality of the area is viticulture, particularly around Asti and Alba. Here there are nearly 150 000 hectares of specialised vineyards which produce the famous sparkling Moscato D'Asti, and Torinese Vermouth. There is also a large production of red wines notably Barbera, Dolcetto and Barbaresco. In other parts there is a patchwork of cereals, pasture and cattle raising, with some fruit and market gardening, but farms are small, generally two to three hectares, and this part of Piedmont suffers from the drift of population to the cities. As the higher slopes of the Apennines are reached, there are hazel nuts, truffles and chestnuts in valleys such as the upper Tanaro.

The Alpine hill country

To the north and west of Turin the plain merges into the Alpine hill country. At the exits to the Val Di Susa and the Val D'Aosta there are extensive moraines which form irregular hills. The Serra D'Ivrea is one example, its surface covered with heathland and chestnut woods. The Val D'Aosta, which carries the beautiful Dora Baltea river has the classic Alpine succession of pasture and tiny fields of hay, rye and potatoes on the flats, orchards on the lower slopes, followed by terraced vineyards supported on trestles to counteract the effects of damp from the high rainfall (100 cm). Above this are bare slopes and the high pastures used for the spring migration of farm animals. There are also abandoned terraces and farmhouses, indicative of the retreat to lower slopes and more beneficial soil. In the western Italian Alps as a whole only 6 per cent of the land is classed as cultivated. In the upper Po valley to the south-west of Turin are the sheltered valleys of the Saluzzese

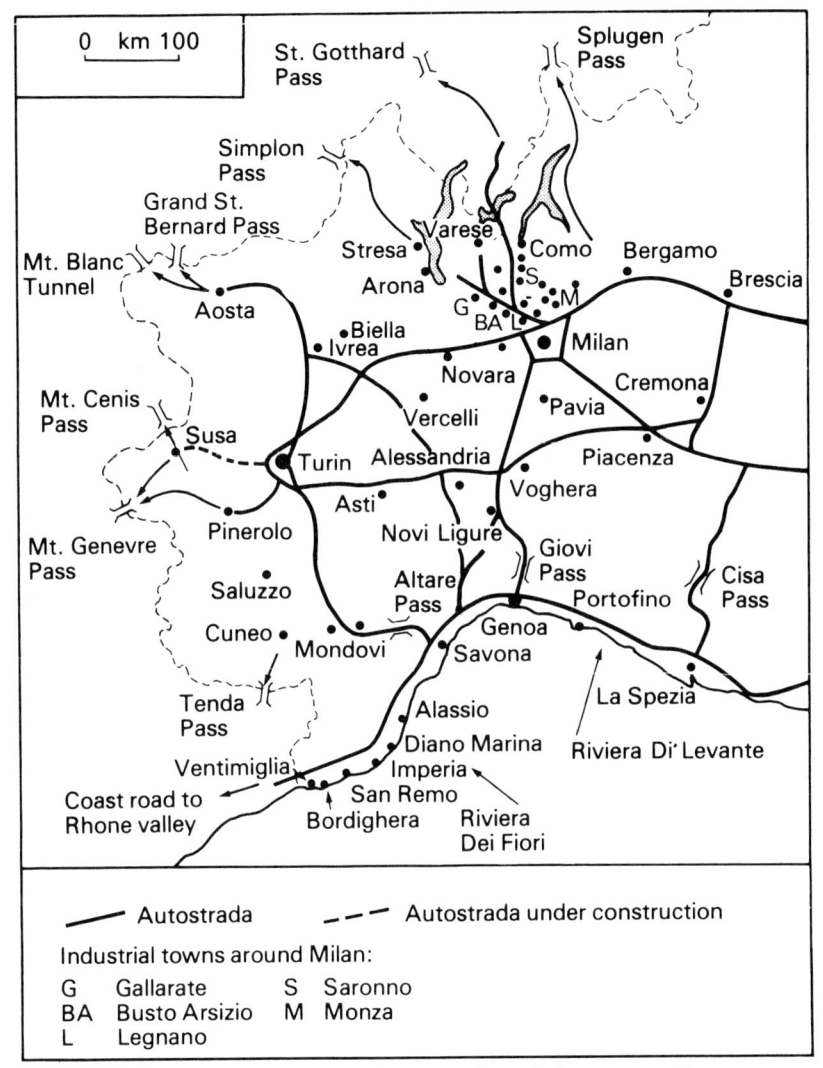

Figure 16.2 Autostrada network and major routes through the western Alps

with orchards, vines, palms and magnolias, and a four-year rotation of cereals with hay and root crops.

The 'Rice Bowl'

Probably the most distinctive farming type is the 'Rice Bowl' of Vercelli and Novara. Half a million hectares of heavy clay soils are irrigated in vast monotonous fields which are carefully terraced and bordered with screens of Lombardy poplar and willow. The farms are large, usually over fifty hectares, and isolated, mostly built around a courtyard and containing grain stores, machinery, sheds and accommodation for seasonal workers. Up to eighty per cent of the land is under rice. The ploughing, cultivation and harvesting of the rice is done by machine, but the planting is done by hand, with seasonal labour. Yields are high, averaging twelve tonnes per hectare. There is heavy use of fertilisers, and although monoculture has traditionally been the pattern, there is now a growing diversification towards rotation fodder and milk production. The Vercellese is an example of the early use of irrigation for intensive production. Fontanili springs were used as early as the thirteenth century, but the real intensification of farming came with the construction of the Cavour canal in 1863. The Vercellese is an excellent example of large-scale, heavily capitalised farming, with irrigation as the basis for intensive rice production.

The Alta and Bassa Pianura around Milan

The part of Lombardy which is centred upon Milan lies mainly between the Ticino and the Adda and exhibits a more straightfoward division than Piedmont, with the water-rich Fontanili spring zone dividing the Alta Pianura, north of Milan, from the Bassa Pianura which stretches south to Piacenza on the river Po. The excellent supplies of water and their control and exploitation are the key to Lombardy's historic agricultural productivity. About sixty per cent of the farmland is irrigated, and of this forty per cent comes from the Fontanili which has been used since the eleventh century. The rest comes from the extensive canal system which utilises the waters of the Ticino, Adda, Lambro, Sesia and Oglio rivers, by crossing the interfluves in an east-west network stretching from Turin to Brescia. The most important of these are the Cavour and Villoresi canals.

Along the line of the Fontanili, the high water table encourages dairy farming. The cattle are reared on water meadows (Marcite) which are irrigated continuously ensuring a large number of fodder cuttings (ten are usual). The cattle are stall-fed in the 'cassini', which is a special variety of the 'Corte', the large Italian courtyard farm. The production of milk is geared to cheese-making, in particular Parmesan and Gorgonzola, and butter.

South of Fontanili, on the lower plain, irrigation, though still important, is more intermittent, and fodder crops are grown in conjunction with wheat and maize. There is often a seven-year rotation, with four years of cereals, followed by three years of meadow, indicating the continuing importance of dairy farming. The main hazard on the low plains is of flooding, creating the necessity for high levees.

The intensive livestock economy thus described is not typical of Italy, nor is the capitalised farm operated with wage labourers generally typical of the Mediterranean area. The medium-sized and large farm holdings with their prosperous 'Corte' are a particular feature of the Lombardy and Piedmont plains. The urban life of the cities of Milan and Turin has been sustained since the Middle Ages and has been a major impetus to the efficient production of food, the utilisation of new techniques and the specialisation associated with commercial agriculture. The administration of these provinces was more enlightened than most, and break-up of the large medieval estates into viable farm holdings was carried out in the nineteenth century.

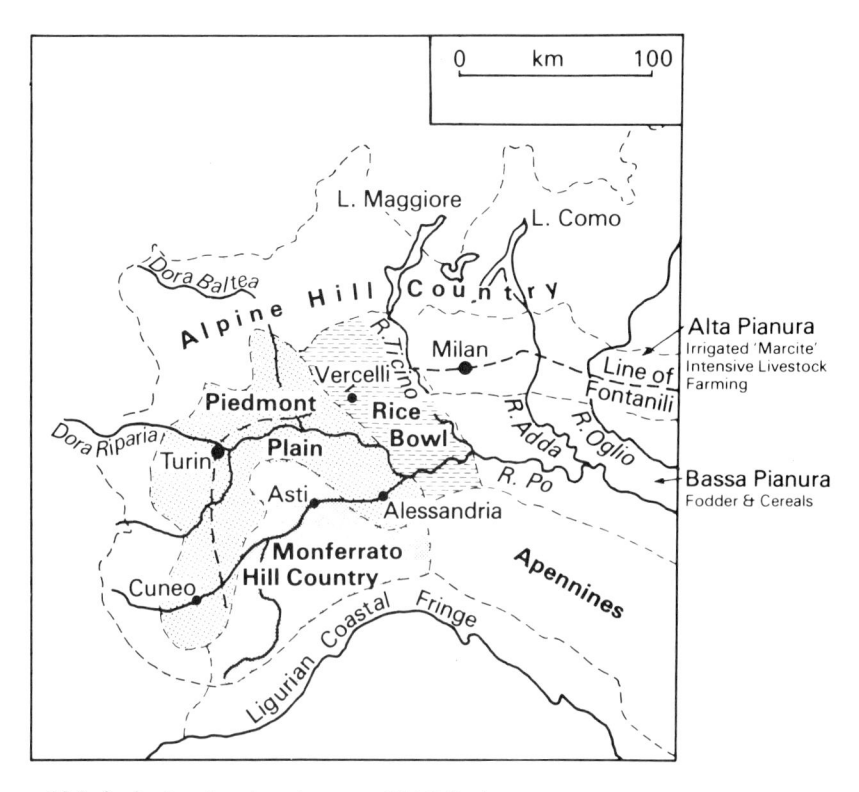

Figure 16.3 Agricultural regions (source – *DS Walker*)

The Ligurian coastal fringe

Liguria forms the maritime facade for the interior plains of Lombardy and Piedmont. It also forms a considerable barrier, with over sixty-five per cent of its land area classed as mountainous and the rest hilly, with only limited alluvial flats. Agriculturally, Liguria is therefore a difficult region, with only eleven per cent of the land cultivated. Tree crops dominate, with olives and vines.

There are three sections to the province, the Riviera Centrale, containing the port of Genoa; the Riviera di Levante, stretching eastwards to La Spezia; the Riviera di Ponente, which stretches from Genoa westwards to the French frontier, and it is this last section which has the most interesting and specialised agriculture. The Riviera Di Ponente is often called the 'Riviera of Flowers' or Riviera Dei Fiori. It faces south-east, and is sheltered from the north by the Apennines. It has drier, milder conditions than the rest of Liguria, and its sheltered bays have many sub-tropical plants, exotic palms, cypresses and bougainvilleas. There is a string of popular holiday resorts from Ventimiglia on the French frontier to Alassio, and the hillsides are heavily terraced with stone walls and covered with glasshouses for the cultivation of garden flowers. Flori-culture is of key importance along this coast, and Ventimiglia has a famous traditional flower market. San Remo in particular is well-known for the cultivation of roses and carnations. Citrus fruits, early vegetables, and orchards of apples and plums are found on flatter alluvial fans of irrigable land, as at Albenga and Diano Marina. Vines, figs, and olives alternate on the lower slopes with chestnut woods, and there is considerable woodland on the upper slopes, petering out into poor pasture.

The Italian industrial triangle

Although there are other heavily industrialised zones in northern Italy, notably at Marghera near Venice, nevertheless the major industrial concentration in Italy lies within Lombardy, Piedmont and Liguria. Here lies the core of Italy's wealth production. Lombardy has little more than one-seventh of Italy's population, but provides one-fifth of the national product, and 40 per cent of industrial exports. It has a quarter of the industrial workers, and provides over 35 per cent of the taxes paid by industrial companies. Its industry is broadly based upon a vast number of small and medium-sized concerns. Around Milan there are over 100 000 industrial concerns, and the largest factory in the city employs 18 000 workers, a small figure when compared with the 70 000 workers of the main Fiat factory in Turin. Piedmont provides one-tenth of the country's national product, and one-seventh of the industrial output, but, unlike Lombardy, its wealth is built around a few large enterprises. Fiat of Turin employs 186 000 workers, and thousands more are directly dependent upon it for a livelihood. Turin is an excellent example of a company city, and certainly its life revolves around the car industry. Liguria's wealth is based first upon the port of Genoa and its subsidiary port, Savona, and the modern steel mills, shipbuilding yards and engineering industries which lie on the coast. The other source of wealth lies in the old established tourist trade along the Riviera Dei Fiori. Many well-known resorts line the coast for 150 kilometres eastwards from the French Riviera around the Gulf of Genoa.

Milan

Milan originated as Roman Mediolanum, and became a famous ecclesiastical centre during the latter part of the Roman Empire. During the revival of city life in the eleventh century it again prospered and was a medieval industrial, commercial and banking centre of great importance. The city typifies the extraordinary tenacity of the cities of the northern Italian plains, as they survived the centuries of stagnation, warfare and misrule, and from the nineteenth century onwards industrial development came steadily, with the concentrated economic advantages which Lombardy could now exploit. These were the accrued agricultural commercial wealth, the enterprising spirit of the population, and hydro-electric power from the Alps. Perhaps most important was the development of modern communications so that the city could assume its natural position as the chief focus of road and rail routes within the plain and across the Alps, and to the Ligurian and Venetian coasts. Food processing (pasta, confectionery, etc.) is concentrated in the Greater

Milan area. In particular, cheese making, with the famous Gorgonzola, Parmesan, Bel Paese and Stracchino cheeses, is found in Milan and the cities to the south - Lodi, Pavia and Piacenza.

The industries which basic are to the prosperity of the area are, however, engineering of all kinds, textiles, and chemicals. The textile industry is located mainly in the upper plain to the north of Milan, and is associated with the growth of a major agglomeration of industrial towns (fig. 16.2) stretching north to the Alpine valleys, which were the original source of power. The cotton industry is represented strongly to the north-west of Busto Arsizio, Legnano, Varese and Gallarate, and to the north-east at Monza and Bergamo. Silk is manufactured in Como, and woollens in Bergamo province, whilst the newer synthetic fibres are present in Milan itself.

The chemical industry is concentrated in Milan and Novara, with many factories of the Montedison group. There are several petrochemical plants, and an important feature is the existence of substantial natural gas deposits in the lower plain. The headquarters of ENI, the state oil and gas agency, is at Milan. Fertilisers, artificial rubber and Pirelli tyres are all important, whilst there is a major pharmaceuticals industry in Milan, based largely upon the high level of demand in the northern cities.

It is in the engineering field, however, that the real industrial vitality of Milan is based. Prior to the post-1945 movement to the integrated coastal steelworks, Italy's steel industry was dominated by Lombardy, and there were steelworks based on scrap at Sesto San Giovanni, Bergamo and Brescia. Output is relatively small, but it has maintained its presence through investment and the large local market of steel-using industries. Foundry and steel-making equipment, presses, lathes and engines, and heavy electrical generating equipment are located at Milan. The vehicle industry is represented by Innocenti and Alfa Romeo. The light engineering section includes machine tools, calculators, precision instruments, textile machinery, motor scooters, motor cycles and sewing machines.

The footwear industry is centred at Vigevano to the south-west of Milan, but Milan itself and Varese are other centres. Paper and furniture manufacture is also represented strongly in the satellite towns of the upper plain from Vercelli to Como.

The great variety of industry in Lombardy, particularly Greater Milan, is a reflection of its mature industrial structure. Milan itself is a city of almost capital rank. It is the financial capital of Italy with major commercial, insurance and banking operations of greater significance than those of Rome. It has the country's most significant trade fair, is a great publishing centre, and its cultural activities are metropolitan in character.

Turin

Turin is the chief city of Piedmont, and stands at the confluence of the Dora Riparia and Po rivers, but the modern significance of its position is the control of the passes emerging from the western end of the Alps. It rose to importance only in the sixteenth century when chosen as the capital city of the

House of Savoy, under whose guidance it became the chief force in the unification of Italy in the mid-nineteenth century. During the twentieth century, with the development of roads, railways and modern autostrada, the importance of its position has again been underlined (fig. 16.2). It controls the important routes south to the Ligurian coast and Riviera, westwards via the Mont Genêvre and Mont Cenis passes, and northwards through the Dora Baltea river valley to the Grand St. Bernard Pass into Switzerland and the Mont Blanc road tunnel into France.

Turin is, however, the second industrial centre of Italy. The rulers of Savoy were instrumental in giving Piedmont a measure of industrialisation which was exceptional for Italy. The Turin arsenal was followed in 1900 by the car companies of Fiat and Lancia, and by Olivetti typewriters. The utilisation of the hydro-electric power of the Alpine valleys and the development of stategic industries under Mussolini in the 1930s firmly established Turin as a leading industrial city. The engineering industry, particularly car manufacturing, has boomed since 1950, and the population of Turin has reached 1.2 million, with the absorption into the city of nearly half a million Italians from the south. Car production is dominated by the large integrated Fiat factory at Mirafiori; Fiat in Turin employs 186 000 workers and produces over a million cars per year. They also produce a vast range of electrical and engineering products such as marine and aero engines, electric motors, railway stock,

The autostrada via the Grand St. Bernard Pass linking Aosta, Italy, with the Rhône valley and Switzerland

tractors and domestic appliances. They are even concerned with the construction of Alpine tunnels and nuclear engineering. Turin (and Ivrea to the north with the Olivetti company) are examples of 'company towns'. Ancillary industries such as sheet steel, machine tools, rubber and ball-bearings, complete the picture of car assembly.

Although Turin does not have a constellation of subsidiary industrial towns like Milan, nevertheless there are a number of important rank. To the south are Alessandria and Cuneo, both route centres, Novi Ligure (steel), and Asti, a famous wine processing town. To the west is the Dora Riparia valley to the Mont Cenis pass, with light industries at Susa and Bussoleno. To the north-east lies the woollen textile town of Biella.

Perhaps the most interesting area of development is the Dora Baltea valley. Along it runs the autostrada to the Mont Blanc tunnel and Grand St. Bernard Pass, the most important routes through the western Alps (fig. 16.2). Hydro-electric power is well developed and aluminium works at Borgofranco and rayon at Chatillon are based upon this. At Aosta itself there is a small steelworks, based upon local supplies of magnetite. Another source of wealth is the rapidly developing tourist industry, of both winter and summer resorts. Courmayeur and St. Vincent in the Val D'Aosta now have a greatly increased accessibility. Before the construction of the modern roads and tunnels in the 1960s, the Val D'Aosta was isolated from France and Switzerland for seven months of the year. This change typifies northern Italy's new close relationship with north-west Europe.

Genoa

Genoa was a flourishing port in the Middle Ages, when it rivalled Venice in the Levantine trade, but its importance shrank from the sixteenth century onwards with the growth of Atlantic shipping. Cavour, the Piedmontese statesman, saw its potentialities as an outlet for the whole of northern Italy and the Alpine regions. Under his guidance the port facilities were considerably improved and a railway link from Turin established. At the same time the Suez canal had a major regenerative effect upon Mediterranean shipping.

Although the port has a major disadvantage stemming from its mountainous coastline and consequent lack of space for development, nevertheless, there is a transport factor of the greatest significance. The mountains are at their lowest and narrowest point behind Genoa, and this funnels traffic from the interior into the Giovi pass (fig. 16.2), which is traversed by two railways and the autostrada. There is also the higher Bochetta road pass close by, and the Turchino railway route. These are important not only in allowing Genoa easy access to Piedmont and Lombardy, but in addition the trans-Alpine passes have facilitated the extension of its hinterland to Switzerland and beyond. It is thus the natural outlet for Turin and Milan, and is an important corner of the Italian industrial triangle.

Genoa is now second in importance only to Marseilles on the European Community southern flank (chapter 8, fig. 8.8). There are seventeen miles of quays and industrial suburbs, Sampierdarena, Cornigliano, Sestri, Pegli and

The coastal autostrada along the intensively cultivated Ligurian Riviera near Savona

Voltri, extending to the west of the city. Genoa deals mainly with imports, ninety per cent by tonnage of its total trade. Oil accounts for half the total tonnage, coal, mineral ores, grain, tropical foods, chemicals and textile fibres also being important. It is thus a major input point for raw materials, and associated with the oil terminals are the pipelines which carry oil from Genoa to Milan, Aigle in Switzerland and Ingolstadt in southern Germany.

Heavy industry has developed along the coast, particularly shipyards at Sestri and Voltri. At Cornigliano is one of the coastal integrated steelworks developed since 1945, entirely dependent upon imports of scrap iron, West African iron-ore and coal. There is also oil refining, metal smelting, machine tools, marine, electrical and railway equipment. Food processing based upon imports is also an important activity, with vegetable canning, grain-milling, soap manufacture, sugar refining, paper and pottery.

The other two Ligurian ports illustrate the importance of hinterlands. La Spezia naval base to the east has an excellent harbour, but is handicapped by difficult passes through the Apennines. Savona by contrast, on the western side, has developed a special relationship with Turin, to which it has easy access via the Altare Pass. These links have increased rapidly since 1972 with the development of the autostrada from Turin, a spectacular example of modern engineering which joins the coastal autostrada at Savona.

Tourism along the Italian Riviera

The Riviera Di Levante, eastwards of Genoa, has a rugged coast of great beauty, with famous resorts like Portofino, Rapallo and Santa Margarita. The section of the coast west of Genoa has already been described as the Riviera Dei Fiori for its associations with fruit and flowers. It too, has a beautiful rugged coastline and many seaside resorts including Ventimiglia, San Remo, Diano Marina and Alassio, which nestle in sheltered bays backed by wooded hills. The whole riviera coast has begun to feel considerable competition from newer tourist areas around the Mediterranean sea. It suffers from overcrowding due partially to its longstanding popularity, and partially to the difficulty of access by the inadequate roads. The completion of the autostrada along the coast from Avignon to Genoa and La Spezia, with interior links to Milan and Turin may well have largely removed this disadvantage (fig. 16.2). The revitalising effects of fast road transport are important to the whole region, but perhaps most of all to this beautiful Ligurian coast.

17

The Mezzogiorno: Italy's problem region

The poor south

The contrasts between northern and southern Italy are dramatic. The well-watered enviroment of the Plain of Lombardy, with its great traditions of urban life, commercial agriculture, bustling industry and busy transport networks, has no counterpart in the south. Italy is a land of two nations. South of Rome there is a clearly indentifiable atmosphere of desiccation, impoverishment, lack of activity and an indefinable impression that this part of Europe is very different. To be more specific, Italians refer to the area as 'Il

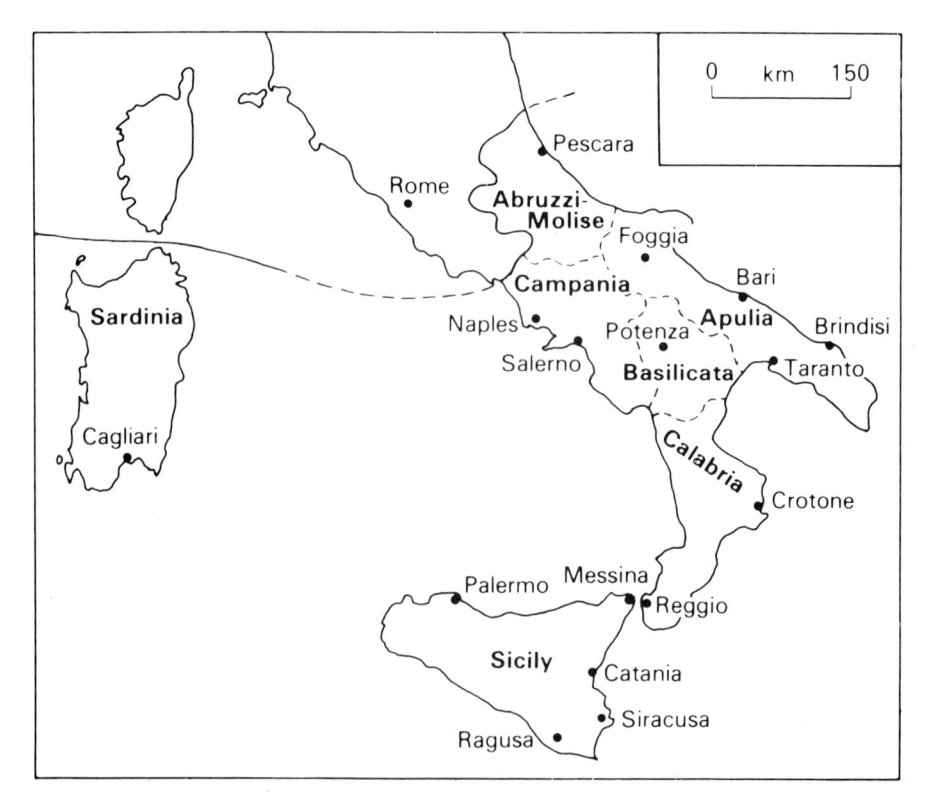

Figure 17.1 The seven provinces of the Mezzogiorno

Mezzogiorno', the land of the noon-day sun. The Mezzogiorno comprises seven of Italy's nineteen administrative regions (fig. 17.1) and contains over 20 million people or 36 per cent of the population. It still produces only about twenty five per cent of the country's gross domestic product (Fig. 17.2). Its population of 20 million people is twice that of Greece, which has much the same total area, indicating a major level of overpopulation. Emigration from the Mezzogiorno has been persistently high. Since 1900 over eight million Italians have emigrated from the south, often to the northern cities, but in very large numbers to North America and other parts of the New World. Real perception of the problem began after the Second World War as illustrated by the following data. In the 1950s more than half the houses in the south were without drinking water, and forty per cent without sanitary facilities. Most significant of all, because it underlines the basic problem, was the picture of agricultural employment. In 1950 the south had 57 per cent of its population employed in agriculture and only 20 per cent in industry, demonstrating its basically under-developed structure. The problems could be summarised as: a low level of economic activity; income levels less than half those of northern Italy; poor living conditions; an overwhelming dependence upon agriculture with resulting rural overpopulation. Persistent outmigration has been one traditional answer to the situation, but it has not been enough. The population has continued to increase steadily and has remained stubbornly too high for the resources of the land to sustain. In the immediate post-1945 period the imbalance between Italy's south and north emerged as one of Europe's most intractable problems.

Gran Sasso mountains, Abruzzi - Molise. Droving sheep near Aquila.

The disadvantages of the Mezzogiorno: physical, historical and economic

Water shortage

The small amount of rainfall and its seasonal occurrence are a major problem. Many parts of the south have only 500 mm of rainfall per annum, and in addition the summer drought, high rate of evaporation and considerable unreliability of rainfall create desiccated conditions for up to five months during the summer. Yet it is precisely during the summer period that the high sub-tropical temperatures would permit the growth of crops not easily cultivable anywhere else in the Community. Unless irrigation water is available, and this is not easy because most rivers dry up in summer, intensive summer cropping is impossible, and the common answer to the drought is a wasteful fallow. There is generally under-utilisation of the land with extensive farming and heavy reliance upon wheat, olives and livestock grazing. Even grazing is limited owing to the absence of good year-round pasture, and the numbers of sheep and cattle are low compared to those in the north. The whole agricultural enviroment of the south is harsh, inferior, and marked by low productivity.

A mountainous landscape

Southern Italy is dominated by the Apennine mountains, and it has been estimated that over forty per cent of the land area is mountainous and too steep for any form of cultivation. Another forty five per cent is classified as hill country which is prone to considerable soil exhaustion. Most of the southern Apennines are formed of limestone, much of which is dolomite. This hard dolomitic limestone gives a landscape of sharp peaks and bare slopes with the smallest vestiges of garigue or scrub vegetation. The highest peak, Corno Grande in the Abruzzi is (2915 metres). In other areas such as Apulia, the limestone gives bare karstic conditions. By contrast much of Calabria is granitic with the Pollino massif reaching 2275 metres. Extensive plains are limited, and lowlands are confined to Foggia, the Naples-Salerno plain, and the 'heel' around Taranto. There is a basic poverty in the Mezzogiorno with its high proportion of uncultivable land, rugged and eroded slopes, dried-up river beds and desiccated landscape.

The legacy of the past

The history of the area has had a dramatic effect upon this landscape. During the period of the Greek and Roman civilisations, the south was a major granary with widespread cultivation of vines, cereals and olives. Mediterranean pines and other woodland covered the hillslopes. With the collapse of Roman power from the fifth century onwards, there was a prolonged period of political chaos and insecurity, with a succession of invasions from the north and piracy from the sea. Settlement moved inland and tended to concentrate in closely-knit villages, often of great size, clustered on hilltops, where they could be more easily defended. The plains were abandoned and the foothills and mountain slopes became over-grazed and deforested, with ensuing soil erosion, extensive gullying on the hillsides, and lowland flooding leading to swamps and malarial infestation. From the eleventh century onwards the whole area, as the Kingdom of Naples, came under the corrupt rule of the Spanish House of Bourbon. Feudal serfdom with large absentee landlord estates called 'latifundia' persisted into the middle of the twentieth century. These are generally large estates on the plains which practise monoculture based upon wheat alternated with fallow. This is extensive farming and under-utilises the resources of both the land and the people. Most of the south's sheep and cattle are grazed on the latifundia and transhumance is practised with livestock being taken to the nearby mountain slopes. The blight which they have brought to the south is seen in the under-employment of the farm labourers. Many peasants would have five or six weeks working time on their own plots of land, and then would look for work on the latifundia, some distance away. The peasants would wait, early in the mornings, in the piazzas of the towns and villages for the farm overseer to hire them for the day. Generally they would average only one hundred days work per year, illustrating the low economic level at which the south existed well into the mid-twentieth century.

Too many people on the land

Pressure of population led to increasing subdivision and fragmentation of land holdings. As late as 1950 over seventy per cent of holdings in the south

Villalago, near Aquila, Abruzzi-Molise. A hill-top village in the rugged Gran Sasso mountains, surrounded by abandoned terraces.

were of less than three hectares. These are the other side of the picture, known as 'minifundia'. They usually exised on the poorer hilly mountainous land which experienced considerable sub-division amongst tenants and share-croppers. In 1950, 45 per cent of all agricultural workers in the south owned no land at all; 28 per cent were share-croppers who surrendered up to sixty per cent of their crop to the landowner as rent; only 27 per cent owned their own land. Fragmentation of farm holdings was the norm, with peasant farmers cultivating anything up to ten widely scattered plots of land. Such a system of tiny holdings, particularly with a proportion being share-cropped, was inadequate to support a family. Crop specialisation was almost impossible because of the need to maintain the family's food supply, and the pattern of farming was subsistence with very little entering commercial channels. Methods of cultivation were antiquated and labour intensive; insecurity of tenure was damaging to morale, and not conductive to mechanisation and capitalisation. Economic feudalism was the most distinguishing feature of the Mezzogiorno.

Isolation and a lack of raw materials

The essential bases for industrial development were lacking, as there was no coal, or other raw materials of any significance, and the development of hydro-electric power, so important in the north, was retarded in the south by the throughout the south industry was mainly small-scale and artisan in character. Perhaps most important of all was the isolation of the south from the mainstream of European development during the nineteenth century, whilst the rest of Europe was industrialising rapidly. The great length of peninsular Italy (1000 km) and its physical nature was an inhibiting factor in communications. The backwardness associated with the corrupt rule of the Bourbon kings of Naples until 1861 meant that there was hardly any infrastructure of roads and railways. In 1861, the Mezzogiorno had only 99

Region	Gross Domestic Product Lire ($\times 10^9$) 1960	1981	Gross Domestic Product as a % EUR 12 1986
Italy	19286	261638	14.3
Lombardy	4147	54040	2.9
Apulia	823	12744	0.7
Basilicata	111	2035	0.1
Calabria	374	5337	0.3
Abruzzi-Molise	370	5626	0.3
Campania	1225	16735	1.0

Figure 17.2 A comparison of Gross Domestic Product (source – *Eurostat*)

km of railway, whilst in Italy as a whole there were 1798 km. All this contrasted markedly with the north of Italy (chapter 16) whose most important foundation for industrial wealth lay in its relatively easy access across the Alpine mountain passes into the core area of Europe. Furthermore, after the unification of Italy in 1861, the contrasts between north and south increased. The economic disadvantages of the south, remoteness form markets, higher fuel costs, and a feudal system of agriculture, were compounded by the illiteracy and unskilled nature of the population. In the newly unified Italy, the north held all the basic advantages, and as it rapidly became industrialised from the 1870s onwards, the gap between the two Italys steadily widened.

The Cassa per il Mezzogiorno

There have been piecemeal attempts to develop the economy and improve the enviroment of the south. Various projects to reclaim marshland and eradicate malaria, and to improve river channels and combat soil erosion were made during the nineteenth century. During the inter-war period, the Fascist government introduced a programme to drain and irrigate the land and improve agriculture. In the period immediately after 1945, land hunger and large-scale unemployment caused considerable unrest, the problems of the south became really acute, and were brought to the attention of the government and public opinion. The new democratic government of Italy realised that from political necessity alone, a major effort was needed to develop southern Italy. Furthermore, statistics were more readily available, which allowed assessments to be made and remedies to be suggested. A long-term strategy was favoured and a supra-regional body was proposed with direct powers and massive investment backing which could see the south as a whole and which would be able to plan a coordinated and vigorous development policy.

In March 1950 the Cassa per il Mezzogiorno was established by the Italian Parliment as an executive body with powers to carry out initially a ten-year basic development plan. It was to operate in the seven regions of the south and also a small part of Latium near Rome. Considerable power was given to the Cassa. In addition to the power to initiate development projects, it also had coordinating power development in every sector of the south's economy. The investment of government money was supplemented by private investment, both Italian and external, and by loans from the World Bank. Since the development of the European Community, the European Investment Bank has made a majority of its loans to the south, 25 per cent of the total as late as 1985 (fig. 17.3). The Agricultural Guidance fund has been very important, and more recently the Social and Regional funds have made significant contributions (fig. 10.8).

Agricultural improvement

Agriculture initially accounted for the major efforts of the Cassa, and the

Province	NCI	Own resources			Total	Percentage share of total EIB loans
		Agriculture	Industry	Infrastructure and services		
Campania	58	–	147	148	353	8.8
Abruzzi-Molise	–	12	10	–	22	0.5
Sud (Apulia, Basilicata, Calabria)	1	–	36	116	153	3.5
Sicily	15	–	23	38	76	1.4
Sardinia	–	–	29	–	29	0.7
Multi-regional projects	240	–	360	102	702	10.9
Total Mezzogiorno	314	12	605	404	1335	25.8

Figure 17.3 EIB loans to the Mezzogiorno, 1985 (million ECUs)

original planned investment earmarked 77 per cent of investment for agriculture (fig. 17.4). Industry was not included at all in the original plan. There were two principal methods of agricultural change: land reform; and the modernisation of farming techniques. Special agencies 'Ente Di Riforma' were set up to administer eight areas of land reform (fig. 17.5). These areas were set up by government decree and were not directly under the jurisdiction of the Cassa, and not all of them were in the Mezzogiorno as previously defined. Nevertheless they performed a valuable function, thereby providing landless peasants with holdings by expropriating large and inefficiently run latifundia. The large estates which were expropriated were those which generally had been badly managed and which had received virtually no investment. Particularly well run farms were termed 'model farms' and exempted from expropriation. The whole object was to carry out close

	Expenditure (%)			Estimated overall expenditure (%)
	1950 (*proposals*)	1950–65	1966–69	1950–80
Agriculture	77.0	56.1	24.7	12.7
Infrastructure	20.5	22.4	19.6	20.2
Industry	–	6.9	36.0	49.9
Tourism and special projects	2.5	14.6	19.7	10.6

Figure 17.4 The Cassa per il Mezzogiorno (source – *Allen and McLennan*)

settlement, to provide for new villages and to increase productivity with irrigation and land improvement schemes, so that intensive cash-crop farming on small holdings would replace extensive wheat cultivation. Viable family holdings (poderi) were created; in irrigated areas the minimum size was five hectares, whilst on hillsides ti could be up to fifty hectares. The typical pattern of change may be seen in the work of the Ente di Riforma Apulia-Basilicata, covering the 'heel' of Italy. An area of 200 000 hectares was expropriated from 1500 land-owners, and assigned to 31 000 families. The change to this much closer pattern of settlement involved the construction of roads, power and water supply, land reclamation, new farmhouses and Villages. A complete new infrastructure was developed, with 15 000 farmouses, 50 service centres and villages, 1700 kilometres of roads, and 7500 wells. A large programme of irrigation transformed the lowland areas of the Metapontino, the Bari-Brindisi coastal strip (Salentine peninsula) and the Tavoliere Di Apulia (fig. 17.6). The 'wheat and olive' landscape has changed to one of citrus fruits, vegetable and industrial crops, and livestock numbers have increased dramatically. The Metapontino illustrates the transformation of the landscape. Formerly a malarial coastal plain, it was drained, reclaimed, and

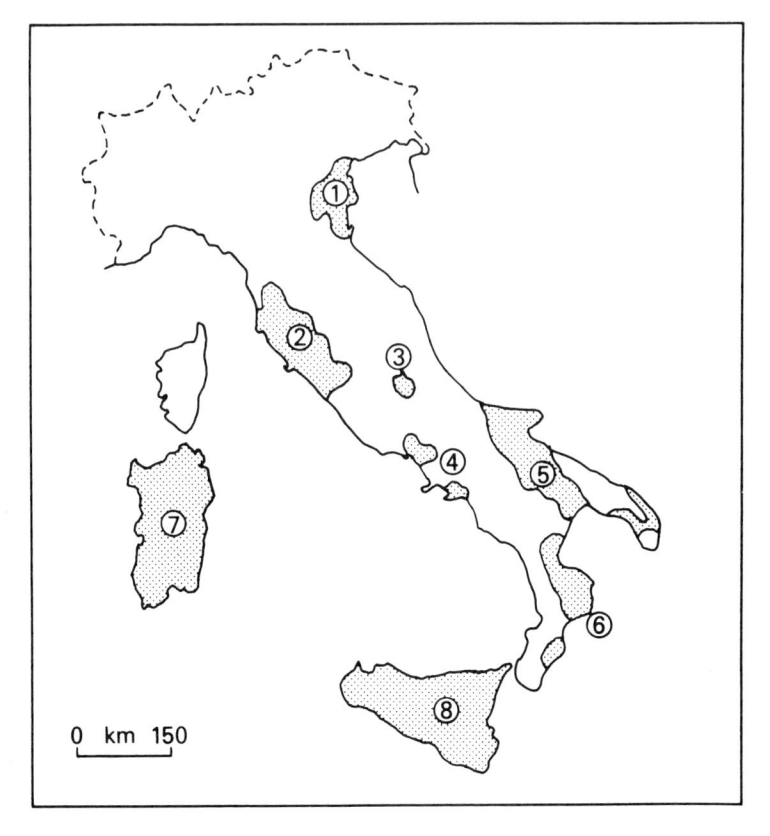

Figure 17.5 Land reform agencies (source – *Mountjoy*) 1. Delta Padano 2. Tuscany - Lazio 3. Fucino 4. Volturno 5. Apulia - Basilicata 6. Sila 7. Sardinia 8. Sicily

the mosquito eradicated. The five rivers irrigate the entire coastal strip and are instrumental in the cultivation of vines, citrus and other fruit trees. Some 750 000 pine trees act as windbreaks. Oranges, peaches, apricots, pears, salad vegetables, sugar-beet, tobacco and tomatoes are the principal high-yielding crops.

The transformation of large areas has been impressive. The largest and least efficient Latifundia have disappeared, and much of the day labour system with them. There has been a notable increase in intensive farming in fruit and vegetable crops. It must be remembered that land reform has only affected a small proportion of the cultivated land in the south, but it was a significant movement in the right direction.

Infrastructure

The Cassa per il Mezzogiorno has been the instrucment of financial assistance for many modernisation and improvement schemes in the areas of land reform. As well as irrigation schemes and new farm houses, it has financed crop and stock improvements and begun the task of consolidating fragmented holdins into viable units. A significant contribution to the rural infrastructure has been the establishment of packing, processing, and refrigeration units. In

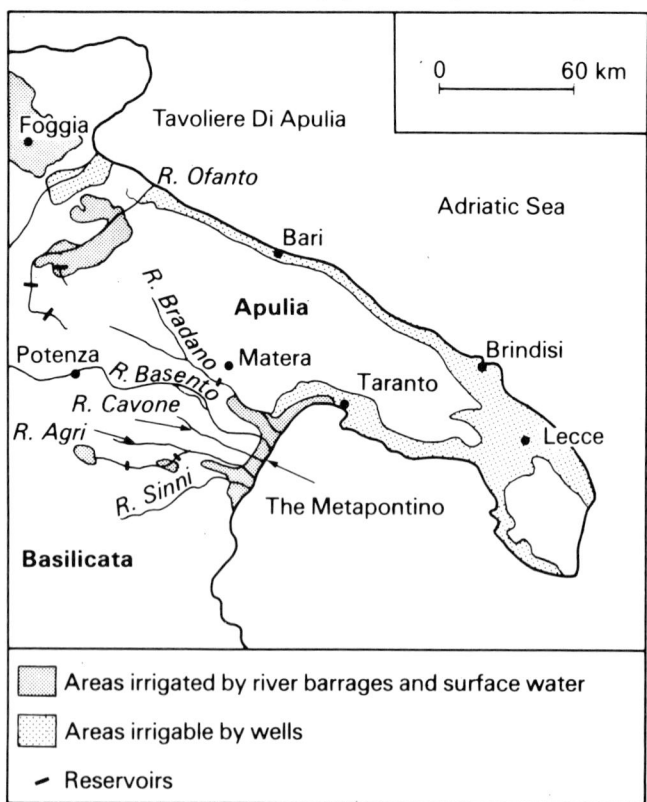

Figure 17.6 Irrigation in Apulia and Basilicata (source – *Mountjoy*)

addition, because the most permanent weakness of the south is its need to overcome the relatively long distances from the markets of western Europe, the Cassa has been very active in developing transport, distribution, and marketing facilities for agricultural products. Aqueducts, reservoirs and the draining of marshes also contribute to the improvement of village life. Schools, hospitals, and training centres are directly associated with the social infrastructure. The Cassa has installed main water supplies for millions of people, and aided local authorities to build the systems for distributing the water to villages. The construction of hydro-electric power stations, telephone systems, new and improved roads, and the railway system from Campania into the toe of Italy at Reggio Calabria, have made large areas of the south more accessible and have improved living conditions immensely.

Industry

During the early period, up until 1957, the main work of the Cassa was to inject new vitality into agriculture, village life, and the rural environment, hoping that this would stimulate demand, and provide conditions favourable to the growth of industry. However, by 1957 it was realised that there would have to be more definite intervention in the industrial sector to provide new employment and to ease the burden on the agricultural sector. The passing of the Industrial Areas Law in 1957 empowered the Cassa to support the establishment of industrial zones in the south, and figure 17.4 illustrates the shift of resources into the industrial sector after 1965. There were various encouragements to industrialists. The Cassa itself provided capital for new projects and for modernisation of existing concerns, up to as much as eighty five per cent of the cost. There was also fiscal exemption, rail-freight concessions and exemption from customs duties on imported raw materials. The European Investment Bank (EIB) has been of great importance here and has helped to establish a large number of factories of various kinds (fig. 17.3). The largest single contribution has come from the government-controlled companies such as ENI and IRI; they have been required by law to place forty per cent of their investment in the south. The state sector accounted for about thirty five per cent of the growth in manufacturing employment initially. Much of the growth has been in heavy industry, mainly in iron and steel, shipbuilding, heavy engineering, cement, oil-refining, and petrochemicals, because of the nature of the state controlled companies. Major projects of international significance include the Taranto and Bagnoli steelworks, and the Montedison petrochemical complexes at Brindisi and Siracusa-Augusta in Sicily. The continuing problem has been that these are mainly capital-intensive heavy industries and they do not provide very large employment possibilities, nor do they necessarily stimulate the development of lighter consumer goods industries, which are essential to balanced growth. Such prestige projects as the Taranto steelworks have been called 'cathedrals in the desert'. Another problem was the dissipation of investment aid over the whole of the Mezzogiorno, which was often spread too thinly to have a significant effect.

Growth poles

The development of a limited number of centres which were individually capable of faster growth was adopted during the latter half of the 1960s. The idea is based upon the economies of scale which accrue when investments infrastructure, industrial linkages, and trained labour supply are concentrated into a smaller cohesive area. Altogether 48 nuclei of industrial development were designated, and subsequent ex perience has shown that five major areas were of greatest significance. These have become 'growth poles' (fig. 17.7) into which most investment has been channelled. Three of the most important of these are the Naples - Salerno pole, the Bari-Brindisi-Taranto triangle, and the Siracusa-Augusta pole.

1. **The port of Naples** has declined considerably in relation to Italy's other ports and now occupies fifth position after Genoa, Trieste, Augusta, and Venice. Nevertheless its shipbuilding and repairing industry has been supplemented by the Bagnoli steel plant, cotton textiles, Alfa-Romeo cars, Pirelli cables, Olivetti office machinery, and Montedison petrochemical works. The Italian aircraft industry is based upon a joint Fiat/IRI venture, 'Aeritalia'. There appears to be the basis of a fairly diversified industrial structure in the Naples area.

2. **Siracusa-Augusta.** The development here is based largely upon petrochemicals, for two reasons. There is a small oilfield at Ragusa, and considerable sulphur and potash deposits in eatern Sicily. More important are the deep water facilities between Augusta and Siracusa which have led to the development of major port installations, which are able to accomodate 250 000-tonne tankers. This is one of the largest oil refinery, chemical and petrochemical complexes in Western Europe. Cement and the refining of non-ferrous metals are other activities.

3. **Bari-Brindisi-Taranto.** The integrated iron and steel works at Taranto, built in 1960, dominates this area. It is one of the largest steelworks in Europe with a capacity of ten million tonnes per annum. Its coastal site and the facility for low cost imports have been instrumental in its success, and it has been accompanied by a large industrial estate with engineering and machine-tool factories, oil refinery, cement works, agricultural machinery and consumer goods industries. At Brindisi is the Montedison petrochemical factory, producing plastics and ethylene. The European Commission has drawn up a detailed development plan for this pole of regional development, which it considers to be of major importance to the Community.

Tourism

The recognition by the Cassa that tourism can play a major part in economic development has meant the increasing allocation of funds for hotel building and modernisation and other tourist amenities. The south has much to offer the tourist: beautiful empty beaches; dramatic mountainous scenery; historic cities and architecture of every type and period. It is the more attractive by comparison with the overcrowded northern Italian resorts. Hitherto there was one major drawback - inaccessibility. The Cassa has assisted in the improvement of construction of hotels and has constructed local roads for sightseeing. It is surprising that the Mezzogiorno has still not attracted the tourist on a large scale. There is an immense tourist potential.

Autostrada

The construction of autostrada has now diminished travelling time quite dramatically. The Autostrada Del Sole, running from Bologna through Rome and Naples to Reggio Calabria is the most well-known, but there are autostrada down both west and east coasts, with cross-Apennine links from Naples to Bari and from Rome to Pescara (fig.17.7). The remoteness of the south is now less critical both for the tourist and the industrialist, and the autostrada may well prove in the long run to be the most important single factor in its development.

The Mezzogiorno in the 1990s

The problem has proved deep rooted and complex: the Cassa has been in existence for over thirty years, during which it has made massive investments. In 1984 there was a financial crisis and its role has been reduced since then, with its powers transferred to the regional governments. It is difficult to see whether the economy has reached the 'take-off point' or whether the 'multiplier effect' is operating, and capable of sustaining long-term growth. Yet there is no doubt that there are the beginnings of sustancial industrialisation, and the employment structure has shown quite dramatic change as seen in fig. 17.8. The infrastructure is much sounder, and agriculture has lost two million workers.

During the life of the Cassa new factories, employing a total of nearly 400 000 workers have been constructed. Most have been in four districts: Latina, south of Rome; Naples - Salerno; the Bari-Brindisi-Taranto triangle and Catania - Siracuse in Sicily. The key problem is the concentration upon large, capital-intensive units based upon steel, metallurgy and chemicals. There have been relatively few of the labour-intensive industries in traditional industries which are necessary for sustained improvement in employment.

Agricultural productivity has improved considerably in the land reform areas with a much greater emphasis on intensive crops and increased numbers of livestock. But has to be taken in context, noting that land reform has so far affected only a small proportion of the cultivated land in the south. The largest remaining problem is that of rehabilitating the eroded mountain slopes and bringing the hill farmers into a modern farming system.

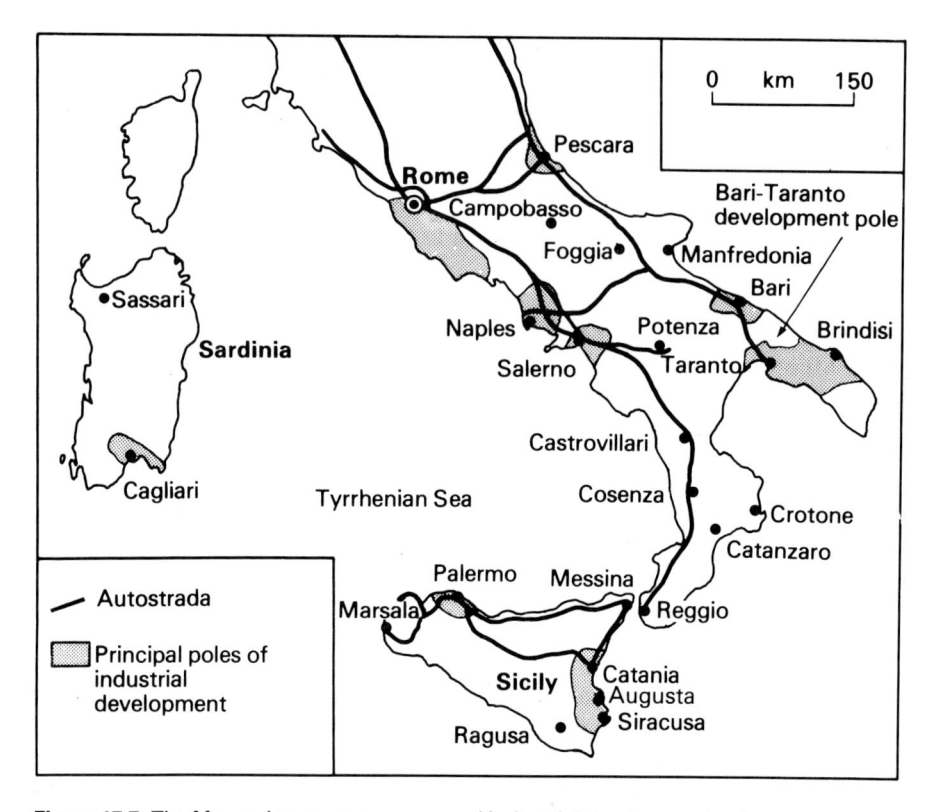

Figure 17.7 The Mezzogiorno: motorways and industrial development poles

The gap between the south and north of Italy remains, because, in spite of considerable advances in per capita income in the south, the north experienced such boom conditions in the decade up to 1975 that the gap has closed only partially as fig. 17.9 shows. The per capita income in Calabria is still below half that of Lombardy and Liguria. Relatively little development has taken place in Basilicata, Calabria and Molise, and these regions have fallen behind the rest. Clearly the improvement has been very uneven and within the South itself major differences are now emerging.

Since 1950 population loss has been over four million people (fig. 17.10). This has been of major significance in siphoning off the excess rural population. Most of the emigration is inter-regional, to the northern cities,

	Agriculture	*Industry*	*Services*
1950	57	20	23
1981	24	28	48
1986	19	29	52

Figure 17.8 The percentage employment structure in the Mezzogiorno (1950–86)

Per capita income index	1950	1970	1981	1985
Italy (average)	100	100	100	100
Lombardy	146	125	132	129
Liguria	151	130	126	126
Apulia	59	79	71	73
Basilicata	51	62	64	66
Campania	56	80	67	70
Calabria	42	61	56	64

Figure 17.9 Changes in per capita income index (source – *Eurostat*)

but emigration to other European Community countries, notably West Germany, also occurs. Since 1975, in the changed economic climate, there has been a reduced rate of intake by the northern parts of the European Community, and a return of migrants to their homeland. Migration from the Mezzogiorno appears to have been much reduced and even reversed, in Sicily, for instance (fig. 17.10).

Global factors have had a major limitation effect. The recession of 1974 damaged the partial success of southern industrialisation, as peripheral regions were badly affected by the fall in demand and by transfer of production by multi-national companies to Third World countries. In the agricultural sector citrus fruits and horticultural crops have suffered fierce competition in the European market from Spain and other mediterranean countries.

Tourism is now the principal avenue through which fuller prosperity could

	Net emigration (in thousands per annum)				
	1966	1970	1975	1980	1985
Campania	26.4	34.9	14.5	–	1.7+
Abruzzi/Molise	15.0	6.5	2.3	–	7.4+
Apulia	27.5	26.4	6.7	5.5	0.3+
Basilicata	8.6	10.7	3.7	3.2	0.2–
Calabria	23.3	26.1	10.4	6.2	4.6+
Sicily	33.5	36.8	11.5	6.0	16.1+
Sardinia	6.5	8.9	1.6	1.6	4.5+
Total Mezzogiorno	140.8	150.3	50.7	22.5	34.4+

Figure 17.10 Net emigration from the Mezzogiorno (source – *Eurostat*)

come. The motorway system has reduced the isolation of the south and as yet, tourism has not touched the Mezzogiorno on any scale outside a few traditional areas such as Sorrento, Capri and Amalfi.

Important questions remain unresolved. Unemployment is amongst the highest in the European Community, and is particularly serious in rural areas. Rural povety is very marked, and the lack of rural development contrasts strongly with the established growth centres. Agriculture dominated the initial stages of the Cassa's work, but only limited areas have been affected. Large state enterprises have moved into the area, but there is a lack of generative smaller scale private industry and coordinated planning. A strong urban tradition does not exist here, as it does in northern Italy. The essential differences between north and south remain with the Mezzogiorno remaining dependent upon the north. The Cassa has achieved a qualified 'economic miracle', and has, it is to be hoped, created the infrastructure which will allow 'take-off' into sustained growth during the 1990s. Only then will the process of 'modernisation' which has been operating for over thirty years, be accompanied by real 'development' of the South.

18

Greece and Ireland: the periphery

Greece and Ireland are two peripheral countries of the European Community, both with aspects of semi-developed and with a markedly lower GDP than the Community average. Their geographical characteristics illustrate the complexity of development problems, as in some ways they are similar, but in others they are unique (fig. 18.1).

Their environments are quite different, with Ireland having a maritime climate with its western mountains and plateaux open to excessive exposure from Atlantic wind and rain. Greece, by contrast, has a desiccated mediterranean climate giving rise to aridity particularly in the mountain chain. Both climates are adverse in terms of the agricultural response. Inaccessibility affects both countries adversely. They are remote from the European core, with air or sea journeys necessary for physical contact. In Greece's case it is

1986	*Greece*	*Ireland*
Area km^2	132 000	68 894
Population (millions)	9.9	3.5
Population density (EUR 12:143) per km^2	75.0	51.0
GDP per capita index (EUR 12 = 100)	56.8	65.9
Percentage workforce in:		
agriculture	28.5	15.8
industry	28.1	28.3
services	43.4	55.8
Car production	nil	nil
Number of passenger cars per 100 inhabitants (EUR 12 = 35)	13	21
Number of telephones per 100 inhabitants (EUR 12 = 48)	33	23
Energy consumption per head (EUR 12 = 3232 toe)	1704	2587

toe = tonnes of oil equivalent

Figure 18.1 Greece and Ireland: an economic profile

possible to travel overland through Yugoslavia, a different journey through similar mountainous country. Marginality to the core is therefore a shared characteristic.

The economic characteristics of this marginality may be expressed in a variety of ways. These include persistent and prolonged emigration, an excess of population in agricultural occupations, a rural subsistence economy with accompanying poverty, and an imbalanced economic structure with low levels of industrialisation, and excessive concentration of population and economic activity in the primate city and its region. Dublin and Athens are both cases in point.

In addition, peripherality is often affected by global factors where interdependent relationships with other countries have had adverse effects. These may be classed as political dependency, cultural disintegration and economic exploitation. Over a long historical period, both countries have been affected by these external factors. Ireland only became independent of Great Britain in 1926 and Greece was a dependency of the Ottoman Empire until 1830. In Ireland the Gaelic language has largely been replaced by English, although Greece has maintained its language and distinctive alphabet intact, probably because of the strength of its ancient classical culture. However, both countries have a dependence upon agriculture and the export of primary products. This is indicative of their general lack of energy resources and their failure to industrialise when other parts of Western Europe were so doing.

Ireland

Emigration

The problem of emigration has been central to the economic geography of Ireland. Up to 1840 the population of the island as a whole (including Northern Ireland) had expanded to approximately eight million. The countryside was relatively prosperous and had sustained the large-scale cultivation of the potato, which was the staple food of the rural population. The failure of successive harvests due to potato blight during the 1840s, and the resulting famine and starvation led to large-scale emigration, many of the emigrants going to the New World. The country never recovered, and the whole basis of life was undermined, resulting in a halving of the population in the century up to 1940. Ireland's population in 1986 was 3 540 000, the lowest density in the European Community. It has been estimated that people of Irish descent living in America, the UK and elsewhere total something like sixteen million, five times the number remaining in Ireland itself.

The question of the great emigration needs a little more analysis in depth. The potato famines lasted for a very short period, but they triggered off a century of decline at the same period when most of Western Europe was growing substantially in population. The answer is probably to be found in the difficult nature of the rural environment, the subsistence character of farming, and low income levels and depressed spending capacity of the population. This must then be placed in the comparative context of the rest of

Europe and the New World. The decline of population began before the start of the nineteenth century, with the commencement of the industrial revolution in Europe. In Ireland there was no industrial base to absorb the movement off the land because there were almost none of the resources to sustain the initial industrial growth which in countries like Britain were supplied by coal and iron-ore. The attraction of employment, high wages and rising standards of living lay abroad, and there was little alternative for the Irish but to emigrate. The potato famines gave a large impetus to this movement, but the underlying cause of Ireland's depopulation lay in its lack of industrial resources at the critical period of the industrial revolution. The fact that Ireland was politically part of Great Britain at this time was also a contributing factor. The close links and ease of access produced a positive inducement to emigrate to more advantageous environments outside Ireland.

Ireland today

Dublin, the capital and primate city, with its port, Dun Laoghaire, has 915 000 people, over a quarter of the total national population. Cork (140 000), Limerick (60 000), Waterford (35 000), and Galway (30 000) are the only other regional centres of any size and importance. Agriculture accounts for sixteen per cent of the labour force. Ireland is thus characterised by an underdeveloped urban hierarchy and an enlarged agricultural sector.

In the west these characteristics are even more extreme. Substantial amounts of land are uninhabited mountains or ill-drained boglands. Settlement is dispersed and villages are rare, the main unit of rural settlement being the single farm. Towns are generally very small, but have the range of social and economic functions normally associated with a town. In many western

Planning regions	1986 Population (1000s)	1986 Population (Percentage)	1979–1981 Net annual migration (rates per 1000)
East	1332	37.6	−0.5
South-east	383	10.8	−1.4
South-west	538	15.2	−1.7
North-east	198	5.6	−3.1
Mid-west	320	9.0	+0.7
Donegal	127	3.6	+2.9
Midlands	262	7.4	−1.1
West	294	8.3	−0.4
North-west	85	2.4	−0.7
Total	3539000	100	−0.7

Figure 18.2 Ireland: planning regions (source – *H Clout*)

countries the proportion of the population engaged in agriculture rises to forty per cent. Poor soils, small and fragmented farms, lack of capitalisation and cooperative organisation, characterise a subsistence-oriented rural economy. Although total population decline has now been arrested in Ireland overall, many parts of the west are still threatened with decline. Selective migration of young people adversely affects the birth rate and causes stagnation in many rural communities. The farm population is relatively old, with many of the farmers over fifty years of age, representing a major obstacle to change.

The north-western regions

The north-west of Ireland is agriculturally poor, and very sparsely populated. Donegal has granite mountains rising to 650 metres, is glaciated and often bare of soil, with the lower lands containing waterlogged areas with extensive blanket peat-bogs. Only about forty per cent of the total area is improved farmland and even good pasture-land is scarce. On the coast are rocky headlands and deep inlets. Rainfall is heavy (up to 1500 mm per year) and the land is bleak and windswept with severe exposure to Atlantic gales. Tree growth is impossible in many areas. Subsistence farming is the norm with cottages usually situated on the sheltered lee of the mountains. Potatoes, hay and occasionally oats, are the only crops possible. Sheep are more numerous than cattle and their wool serves as the basis for the manufacture of homespun cloth, knitwear and Donegal Tweeds. The only town of any size is Donegal (1500 people) which acts as county town and market centre, and exemplifies the restricted development of most Irish towns.

The mountains of Mayo and Connemara are scenically beautiful, but are also regions of difficulty, rising to 800 metres. There is a combination of bare ice-scoured rocks and peatbogs, and the mountains are practically uninhabited. Where there is settlement, it is characterised by isolated farms and cottages, with low standards of living. These north-western coastlands lie on the remotest fringes of Europe and illustrate both the worst extremes of the Atlantic mountain environment and some of the most marginal economic conditions in the whole of the European Community.

The South and East

The southern and eastern regions, including particularly the Limerick-Shannon lowland, the south coast from Cork to Wexford, and the East Central Plain centred upon Dublin, have a kinder physical enviroment and a greater accessibility to the United Kingdom and the rest of Europe. About ninety per cent of the East Central Plain is improved farmland. The generally humid atmosphere encourages a thick growth of grass on limy glacial drift soils. Dairy farming and market gardening take place around Dublin to supply the large urban market, but beef cattle are also of great importance. Live cattle, both store and fatstock, are exported. To the south-west the region merges into the dairy farming zone of Limerick and the Golden Vale.

The southern coastlands around Waterford and Wexford have the sunniest climate in Ireland and the proportions of arable land are relatively high, rising to one-third of the total farmland on the Wexford plain. The Wexford plain is known as 'The Garden of Ireland' with fertile soils, based on weathered sands, clays and marls supporting crops of oats, barley, potatoes, wheat and sugar beet. Cork, with its outport Cobh, is a port of call for transatlantic liners. It has good rail links with Dublin, has an excellent ria harbour, and is the second largest city. It has a substantial manufacturing base with a small steelworks, vehicle assembly, agricultural machinery, rubber, clothing and footwear, and food processing of various types.

The growth of Dublin in particular has been a feature of the last forty years. Although a small city by European standards, Dublin and its outport Dun Laoghaire have increased their share of the country's total population to twenty six per cent. It is seven times as populous as the next largest city, Cork. It accounts for half the national industrial output, and employs over two-fifths of the total industrial labour force of Ireland. Its industries include engineering, clothing, footwear, meat canning and bacon processing, brewing, distilling, biscuit and jam making, tobacco, and fertilisers. Dun Laoghaire is the country's principal passenger port and Dublin city dominates Ireland's economy to an increasing extent, giving rise to a major imbalance between the western underdeveloped region on the one hand, and the richer south and east on the other (fig. 18.2).

The western underdeveloped region

Although the whole of Ireland qualifies for aid under the European Community Regional Policy, the western part of the country is markedly poorer, more marginal and more isolated than the east. Its dependence upon subsistence agriculture is such that purchasing power is often minimal. Average income per head is a useful indication of the differences. In the west and Donegal, the average income per head is only three-fifths of that in Dublin. Ireland had a per capita income of only 62 per cent of that of the UK in 1986. These are highly significant figures when placed into the context of the European Community as a whole.

In 1949 the Irish Development Authority (IDA) was established with two important principles: (1) the encouragement of industrial investment (particularly for export-oriented industries) by means of incentives; (2) the ending of restrictions on foreign ownership of Irish companies. Foreign capital investment has risen dramatically with the creation of 75 000 new jobs. The problem was that much of this investment continued to go into the south-east and Dublin area.

The Underdeveloped Areas Act 1952 designated the 'Western Underdeveloped Region' (fig. 18.3). It comprises twelve out of the twenty-six counties of the Republic, and includes one-third of the total population. Industries established within the region are eligible for a higher rate of aid than in the rest of the country. Aid falls into four groups: capital grants of up to two-thirds; long-term loans at favourable interest rates; tax relief of twenty per cent on buildings; and grants towards the training of workers. The launching of the 'Small Industries Act' of 1969 was another attempt to assist the modernisation and enlargement of many existing local craft industries scattered throughout the small towns in the west: the Irish linen industry is an example.

In 1968 Professor Colin Buchanan recommended to the Irish Government the strengthening of the urban hierarchy by means of the designation of

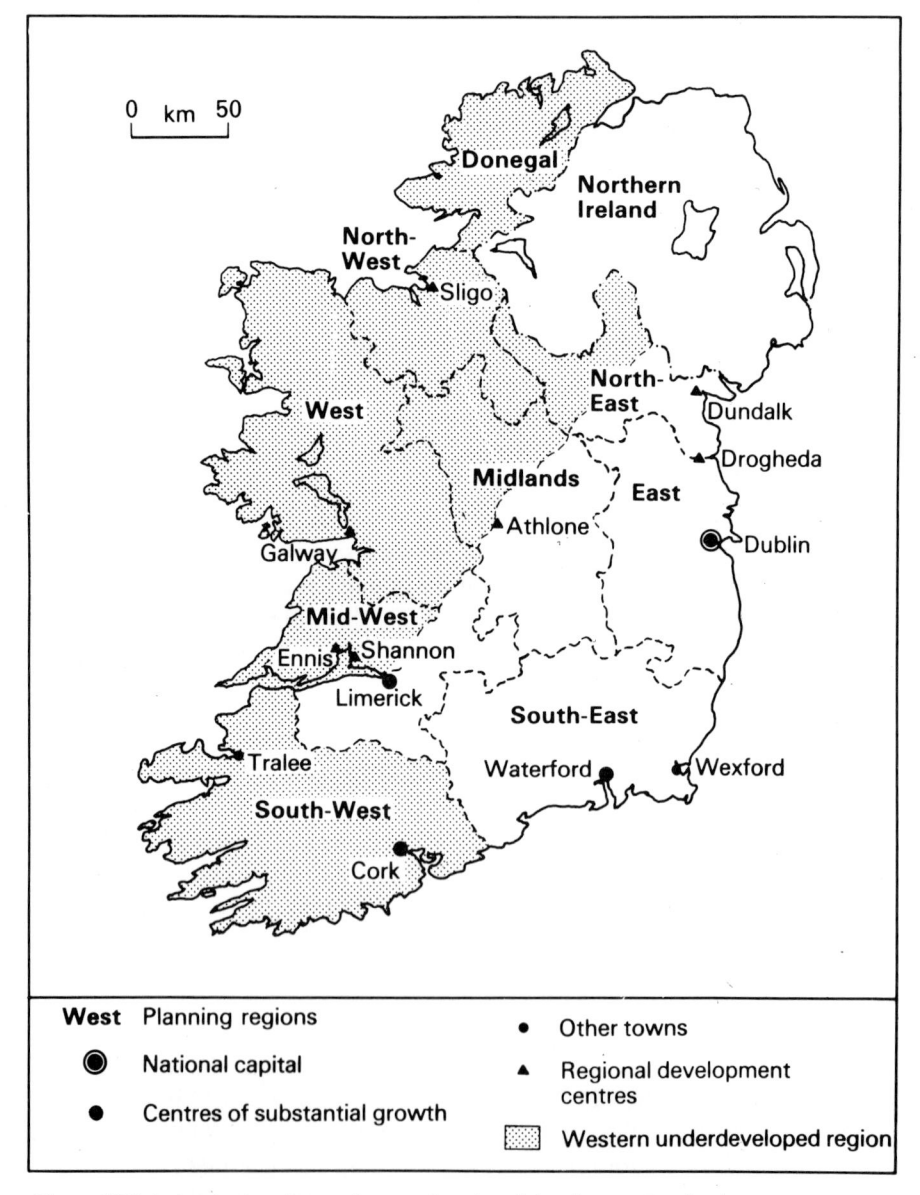

Figure 18.3 Ireland: planning regions and regional development poles (source – Buchanan)

regional development centres (fig. 18.3). The IDA was reorganised and given wide-ranging power, and progress has been made to attract investment into the three large regional centres of Cork, Waterford and Limerick, which have an existing industrial base. Perhaps more significant are projects established in areas such as Donegal. Two examples of textile factories are Snia Viscosa and Courtaulds. The lower wage levels and the reserve of under-employed female labour are two significant factors. Most industry is associated either with the agricultural processing section (such as milk-processing and fertilisers) or else is classed as 'footloose', covering a wide range of light industries and consumer products ranging from bathroom scales to Sligo to ballbearings at Tralee.

The Mid-west regional plan

The Mid-west comprises the counties of Clare, Limerick and part of Tipperary, 308 000 people in all. The regional plan focuses upon the growth pole of Limerick-Ennis-Shannon. The development of the Shannon Industrial Estate since the 1960s has provided the basis for the country's third largest industrial concentration after Dublin and Cork. The Airport Act of 1947 established Shannon as the first customs-free airport in the world. The Airport Development Authority promotes the use of the airport — for freight handling and warehousing — and the industrial estate, and operates the

Shannon airport, with the adjoining industrial estate

and fiscal incentives referred to previously. Factories can be rented or purchased, and now over 400 people are employed by thirty companies. This industrial estate produces thirty per cent of Ireland's manufactured export goods. There is a wide variety of light and specialised products, predominantly high value in relation to weight and readily adaptable to air transport. The most important single type is electrical and electronic equipment. Shannon Industrial Estate has provided a singularly successful growth point in the west and population growth is now substantial around Limerick and Ennis.

The continuing problem

The work of the IDA has to be analysed in a complex global context. Although it has successfully created 75 000 new jobs by the mid 1980s, the Irish economy also shed nearly 60 000 jobs owing to international recession and the inability of Irish industry to withstand European competition. Levels of unemployment (at eighteen per cent) are amongst the highest in the European Community.

The problem of emigration, which has been very substantial during most of the twentieth century, has not really been solved either. Emigration was increased by the attractions of the British economy as it recovered from the Second World War. It was not until the 1971-81 period that the total Irish population began to increase and there was a substantial economic boom during that period. However, there are signs, since 1981, that out migration is beginning to assert itself again. Although the total population continues to increase because of substantial natural increase, nevertheless the selective drain of manpower to external attractions continues. Ireland continues to suffer because of its peripheral position *vis à vis* the rest of the European Community.

Greece

Greece lies at the southern end of the Balkan Peninsula and in 1986 had 9.9 million inhabitants. In area the country is almost the same size as England and Wales. Greece has had an association agreement with the European Community since 1962. From 1967 to 1974 the monarchy was replaced by a military dictatorship. During this period relations with the European Community were suspended and not resumed until 1975 when Greece, having restored a democratic government, applied to join as a full member. She acceded to membership in January 1981.

Greece has traditionally been pictured as a poor semi-developed country. At the end of the Second World War nearly two-thirds of the labour force were still engaged in primary activity. Even today thirty per cent of the population live in villages of less than 2000 people. Over twenty-eight per cent of the population are employed in agriculture (compared with the European Community average of eight per cent), contributing seventeen per cent to the Gross Domestic Product. The principal agricultural products are cereals, citrus fruits and vegetables, raisins, wine, tobacco, olive oil and cotton.

Pendeli monastery near Athens, with the rugged Pendeli mountains, where marble is quarried, in the background

The environment is extremely difficult, being mountainous with a desiccated semi-arid Mediterranean climate. Mount Olympus, 3200 metres, is the highest mountain. There are three physical regions, the rugged Pindus Mountain chain which covers some eighty per cent of the country and several small areas of lowlands principally around Athens, thessaloniki and in Thessaly. Finally, there are the Greek Islands in the Aegean Sea, a complex group which includes Crete, Rhodes and the Cyclades archipelago. (fig. 18.4)

Part of the problem of Greece lay in its history. Having been one of the pillars of Western civilisation in classical times with a great Empire and the home of great philosophers and architects, Greece fell into a long decline and for several centuries was ruled as part of the Ottoman Empire in the Balkans. It emerged in the early twentieth century as a semi-developed country with an economy dependent upon agricultural exports.

Prior to 1960 national development was concerned with recovery from damage inflicted in the Second World War Nazi occupation, but during the 1960s great changes began to take place. These may be examined in four key

Kalymnos island in the Dodecanese group in the Aegean Sea

sectors: urbanisation; population migration; industrialisation and tourism.

Urbanisation and the primate city strikes the visitor to Greece most clearly and substantially. The overwhelming importance of Greater Athens is very marked. The city, including the Port of Piraeus, has grown dramatically so that, with over 3 000 000 people it contains one-third of the national population total. Thessaloniki with 164 000 are the next largest cities.

The rapid growth of Athens, has sadly damaged its historic character and the summer heat poses a particular problem where large quantities of water and electric energy are needed to supply the growing urban population. Conditions in the summer in cities such as Athens are less than pleasant. Urbanisation in Greece, and Athens in particular, has been less influenced by industrialisation than by political causes. The civil war immediately following the Second War World between the monarchy and the communists produced major population movements to the cities for protection, and few have returned to the land since. The growth of the city has been therefore singularly affected by the rural 'push' factor for other than economic reasons.

Industrialisation has in a sense, followed urbanisation in Greece. Manufacturing traditionally suffered from a lack of raw materials, shortage of capital and limited markets, and cottage industry has remained important for sectors such as shoe and clothing manufacture. The economy is now developing fast, however, and new heavy industries have been established such as aluminium, chemicals, metallurgy and shipbuilding. Textiles, paper making and food processing are also important. Electricity generation has been a major factor

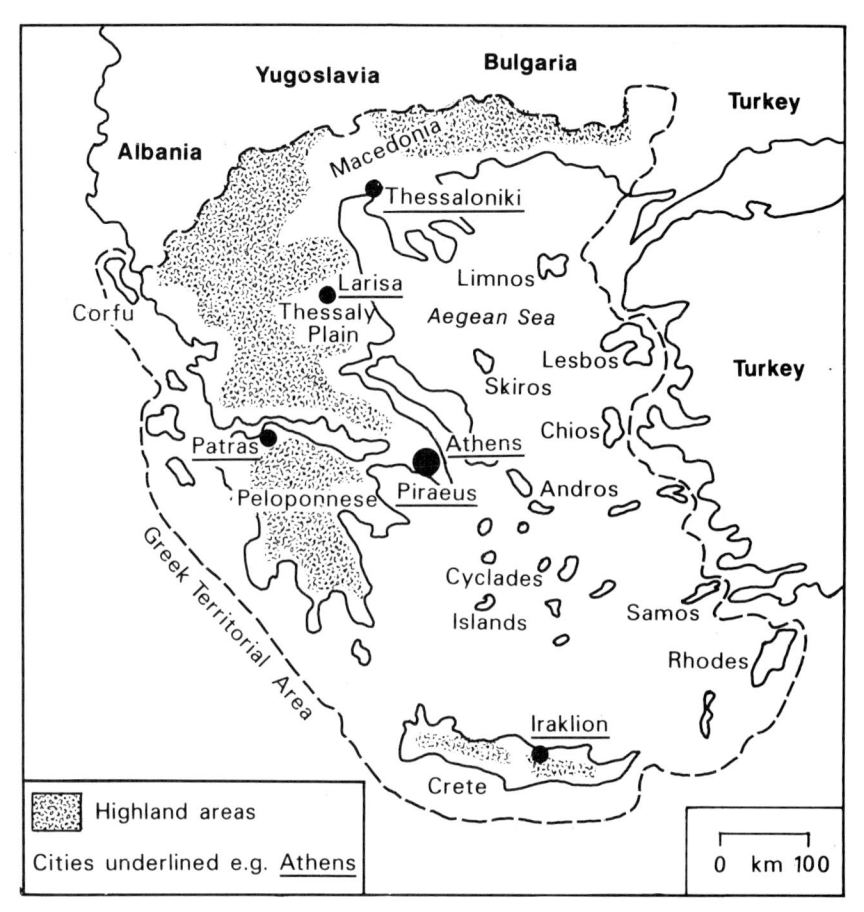

Figure 18.4 Greece: main features

in stimulating Greek manufacturing since the mid 1960s. Greece is highly dependent upon imported oil for nearly seventy per cent of its total primary energy requirements, but new hydro-electric power projects have been of major importance. Most economic activity is concentrated in the Athens Piraeus lowlands which has one-third of the country's population and over fifty per cent of the total industrial employment. The Athens conurbation provides a large labour force, a ready market, and is the country's most important transport focus. There is a steelworks at Piraeus producing one million tonnes a year, based on imports of coal and scrap.

One of the most important features of the Greek economy is its merchant fleet, a feature of the trading capacity for which Greece is famous. The merchant fleet, mainly based at Piraeus, is thirty five per cent of the European Community total and a major prop to the economy. Oil tankers form a substantial part of this merchant fleet. Cruise liners are also important and Greece is the world's fifth greatest ocean passenger-carrying country.

The decentralisation of manufacturing to the provinces has only been partially successful because of high transport costs. However, mining activity is increasingly important in the North-east of the country with lignites, asbestos, nickel, bauxite and manganese being particularly important. In addition, oil has been discovered in the Aegean sea.

Migration, both internal and external, has formed an important part of the economic changes in the country. During the late 1960s external migration reached a rate of 160 000 per year with many going to the boom economies of West Germany, Australia and the United States. However, since 1974 there has been a net return of people, and their skills and capital.

Internal migration is of the greatest geographical importance. Migration between rural areas is significant, motivated often by a desire amongst people in mountainous areas to settle on the plains. Attractive reception areas for rural migrants include the plains of Macedonia, the Soufi plain and the lower Thiamis Valley opposite Corfu. However, most internal migration is destined for Greater Athens, and Thessaloniki. The reasons for migration are complex, but include both the fundamental problems of farming as well as social phenomena. Inadequacies in Greek agriculture include small fragmented farms, obsolete techniques and a shortage of capital. Strict rules forbid marriage between blood relations, hence marriage partners often have to be sought outside the home community. Many rural migrants are women who have no particular attachment to their family farms. Marriage is also one of the major reasons why so many women migrate to Athens.

The fourth key factor in the Greek economy is tourism, originally based

The Parthenon on the Acropolis, surrounded by the modern city of Athens

upon the classical sites such as Delphi and the Acropolis. Numbers of tourists have risen strongly since the 1950s from 33 000 in 1950 to over one million in 1965 and nearly six million visitors to the country in 1982. Greece has now over twenty per cent of the air package holiday market in Europe. Over three-quarters of these are from Europe of whom most are attracted by the hot sunshine, cheap food and wine. The availability of the package holiday, charter flights and the development of airstrips in all but the most inaccessible islands have been of major revitalising importance to the Aegean islands, including Rhodes and Crete. Most holiday makers come in the summer season and efforts are being made to make the low season, from November to February, more attractive. Only one-tenth of visitors arrive during this period.

Significant spatial disparities exist in Greece between industrialised and agricultural regions, between contrasting physical areas and between the mainland and the islands. Environment is clearly important but probably the most important single reason for regional disparities is the distance from consumer markets. Accessibility problems are at the root of the backwardness of areas such as Epirus and the mountains and islands, by comparison with Greater Athens or the Thessaly and Macedonian plains. Agricultural contrasts are the most obvious sign of difference in prosperity. In the plains of Thessaly and Macedonia there is a relatively properous look to the countryside with average farm sizes larger and mechanisation and irrigation more widespread. These two regions have over half the total irrigated land and produce much of the country's tobacco, wheat and cotton. By contrast the olive and vine are the leading crops on fragmented farms in much of the mountainous hinterland and the Peloponnese, and on islands such as Crete. Out-migration is the usual reaction from these regions of difficulty.

The Greek economy is already closely tied to the European Community with about sixty per cent of total trade with other Community countries. Its economic development has been accompanied by a growing trade deficit caused by the country's dependence upon imported oil and capital goods. The trade balace is heavily in deficit and only partially covered by earning from shipping and tourism. The economy is structurally weak with low levels of industrialisation and an over-developed tertiary sector, over populated rural areas with high levels of unemployment and small fragmented farm holdings.

In 1983 a five year development plan aimed at modernising industry and promoting more work on the land was established. It asks for exemption from European Community competition and industrial subsidy rules, and the allocation of special funds to assist industrial, commercial and regional development. Greek regional development should benefit from the 1.7 billion ECU investment loans approved by the European Community in 1985 as part of the Integrated Mediterranean Programmes. (IMPs) which are operating up to 1990. The funding is earmarked for regional and social schemes, agricultural modernisation, forestry, rural diversification and tourism, and training, and for helping Greece prepare to cope with the increased competition, in particular from Spanish and Portuguese membership of the Community.

19

The Paris Region: a problem of definition

What is the Paris region?

The Paris region poses an important problem, that of definition. What is the Paris region? Is it the city of Paris, the Paris agglomeration with over eight million people, the Ile de France Planning Region which includes four départements and extends up to sixty miles away from the city, or the Paris

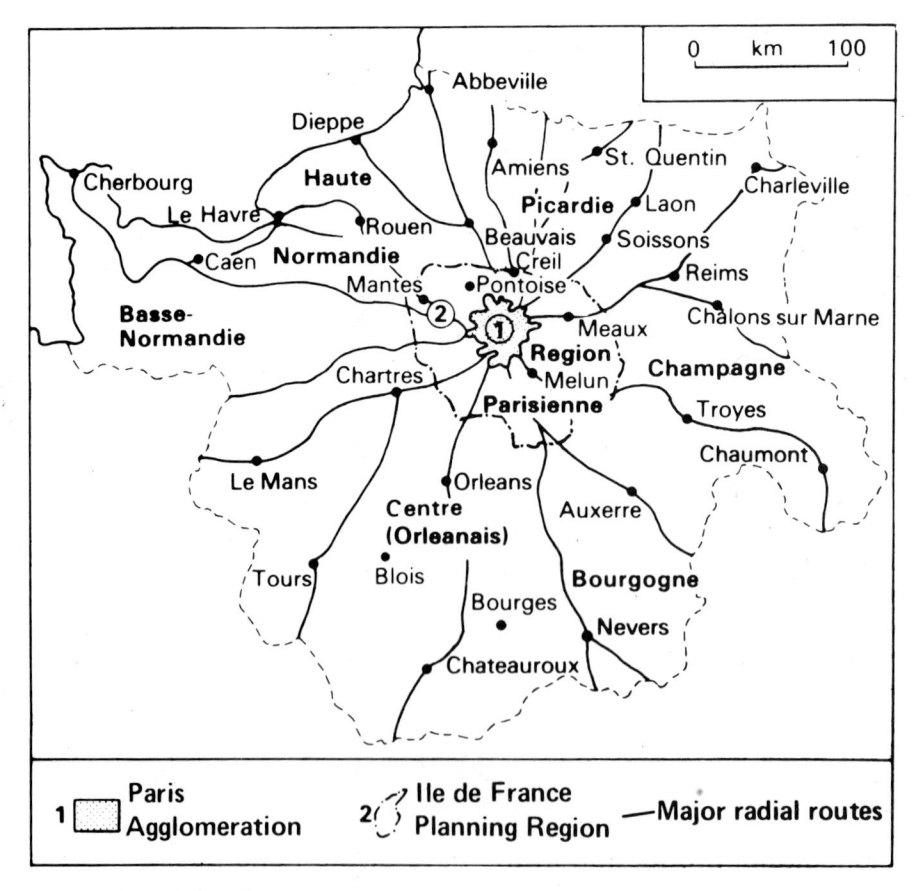

Figure 19.1 The Paris Basin Planning Unit

Basin planning unit (fig. 19.1)? Alternatively, there is the traditional view of the Paris Basin in which the convergence of the Seine and its tributaries is the centre of a key agricultural lowland extending from the Loire to Normandy and from the Champagne scarplands to Picardie (fig.19.2). This problem of definition may be examined by means of two themes which stress the economic value of the area in different ways: the scale and variety of agricultural production on the fertile 'pays' of the Paris Basin; the power of the city in attracting population and resources, and the planning problems which this has involved.

Agricultural production in the pays of the Paris Basin

The land-use diversity of the Paris Basin depends essentially upon geological characteristics and their effect upon landscape (fig. 19.2). The four major elements are:

1. The Tertiary Basin of the Ile de France. The central area of limestones, clays and sandstones overlain by superficial deposits of limon and alluvium and divided into separate 'pays' by the tributaries of the Seine.

2. The scarplands of Champagne and Bourgogne to the east and south-east.

3. The chalk plateau of Artois and Picardie to the north.

4. Haute Normandie. The chalk plateau and lowlands of the Lower Seine.

The Ile de France Tertiary Basin

One of the most distinctive 'pays' is the **limestone plateau of Beauce**, which has a thick covering of porous limon and a lower water-table than most other parts of northern France and, most importantly, is almost entirely flat. The outstanding feature is the almost complete utilisation of the land for agriculture; in many areas cereals occupy eighty to ninety per cent of the land. Beauce is the foremost wheat-growing area of France, with barley and maize as other significant crops. Maize in particular has grown in popularity because of the ease of mechanised harvesting techniques. The uniform relief has helped efficiency and mechanisation, and farms are large with consolidated holdings and large expanses of fields in a prairie-type landscape. Settlement is concentrated in large villages, and there is nucleation around wet points because of the relative lack of surface water.

A significant contrast is provided by **le Pays de Brie** which lies between the rivers Seine and Marne, south-east of Paris. Here the limestone contains bands of clay overlying impervious marls, there is abundant surface water, and the water table is much higher than in Beauce. In former times, Brie was forested and marshy, and even now there is abundant woodland and a more varied land-use. Agriculture has a more mixed character, cereals comprising often one third, and permanent pasture another third of the agricultural land.

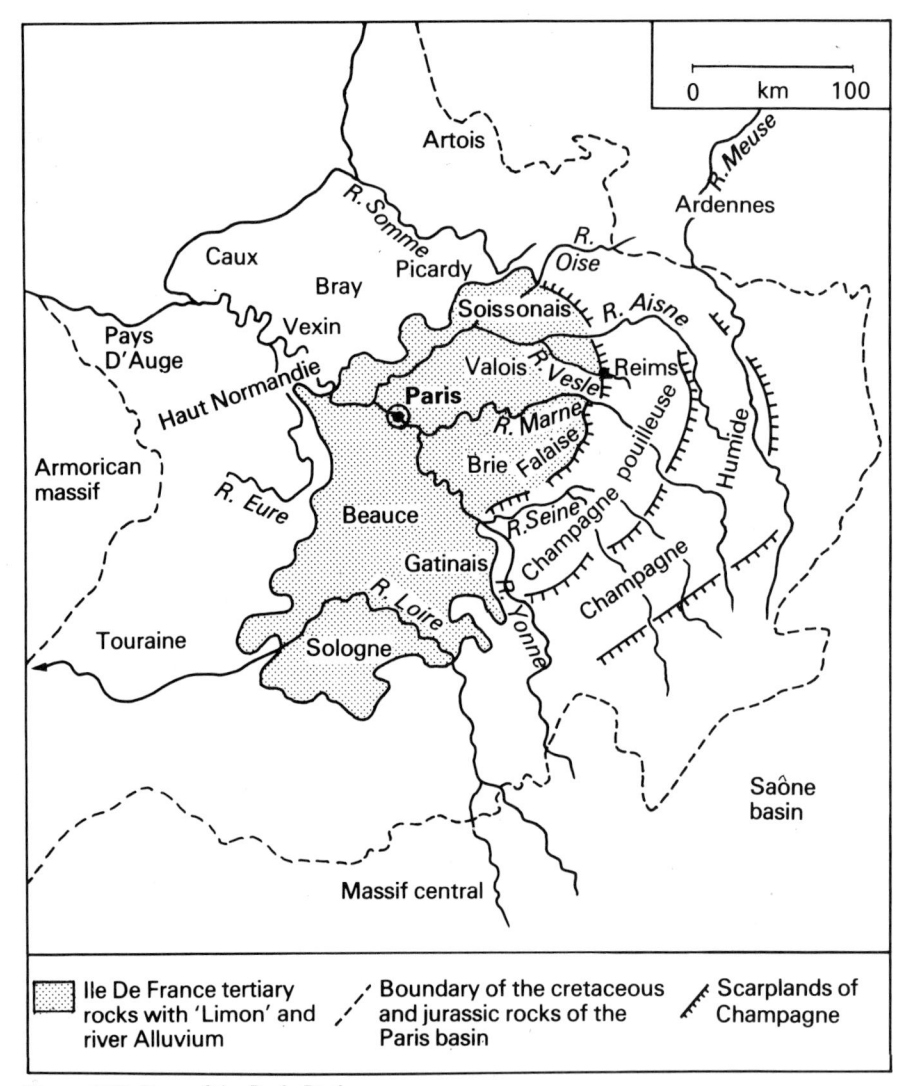

Figure 19.2 Pays of the Paris Basin

Fodder crops are important, with rotation pasture and sugar beet cultivated as cash crops and fodder, as cattle and sheep provide much of the income. Brie cheese and butter are two significant products. The smaller average size of farm and greater dispersal of settlement is more typical of dairy-farming country. Hamlets and isolated farms are common. The contrasts between these two 'pays' of the Ile de France are almost absolute.

The Champagne Scarplands

To the east and south-east of Brie lie the succession of chalk scarps and clay vales which are formed by the south-east facing escarpments of the Falaise, the 'Champagne Pouilleuse' and the 'Champagne Humide'. The variety of landscape is compounded by the river gaps of the Seine, Yonne, Aube, Marne, Vesle and Aisne which cut large embayments through the scarps.

The tertiary scarp, known as the Falaise de L'Ile de France, stretches from the Oise to the Seine but is best known in the region of Reims and Epernay for the vineyards of the Champagne wine district. There is a conjunction of physical factors. The chalk scarp has a generous covering of loam which promotes both drainage and aeration; the marginal climatic conditions are modified by the south-east facing scarp slope which gives maximum insolation and which provides frost drainage; chalk bedrock reflects light on to the plants and allows warmth to penetrate the soil. The prosperity of viticulture in these marginal climatic conditions, however, is largely due to human factors. The medieval trade fairs, the commercial acumen of the Bishops of Reims and Châlons-sur-Marne, expertise in blending, and specialisation in sparkling wines, has given the wines of Champagne an international reputation. There is also considerable capital needed to sustain the blending processes, manufacturing, storage, and maturing. This was provided originally by the Benedictine Abbey of Hautvilliers, and now by the 'Maisons de Champagne', an association of manufacturing firms, including Pommery, Heidseck, Clicquot, Bollinger, in and around Reims and Epernay. The marginal nature of the area climatically, resulting in many poor years, favours the large producer who can carry large stocks. This tends to maintain the system whereby eighty per cent of champagne is produced by the four large companies. The actual farm-holdings are, however, small. There are over sixteen thousand smallholding vine growers, and about ninety per cent have farms of two hectares and below. The 'Maisons de Champagne' buy most of their grapes from these small tenant farmers.

La Champagne Pouilleuse (dry Champagne) succeeds the tertiary scarp, and is a landscape of thin chalk without limon, with consequently little surface vegetation. It carries a poor quality grassland with outcrops of chalk and traditionally was devoted to sheep rearing, with some cereals and large areas of fallow. The area had a history of depopulation, a low population density, and a general air of sterility, with long distances between villages. It was an almost empty land. In the last fifty years this area, under new attitudes and values, has become a valuable agricultural resource, as it provided a large under-used reserve of land for food production. Large-scale remembrement and a uniform open landscape has created large farms ideal for mechanised grain farming, and extensive use of fertilisers gives very profitable farming. There is a mixed farming system, with cereals, sugar beet, potatoes, fodder and root crops, and lucerne. Cattle are kept, giving dairy products for the Paris market. The landscape, still with a deserted appearance, now looks more like a prairie scene, with infrequent villages, and arable fields stretching into the distance. The N44 road between Châlons-sur-Marne and Reims gives a typical view of this landscape.

The dry chalk country, terminated by a second scarpline, is followed by the Champagne Humide. This is lower cretaceous sand and clay, akin to the sub-scarp Wealden and Gault country of Southern England, and comparable in landscape and economy. Woodland and pasture, interspersed with orchards, produce a dairy farming economy, of which the most typical is the Marne valley from Vitry-le Francois to Chaumont. A 'bocage' landscape, well watered and with many villages and hamlets and isolated farms, is the norm.

The chalk plateau of Artois and Picardie

North of the river Oise, the chalk plain of Picardie, with its thick cover of limon, rises gradually across the Somme Valley to the outer rim of the chalk plateau in Artois. Arable farming with cereals, sugar beet, potatoes and fodder crops is the rule. Farms of over fifty hectares, highly mechanised, rationalised by remembrement, and open fields stretching to the horizon, underline the picture of large-scale agriculture. There is considerable variation with several 'pays': the Bas Boulonnais near Boulogne has jurassic clays providing good pasture for cattle and horse rearing; between Arras and Cambrai is an intensive rotation system with the highest crop yields of anywhere in Northern France; the Somme valley, often marshy, is devoted to pasture, with market gardening around Amiens.

Haute Normandie

This comprises the western side of the chalk rim of the Paris Basin, the undulating limon-covered plateau being the dominant feature, with cereal farming on medium to large, efficient farms, especially in the pays de Caux. Further east, in Vexin, there are greater concentrations of sugar beet, barley for brewing, and other industrial crops. Near Caen is the pays d'Auge, which is bocage country, with dairy products, including the famous Livarot and Camembert cheeses. The clays and marls in the anticline of Bray and in the valley of the Eure produce dairy farming and orchards. Cider apples and Calvados are specialities.

Description of these 'pays' is intended to illustrate the traditional picture of the Paris Basin as the 'Granary of Europe', with its vast tracts of fertile sediments overlain by limon. It is the most advanced agricultural province in France, favoured by the physical environment and also by proximity to the industrial and urban populations of Paris and Northern France. This has stimulated food production. Movement off the land towards the cities has also stimulated a greater level of mechanisation, making the area one of the most efficient farming regions in France. Thus the Paris Basin has a much lower density of population and larger farms than other parts of France, providing a marked contrast to the French peasant of the 'Midi' on his smaller and fragmented farm holding. There is a concentration of croplands, particularly cereals, in what is known as the 'intensive grain core' extending from Beauce to Artois. At the same time the widely differing 'pays' landscapes have provided a great range of farm products. The Paris Basin has both scale and variety.

Regional Centre	Population 1986	Regional Centre	Population 1986
Rouen	388 000	Cherbourg	79 000
Le Havre	266 000	Bourges	76 000
Tours	245 000	St. Quentin	70 000
Orleans	209 000	Charleville	64 000
Reims	197 000	Chartres	59 000
Le Mans	192 000	Châlons-sur-Marne	56 000
Caen	181 000	Châteauroux	55 000
Amiens	154 000	Nevers	55 000
Troyes	130 000	Beauvais	50 000

Figure 19.3 Population of regional centres in the Paris Basin

The growth of the Paris agglomeration

Figure 19.1 illustrates the position of Paris as the centre of French communications, and the relatively small size of cities in close proximity to Paris (fig. 19.3). Paris has always been very much the undisputed capital city of France, the administrative, industrial and cultural centre of the country. The kings of France had traditionally centralised power in Paris, and this was underlined by Napoleon, who destroyed the power of the great historic provinces such as Lorraine and Burgundy by creating the smaller départments, and by focussing the system of 'routes nationales' upon Paris. There is a strong historic tradition of central government.

The economic consequences are shown by the employment characteristics. The Ile de France had 10.1 million people in 1982 (19 per cent of the French population) and employed 21 per cent of the French labour force, with over four million workers. Its predominance in specialised functions and skilled labour is even more significant; it has over 65 per cent of French research workers; 48 per cent of the total qualified engineers; and 40 per cent of all professional and managerial grades. It produced 30 per cent of the output of the aircraft industry, 40 per cent of French cars, and about 50 per cent of all precision, electronic, radio and television products. The service sector is even more significant, the city having a virtual monopoly of banking, insurance and company head offices. The large consumer market, pool of skilled labour, and commercial and financial institutions have given Paris technologically advanced industries with a high growth potential, and income levels well above the national average. This economic strength is matched by an equivalent concentration of educational provision, artistic and cultural activity, which gives the city a power and relative importance in French life which is unequalled anywhere else in Europe.

At the last census (1982) there were 8.2 million people in the Paris

310   THE NEW EUROPE

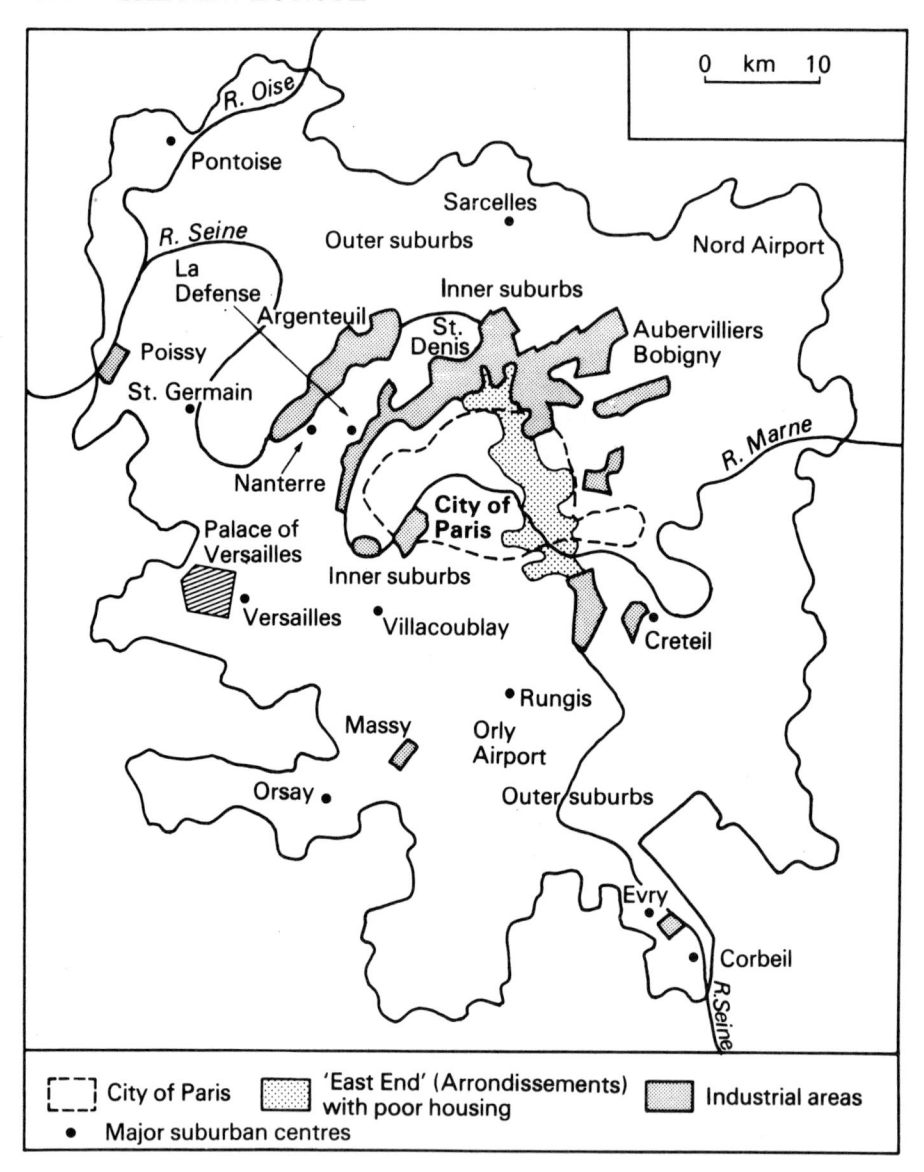

Figure 19.4 The Paris agglomeration

agglomeration, but the component parts of the city and its region need to be examined in more detail.

The Paris agglomeration consists of the city and its suburbs. The administrative, cultural, financial, tourist, retail and commercial centre, together with railway termini, mixed industrial and traditional residential districts such as Montmartre, is surrounded by the framework of the 1798 city wall, along which are famous gates such as the Porte d'Italie and the Porte d'Orleans. Along the approximate line of the city wall runs the Boulevard Peripherique, the inner ring road. In the inner suburbs lie most of the industrial zones (fig.

19.4) including St. Denis and Aubervilliers-Bobigny in the north, and the banks of the Seine around Argenteuil, which together constitute the northern arc of nineteenth century industrialisation.

The two major inner city problems are congestion and over-crowding. The transport system has to cope with three million daily commuters, of which thirty five per cent travel by private car, for whom the main problem is parking. This seems strange when considering the wide boulevards of Haussmann (1853-70) but these have to cope with dense traffic flows, and kerbside parking is therefore highly restricted. Some fifteen per cent of the commuters travel short distances by foot, and about fifty per cent rely upon public transport. The Metro, the underground system, is not as extensive as that of London, and only goes as far as the inner suburban ring, at which point commuters travelling to the suburbs have to transfer to the municipal bus service. Housing in inner Paris is badly overcrowded. The eleventh arrondissement is one of the worst examples, with a substantial proportion of its houses without basic facilities and in a very decayed condition. The problem is worst in the east end arc of the city, from Montmartre in the north through Temple, Le Marais, Popincourt and Bastille to the Gare De Lyons. Industrial decline in the older inner city areas has created considerable economic and social problems. The problem is one of redevelopment of the inner areas maintaining adequate residential employment and recreation facilities, and adequate road systems.

The Basilica of Sacre Coeur, Montmartre, Paris, one of the major tourist attractions of this historic city

The outer suburbs (banlieu) constitute the lower density residential zone, but they also include high density 'Grands Ensembles' such as Sarcelles (fig. 19.5), decentralised industry in the new estates such as Poissy, Massy and Creteil, and the development of public service areas such as the Rungis Food Market, Orly airport and Paris Nord airport. Expansion along the major radial routeways and rapid population growth has incorporated many previously independent settlements such as Versailles, Pontoise, St. Germain and Evry-Corbeil, which are now outer suburbs of the city.

The Ile de France Planning Region

The Ile de France Planning Region (fig. 19.6) includes a number of départements adjacent to Paris which are closely linked to the economic life of the city. There are still separate, but fast-expanding, medium-sized towns surrounded by agricultural land and large areas of woodland, forest, and protected areas, particularly the Forest of Fontainebleau to the south, Rambouillet on the south-west and Chantilly to the north. The area acts as an open lung for the city, a safety valve, where expansion will be directed along carefully planned lines. Figure 19.5 shows the remarkable growth of these départements, Seine et Marne, Yvelines, Essonne and Val d'Oise. Towns in this commuter zone such as Mantes, Creil, Lagny, Etampes, Meaux and Melun are amongst the fastest growing places in France.

The Paris Basin Planning Unit

This is based principally upon the historic provinces of Champagne, Picardie, Normandy and Orleans (fig. 19.1). From the nineteenth century onwards the effect of Paris has been felt in an adverse way in the small farming communities and villages of these provinces. The drainage of manpower and resources to the capital from what has been termed by J F Gravier as 'le desert Francais', has continued to the present day. The urban centres in the Paris Basin such as Reims and Orleans are medium-sized cities and towns which act as natural centres for the agricultural 'pays', but they are considerably smaller than would normally be expected. They have existed under the shadow of Paris for a very long period. There is therefore, a lack of equilibrium within the Paris Basin Planning Unit between the city of Paris and the major regional centres.

Population change 1968–82

The Paris region illustrates the complex nature of population movement during the last forty years. Until 1968 the rapid growth of the city was an integral part of the remarkable demographic and economic growth of France and, indeed, the rest of Western Europe. Urbanisation was the dominant force and by 1961 when the Paris planning strategy was begun, the city was acting as a magnet and was growing at an alarming rate.

Since 1968 however, Paris, and France as a whole, appears to be conforming to the process of counter-urbanisation, involving decentralisation from

The Arc de Triomphe, Paris, viewed from the Champs Elysees

cities to suburbs and higher growth rates lower down the urban hierarchy. The net population loss in the city of Paris since 1968 (fig. 19.5) and in the agglomeration since 1975, is matched by dynamic growth in the four départements in the outer suburban ring, of which Essonne achieves the highest growth rates. Population change is paralleled by the decentralisation of employment from the core to the outer suburbs and adjacent rural areas.

Planning for Paris and its Region

The long-term strategy for planning was begun by the PADOG proposals in 1960, followed by the Schema Directeur in 1965. This has been an extremely flexible plan which has been adapted to take into account changing economic and demographic conditions in the city region. Revised versions were introduced in 1969 and 1975. The key feature of the Schema has been the recognition of Paris as a metropolitan world city and that limitation of its growth would be counterproductive. It therefore set out to regulate rather than contain growth, and accepted that the population of the Paris region could grow to fourteen million by AD 2000. This estimate was subsequently revised downwards in the light of the 1973 recession and the slowing down of population growth. The principal aims were the reinvigoration of the central areas of the city, the improvement of the transport infrastructure, the creation of new service centres in the suburbs, planned growth in the lower Seine valley, and balanced development in the Paris Basin as a whole with the strengthening of regional centres such as Orleans and Reims. Key aspects of the plan are as follows:

(a) The 'La Defense' redevelopment scheme creates a new inner suburban node with commercial, cultural, administrative and public buildings, adjacent to the University of Nanterre. Its purpose is to establish a growth centre to the west of the existing central area of Paris. There are

314   THE NEW EUROPE

Zone	Population total (1000) 1968	1975	1982	Percentage annual change 1968–75	1975–82
Paris city (1)	2591	2290	2162	−1.69	−0.8
Inner departments (2)	3823	4129	4119	+0.4	−0.3
Paris agglomeration	8197	8424	8247	+0.4	−0.3
Seine et Marne	604	706	819	+2.4	+2.3
Val d'Oise	693	848	925	+3.2	+1.3
Yvelines	853	978	1074	+2.1	+1.4
Essonne	674	928	993	+5.4	+1.0
Outer Suburban ring (3)	2824	3460	3811	+3.2	+1.5
Ile de France total (1) (2) (3)	9238	9879	10092	+1.0	+0.3

Figure 19.5 Population change in the Paris Region (Ile de France) 1968-82

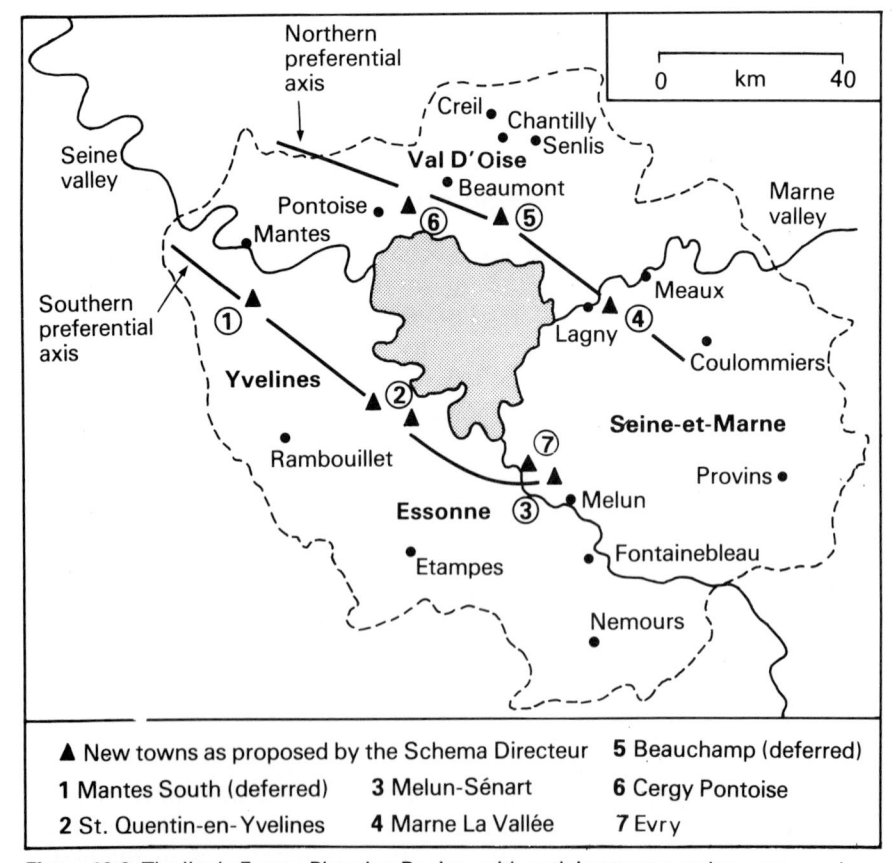

Figure 19.6 The Ile de France Planning Region, with outlying towns and new towns along the preferential axes

similar projects in Villacoublay, Rungis and Creteil to the south, and St. Denis and Bobigny to the north (fig. 19.4). The strategy is to reinvigorate the inner suburban zones. Urban renewal has also taken place in the historic city centre, in Les Halles and Montparnasse where public buildings have often replaced housing.

(b) In addition to the expanded Metro system, a new fast suburban railway 'The Reseau Express Regional' links a major east-west line with two north-south lines (connecting Orly with Paris Nord Airport). Three ring motorways, the Boulevard Peripherique, the Rocade De Banlieu at fifteen kilometres, and the Autoroute Interurbaine de Seine et Oise at twenty kilometres, link with the radiating motorways out of the city.

(c) Growth axes parallel to the river Seine were designed to preserve the open land alongside the river and to provide for up to eight new towns at a distance of up to 35 kilometres from the city centre. These were later reduced to five (fig. 19.6) as population growth slowed after 1968. Marne La Vallée, St. Quentin-en-Yvelines, Cergy-Pontoise, Melun-Sénart and Evry have been the principal foci of population growth in the region, with high levels of natural increase and in-migration levels of five per cent per annum.

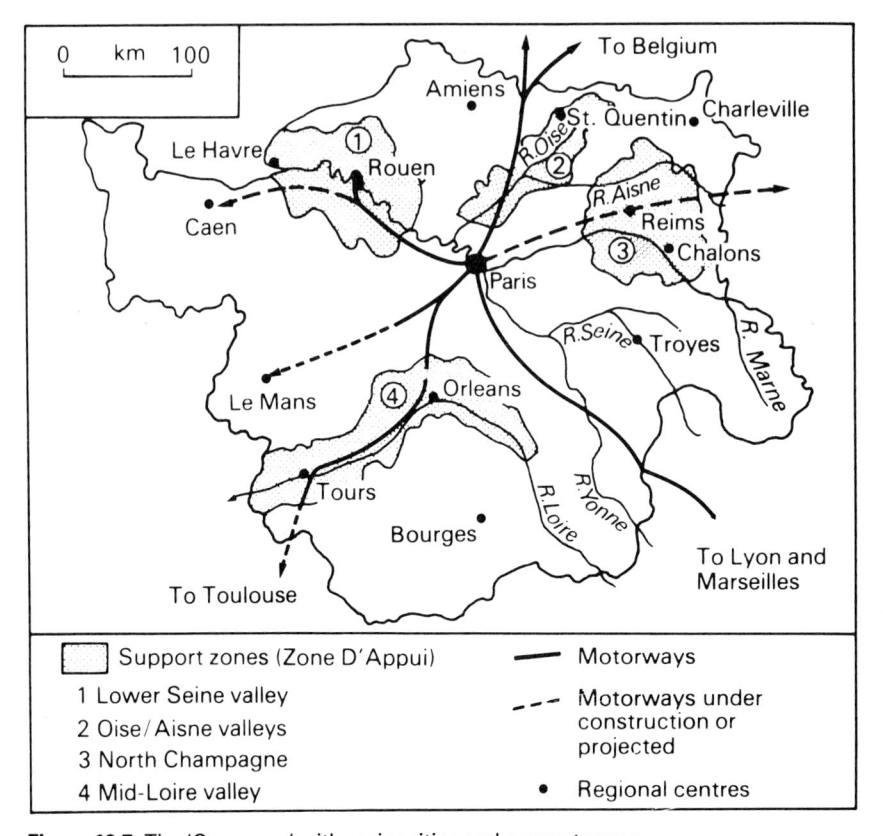

Figure 19.7 The 'Couronne' with major cities and support zones

(d) The 'Couronne' is a ring of historic towns and cities at up to 200 km distance from Paris (fig. 19.7). These cities such as Amiens, Chartres, and Reims lie on the major radial routeways and at key points in the agricultural 'pays'. Four areas were designated as 'support zones' (zones d'appui), major new revitalised centres, with other regional centres expanded where necessary (fig. 19.3).

(e) The most important single area with growth possibilities is the Lower Seine (Rouen-Le Havre) axis. This is a major line of movement via railways, waterways, pipelines and motorways. There are deep water access and port facilities at Le Havre including oil refineries and terminals, and the area also includes the two largest existing urban centres outside Paris itself - Le Harve (270 000), the second port of France, and Rouen (390 000). This axis is to be developed as a strategic growth corridor from Paris to the coast.

The growth strategy (d) and (e) for the outer areas has been revised and reduced since the mid 1970s. The Ile de France planning region has now stabilised with a population of 10.1 million, in line with the general economic and demographic slowdown in France during the 1980s. As a result the 'zones d'appui' have a lower priority as reception areas for decentralisation from the city of Paris.

20

Spain and Portugal

Introduction

For much of this century Spain and Portugal have been on the margins of West European life, but since 1960 both countries have undergone marked economic, social and political change. Spain, in particular, achieved a very rapid economic growth rate during the 1960s whereas Portugal lagged well behind, partly because of its smaller size and fewer resources, but in particular, because of her African colonial struggles which slowed progress condiserably. They now stand at an intermediate level of development between the industrialised North-West Europe and the emergent nations of the Third World (fig. 20.1).

There has also been major political change from the mid-1970s when both countries emerged from the dictatorships of Franco and Salazar, and entered the ranks of the Western European democracies. They have simultaneously, however, had to face major change, in particular the global economic recession in the 1970s and 1980s.

Spain

Spain, with 39 million people, is by far the largest economy, next to Italy, in the Mediterranean region. It is a country of great potential and with significant resources, but one in which industrialisation started late.

It is a country of generally adverse physical character, particularly the interior Meseta plateau, rugged mountain ranges and the semi-arid south. Alpine fold mountains include the Sierra Nevada, Cantabrians and Pyrenees. The Pyrenees reach 3470 metres (fig. 20.2). The Cantabrian mountains isolate the northern coasts of Spain from the semi-arid interior basin of Old Castile. The rugged landscape has given rise to great difficulties of communication, and major cultural differences between Castile, Catalonia, Andalusia, and in particular, the Basque region. Reference to the historic divisions between the 'Eight Spains' illustrates the cultural diversity. Historical and economic problems have a striking resemblance to Italy's Mezzogiorno.

The semi-developed nature of Spain is also illustrated by the contrasts between agriculture and industry. Agriculture involves sixteen per cent of the labour force and is characterised by an absentee landlord system with large

318   THE NEW EUROPE

	Spain	Portugal
Area (sq km)	505 000	92 000
Population (millions)	38.5	10.2
Population density per sq km (Eur 12 = 143)	76	110
Gross Domestic Product index per capita (Eur 12 = 100)	72.1	52.3
Percentage of workforce in agriculture	16.1	21.9
Percentage of workforce in industry	32.1	34.1
Percentage of workforce in services	51.8	44.0
Car production (1000)	1 028	nil
Number of passenger cars per 100 inhabitants (Eur 12 = 35)	23	16
Coal production (million tonnes)	15.9	0.2
Steel production (million tonnes)	14.2	0.6
Energy consumption per head (Eur 12 = 3232 toe)	1 847	1 091
Number of telephones per 100 inhabitants (Eur 12 = 48)	35	17

Figure 20.1 An economic profile, 1986: Spain and Portugal

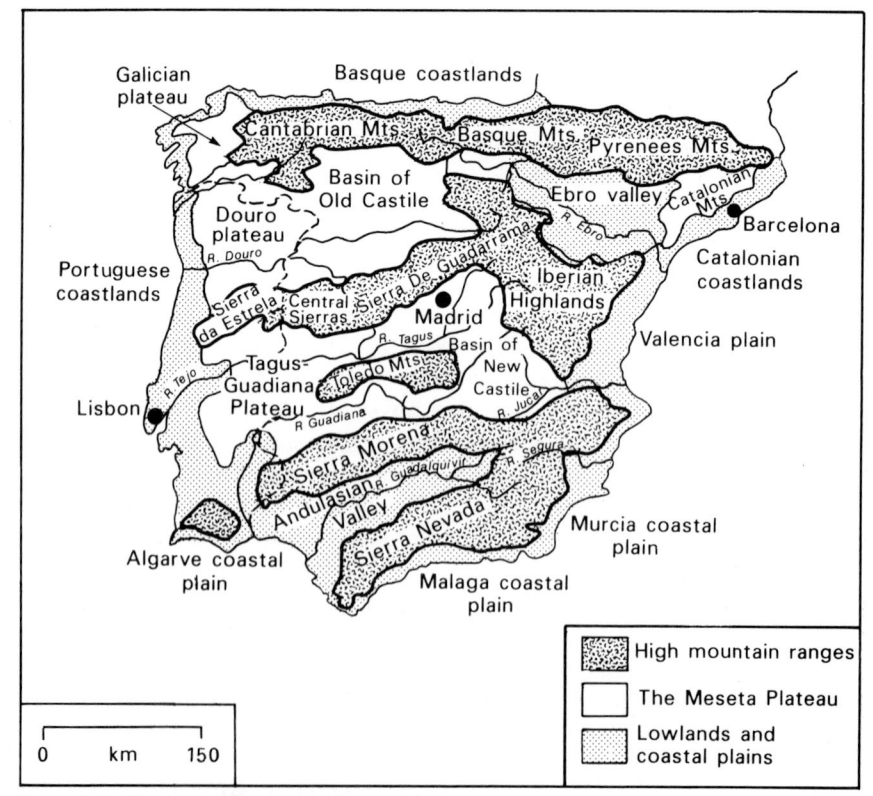

Figure 20.2 Spain and Portugal: physical features

feudal-style estates, fragmented farms, many of which are below two hectares in size, and low productivity which is generally fifty per cent below the European Community average. On the other hand, Spain is also the tenth largest industrial country in the world. On the north coast are coal reserves and Spain produces sixteen million tonnes of coal per year. Bilbao mines high grade iron-ore and is the centre of a major steel-producing area. The Sierra Morena region in particular is a substantial producer of copper, lead, silver, zinc, manganese and mercury. Major rivers such as the Guadalquivir, Tagus and Guadiana are being harnessed for hydro-electric power. As well as irrigation control, and increased agricultural productivity, this is significant for industrial development. Spain now follows France and Italy in HEP production. There are major industries of textiles, steel, shipbuilding, engineering, and car assembly. Nearly half of the economy is based on tourism, and this last fact indicates the great imbalances in an otherwise potentially large and rich economy. Spain, although semi-developed, is thus experiencing substantial transformation from an agricultural into an industrial country. The change has been relatively rapid during the last two decades, for several reasons. These include the opening up of the country to western influence from American air bases and the enormous influx of capital from the tourist trade since the 1960s which has given Spain finance for development projects. Since General Franco's death and the return of constitutional monarchy, Spain has successfully pressed for European Community membership.

Portugal

Portugal has a long Atlantic coastline, with a narrow rugged hinterland rising eastwards towards the Spanish Meseta. Many of the Spanish mountain ridges such as Serra da Estrela (1990 metres) continue into Portugal (fig. 20.2). However, the relief is generally lower than in Spain with river basins including the Douro, Mondego and Tejo (Tagus). The lowland fringes of Portugal face westwards towards the Atlantic Ocean and the physical division from the Spanish plateau helps to explain the traditional political separation and different outlook of the two countries.

Portugal has a different climate from most of the Iberian peninsula and is dominated by maritime influences from the Atlantic Ocean. Most of the country has an equable, warm temperature climate with considerable rainfall, mild winters and hot summers. Natural vegetation reflects the abundance of rainfall. Portugal originally had a forest cover of evergreen oak, mediterranean pines, cork oak and sweet chestnut, and almost one third of the country is still wooded. Agriculture follows traditional lines, with small subsistence farms cultivating wheat, barley, maize and rearing cattle. Tree crops are widespread, including peaches, apples and olives. The special vineyards of the Douro valley are most important as is the Tejo valley around Lisbon, which produces forty per cent of Portuguese wines. The Algarve in the extreme south has semi-arid mediterranean type conditions.

Portugal was one of the first European colonial powers, but since the

Terraced vineyards of the Douro valley near Oporto

seventeenth century has suffered a long decline for several reasons. It is almost totally lacking in mineral resources, and hydro-electric power has been retarded by irregular river regimes, although there are now major HEP schemes on the Tejo and Douro rivers. Industrialisation is limited. The African colonial empire proved a costly drain on scarce resources during the long guerilla war in the 1960s and 1970s, after which the last two major colonies, Angola and Mozambique, became independent. There was a long period of dictatorship under which economic development was neglected, after which Portugal emerged with a democratic government in 1976. The single most significant source of income has been the revenue from the tourist industry in Lisbon-Estoril, and in the Algarve. Portuguese industry is composed largely of small businesses and productivity is low. Agriculture too is in no shape to face European competition. Portugal is semi-developed with the lowest income of any in the European Community, and stands to gain considerably from European Community membership and its structural and regional funds.

Tourism

Tourism is another very significant element in socio-economic change. Over forty million people visited Spain and nearly nine million Portugal in 1986. Spain became important during the 1960s for tourism with the development of the cheap holiday package and charter flights to airports such as Perpignan, Barcelona, Malaga and Palma. In addition the low cost of living and cheap

High-rise development at Benidorm on the Costa Blanca

accommodation during the early years were major consideration. However, most important of all for Northern European tourists was the hot sunshine. 70 per cent of the visitors are from West Germany, the United Kingdom, Scandinavia and the Benelux countries. The tourism involved is largely coastal, for few tourists visit the interior on any scale, although this is now increasing as the historic attractions of interior cities such as Seville and Cordoba became more well known and accessible.

The physical advantages of much of the Spanish Mediterranean coastline include backing mountains giving a sheltered aspect, attractive scenery and sheltered beaches, warm sea with up to ten hours of sunshine daily, rainless summers and extremely mild winters. The first area developed historically was the Costa Brava, the so-called rugged coast, which is closest to the French frontier, easily accessible by road and the initial area of interest during the late 1950s and early 1960s. It has a very attractive rocky coast with resorts such as San Feliu and Tossa da Mar. Further south is the Costa Dorada and around Alicante is the Costa Blanca. The Balearic Islands of Majorca, Minorca and Ibiza and the Canary Islands have also benefited on a large scale from air package tours. Most significant of all, however, has been the enormous influx of tourists to the Costa del Sol. This is the most southerly coastline in Spain and faces North Africa. Malaga Airport alone received over three million visitors in 1986 (fig. 20.3).

The physical effects of tourist development have transformed small fishing villages into a continuous ribbon of developments, with high-rise hotels and apartment blocks stretching along large sections of coastline. The rapid development has involved severe planning problems of water supply, building safety, congestion and so on. Marbella, Estepona, Benidorm and Torremolinos are some of the more well-known resorts. Of most dramatic economic importance has been the significance to the Spanish economy, where the income from tourism has been absolutely crucial. It was tourism that broke the vicious circle of Spanish underdevelopment and this has been the main financing force behind Spanish and Portuguese development. Tourism employs nearly eleven per cent of Spain's working population.

Although tourism has transformed many of the coastal economies, nevertheless it has also had negative effects. There has been a lack of physical planning resulting in high-rise development, anarchic coastal sprawl and considerable congestion and pollution. But most important of all, tourism has been predominantly a coastal phenomenon and the economic over-heating of the coasts is characterised by seasonality in employment. Not even the seasonal income from tourism is as high as might be expected because of the control of the large tour operators who dominate the marketing of Spain and Portugal in Northern Europe and who strongly influence prices and conditions. As a result even tourism along the 'Costas' illustrates a type of dependency status.

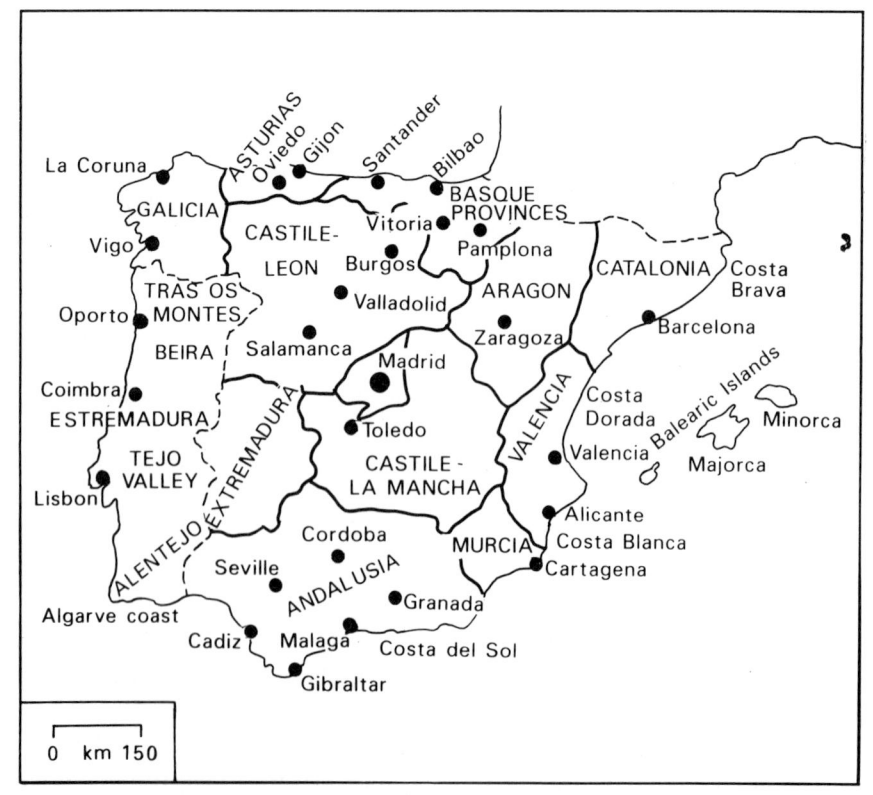

Figure 20.3 Spain and Portugal: autonomous regions and major cities

Regional Imbalance

Both countries have some of the most severe regional problems in Western Europe. Vast areas of Iberia have little or no interaction. There are marked concentrations of wealth in the national capitals, in commercial cities like Barcelona, industrial cities like Valencia, the Basque region in northern Spain, and the tourist coastlands. These areas are bourgeois Spain, middle class areas with an advanced economy of an industrial/service type. By contrast large tracts of underdeveloped regions include much of the north-west of Spain, Galicia, large parts of the central Plateaux, Estremadura, Castile and Southern plain of Andalusia. In Portugal, apart from the coastal areas around Lisbon and Oporto much of the interior is very considerably underdeveloped. In many of the rural areas of Spain and Portugal the essence of the problem is not so much the harshness of the natural environment, the lack of resources and the climate, as much as the human and institutional factors. In Andalusia there are latifundi, the large estates of absentee landlords based upon feudal land-ownership patterns, with low productivity, sporadic employment, poor living conditions, lack of infrastructure, low levels of educational ability and little vertical social mobility.

Spain: Core regions

Madrid lies almost in the geometrical centre of Spain, and became the national capital in 1561 (figs. 20.4 and 20.5). Its importance was underlined permanently as it became the focus of the national road and rail systems. Madrid is a political, administrative and cultural capital. Its population has doubled during the last twenty years. It has benefited from large-scale rural-urban migration providing it with a large labour force. Modern industrial development includes vehicles, electrical equipment, chemicals, food processing and consumer goods.

The northern coastlands include the Basque region of Navarre and Bilbao with high-grade iron ores, coal and manufacturing industry. The Basque provinces are the most densely populated in Spain and have a distinct language and culture. Bilbao and the surrounding towns have a large steel industry with shipbuilding, agricultural machinery and diesel engines. To the west are the ports of Santander and Gijon and the coal-mining centre of Oviedo.

Barcelona is the ancient cultural capital of Catalonia, and is a financial, industrial and commercial metropolis. Much of Spain's hydro-electric power has been developed in the Catalan mountains behind the city. It is an important port with a well established textile industry. Other industries include railway rolling stock, electrical equipment, diesel engines, vehicles, shipbuilding and food processing. Further south, at Valencia, is a modern integrated coastal steelworks, and a variety of industries including textiles, chemicals, shipbuilding, pottery and glass-making.

The Murcia coastlands are one of the richest agricultural areas in Spain. Irrigated terraced coastal lowlands called 'Huertas' cultivate cereals, tobacco, oranges, tomatoes, artichokes, cauliflowers and rice. On the terraced slopes away from the coast are vineyards and olive groves. The whole mediterranean coastline (from the Costa Del Sol to the Costa Brava) has become one of the most important tourist areas in Western Europe.

Spain: Underdeveloped regions (figs. 20.4 and 20.5)

Galicia, on the north-west Atlantic coast, has good ports such as Coruna and Vigo and a pleasant wooded landscape with pasture. There are mixed farms with cattle, pigs, maize and orchards, but holdings are small and incomes are low. There is tunny, anchovy and sardine fishing. However Galicia is one of the remotest parts of Spain. It has a large rural population, a lack of industry, unemployment, and their is a steady drift of population away from the region.

The interior of Spain is dominated by a semi-arid tableland, the Meseta, which has an average height of 500 metres. It is crossed and flanked by high mountain ranges. The basins of Old and New Castile are separated by the Sierra de Guadarrama. The basin of Old Castile has a steppe landscape, with vast estates (latifundi) belonging to the aristocracy and the church. Wheat and barley are grown in open fields with wasteful fallow and low yields. To the south of Madrid lies the basin of New Castile, part of which is known as La Mancha, or 'the desert'. With less than 500 mm of rainfall per year it has dry-farming, large estates and primitive farming methods. The Tagus-Guadiana plateau lies to the south-west with rugged sierras and interior basins. It is known as the Estremadura and large areas are given over to pastoral farming with Merino sheep, and goats.

In the extreme south is Andalusia (the Guadalquivir valley). It has mediterranean winter rainfall and dry dusty summers. Since the medieval period this has been a region of rural mis-management and absentee landlords. Farming is extensive with a poor peasantry, backward methods, extensive fallow, little mechanisation and low yields. Cultivation is based upon the classic Mediterranean combination of wheat, vines and olives, but it is also important for Seville oranges and tobacco and the town of Jerez is the chief centre for the bottling of sherry. There are historic cities like Seville, Granada, Cordoba and Cadiz. Andalusia has great potential for intensive agriculture based upon irrigation projects. The marshes at the mouth of the Guadalquivir have been reclaimed and cotton, rice and other cash crops and cultivated.

Portugal: coastal and interior regions

The basic problem is again the structure of the rural sector. In 1986 twenty two per cent of Portugal's labour force remained in agriculture. Productivity is low and there is considerable rural unemployment. The worst affected area is the interior. Much of it is mountainous with a sparse population. The Spanish frontier hinders communications and trade, and increases the remoteness of Portugal from the rest of Western Europe. Tras os Montes in the north, the Serra da Estrela in the centre and the Alentejo in the south are

The narrow alleyways of the historic centre of Lisbon

Portugal's most under-developed areas. Villages are few, towns are small, and agriculture is largely restricted to pastoral farming with rough grazing for sheep and goats, some cultivated stretches of wheat and vines, and cork oak woodlands. Rural depopulation is a problem, and there is considerable emigration towards Lisbon.

Economic life is centred on two coastal cities, Lisbon and Oporto (fig. 20.3). Lisbon lies at the mouth of the Tejo river and is the national capital and the commercial centre of the country. It is a major Atlantic passenger port, a NATO base and has shipbuilding and oil-refining industries. It is an important entrepôt port and exports cork, wine, sardines and petroleum products. Oporto on the Douro river in the north is Portugal's second city. It is the centre of the port wine trade, and has exported wine to England since the sixteenth century. There are a variety of industries including sardine canning, textiles, pottery, tobacco and electrical engineering.

Other important coastal lowland areas are the densely populated Minho valley with its provincial capital of Braga, and the ancient university of Coimbra in the Mondego valley. On the south coast, the Algarve, with a scenic coast, sandy beaches and hot dry summers has experienced rapid growth as an international tourist resort.

Population Migration

The transformation from predominantly rural-agrarian society into modern urban-industrial society has gone much further in Spain than in Portugal. The principal element of geographical importance is rural to urban migration, the flight from the land to the city.

In Portugal, the coastal areas around Lisbon, and to a lesser extent Oporto and the northern coasts, contrast with a thinly populated interior. In Spain the pattern is more complex with a coastal pattern of settlement dominated by the Basque region, Catalonia and the mediteranean coastal strip, plus the large population centre of Madrid and a hierarchy of other smaller cities such as Zaragoza, Valladolid and Burgos in the interior. The period from 1960 to 1975 saw large-scale migration from the impoverished interior provinces into the core areas of rapidly expanding economic opportunities. The socio-economic picture thus created was one of wide income differences between the well-to-do urban dwellers and the shanty towns on the edge of the city. On the other hand, rural exodus led to the decay of small towns and villages and a lack of infrastructure in the countryside. This pattern of uneven social and economic development showed widening differences between the core industrialised regions and backward agricultural regions: an industrialised

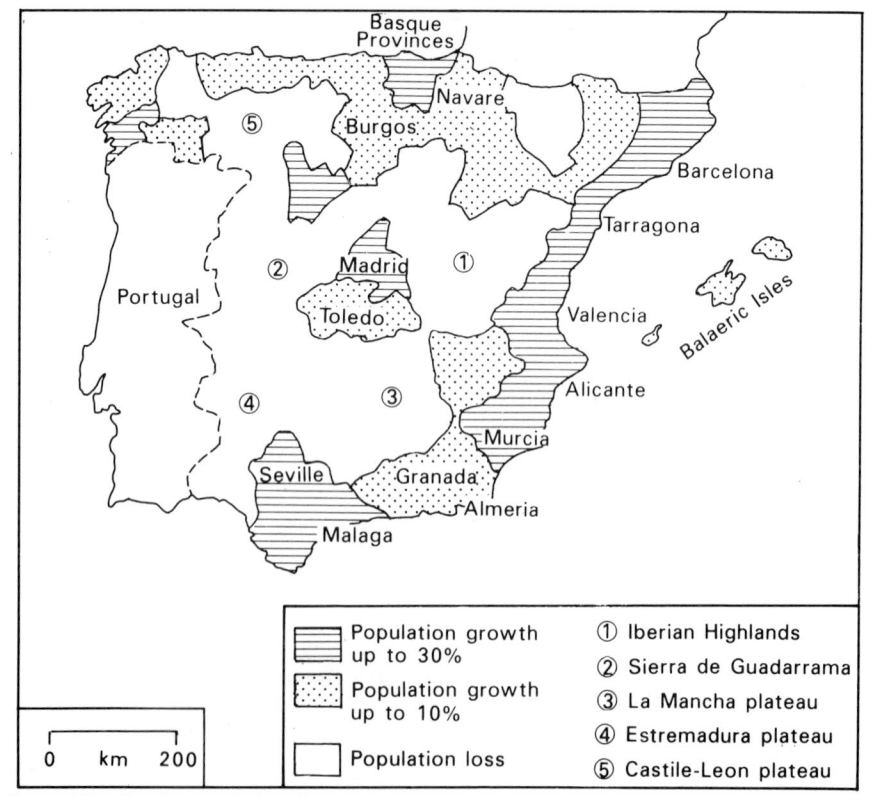

Figure 20.4 Spain population change 1971–81

and advanced Spain co-existing with a rural backward one. The present pattern of population density shows a marked contrast between the core areas in the north and east which contain the major cities, industries and services, and a peripheral south and west which is sparsely populated. (fig. 20.5) Urban primacy is most marked, with the city regions of Madrid and Barcelona accounting for twenty five per cent of Spain's total population.

Since 1975 the pattern of migration has become more complex, whilst the pace has been reduced. The economic recession and the problems of integration into the European Community economy have terminated economic growth in Spain and Portugal. Rates of depopulation of the interior have slackened. Madrid and other major cities, and the tourist coastal areas continue to grow, but a new feature has emerged. The northern provinces of Asturias and the Basque region (Vizcaya and Guipuzcoa) have been affected by economic recession, the decline of coal mining, steel and ship building, high unemployment and political problems, namely the terrorist tactics of ETA. Economic growth and population change in Spain is beginning to show similar patterns to other European Community countries further north.

Intergration into the European Community

There are two other key elements in the political and economic transformation of Spain and Portugal during the 1980s. One of these is the new

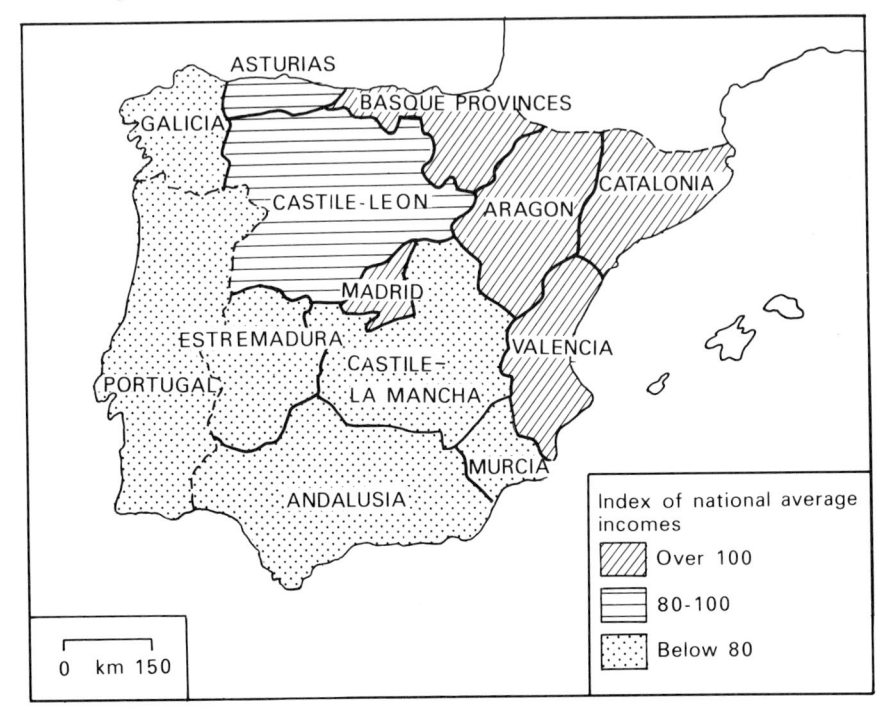

Figure 20.5 Spain and Portugal: regional incomes per head 1984–85

international division of labour, this is the challenge to the industrial supremacy of North-west Europe which is now being made by the newly industrialising countries of the Third World. This has had graver effects upon newly industrialised Mediterranean countries such as Spain, than on other countries in the European core. One reason for this is that the industrialisation which took place during the economic miracle of the 1960s and early 1970s in Spain was characterised by small family firms with low capitalisation and low-level technology, inadequate distribution networks specialising in low-cost goods often for local customers. On the other hand the giant state-enterprises created under the dictatorship of Franco are the older smoke-stack industries of steel, ship building, oil refining, petrochemicals and heavy enginerring which were characteristic of the national status industries of the 1960s. These now face European and worldwide intense competition.

The second step in the political transition of Spain and Portugal has been their accession to the European Community in January 1986. The overriding attraction of the Community for Spain and Portugal is the harmonisation of living standards based upon the European Community principle of convergence, and the alleviation of regional poverty. Per capita incomes in Spain and Portugal are respectively a little more than half and a quarter of the Community average. Spanish and Portuguese membership of the Community has widened regional disparities and substantially increased demands upon Community funds. The hopes of many regions are focused upon funds such as the ERDF and the European Social Fund.

There are bright prospects for agriculture in some regions including some of the poorest regions in Iberia. The provinces of Huelva, Malaga, Seville, Granada, Almeria and the Portuguese Algarve are now capitalising upon their high sunshine and temperature rates and are specialising in the intensive cultivation of a wide variety of fruits, vegetables and flowers. However, the outlook is particularly grim for many farming regions in the interior and the north of the peninsula with traditional mixed farming, agriculture and livestock. Spain's dairy farms will be affected by competition from Northern Europe, but Portuguese agriculture is least able to stand up to Community competition and it will require massive assistance to restructure. The demands upon the Community's farming and regional aid budgets will be substantial. Spain alone has brought into the EC an area of farmland, mountainous and marginal, equivalent to a third of the Community's existing farm land.

In the industrial sector similar heavy demands upon the EC's regional funds will be made as labour costs are rising and with the adoption of the Common External Tariff industries will be open to imports from Third World countries with which the EC has preferential access agreements. The impact of the competition will be particularly severe in Spain and some regions specialising in traditional manufacturing such as steel shipbuilding, textiles, clothing and footwear, will find it hard to compete unless they move up-market. The Spanish textile industry, in particular, is concentrated in Catalonia and has been traditionally accustomed to high tariff protection. Major reductions in steel-making capacity and the accompanying loss of jobs is occuring in the Bilbao and the Basque provinces.

21

The Integration Process: the Single European Act and progress towards the Single European Market

In Western Europe since 1945 there have been three distinct phases of development, all of which indicate a close relationship between geographical, economic and political factors. Immediately after 1945 there occurred the period of post-war reconstruction. This was a period of general economic recovery, and with the help of massive loans from the United States such as Marshall Aid, the basic economic infrastructure was rebuilt. This was carried out in the presence of a general feeling by both governments and population, that there was a need for political and social, as well as economic reconstruction. With the threat of Russian domination, there was a movement to create larger groups, more powerful and protective than the traditional nation state. In the sphere of defence, NATO was an illustration of this.

The 1950s and 1960s were a period of great prosperity. Economic growth was widespread, substantial and sustained, with many countries achieving an average yearly growth rate of over four per cent. Europe's raw materials, particularly oil, were cheap, and the terms of trade lay heavily in favour of industrialised countries. Agricultural production grew so that food supplies were more than adequate, leading eventually to the periodic food surpluses which have been such a feature of the contemporary scene. This widespread economic wealth and confidence was accompanied by rapid progress towards the formation of political institutions and policies which were to lead to a measure of integration in continental Western Europe. The European Economic Community of 'The Six', later to be enlarged, was formed by the fusion of three supranational institutions, the ECSC, Euratom and the European Economic Community. The Customs Union and the removal of trade barriers were accompanied by substantial levels of convergence and harmonisation of the economies of the member states. These include the regulation and integration of the coal and steel industries, increased intra-Community trade, the operation of the Common Agricultural Policy, the construction of the Euro-route system, regulations for fair competition, and investment aid from agencies such as the European Investment Bank.

During the 1970s the position changed dramatically. The increase in the price of oil and other commodities in 1973-74 completely altered the terms of trade, and were major causes of the inflation, loss of confidence, economic stagnation, and recession of the period from 1974-1982. Growth in all the European Community countries was severely curtailed, and their economies began to diverge quite seriously.

The supply and cost of energy became a major problem. The Community now has a wide range of energy sources, but relatively few of these are indigenous. There are a number of choices: to end its dependence upon imported oil from politically unstable sources; to enter a period of energy conservation; greater utilisation of home energy resources; research into new non-fossil sources of energy. The lack of a coherent energy policy probably reflects the very divergent positions, requirements and priorities of the member states.

Another serious problem is economic divergence both at a national and regional level. Not only has West Germany become the most economically powerful state in the Community, but on a European scale the central regions of the Community have grown at the expense of the periphery. In a sense this is the result of the free market policies of fair competition pursued during the early years of the European Community. In conditions of normal competition, the natural advantages of the Rhineland axis, with its dense population, economic resources and highlevels of mobility, would have emerged in any case. The success of the European Community in lowering tariff barriers, the ECSC in restructuring the coal and steel industry of the Heavy Industrial Triangle, the effect of the CAP in favouring the efficient food producers of the Paris Basin and the Netherlands, and the concentration of motorway linkages between Belgium, the Netherlands, north-eastern France and the Rhinelands, have all enhanced these advantages. The north-south divide in the UK is related to the greater accessibility of the south-east to the European core. It may well be that the economic effect of the Channel Tunnel will accentuate the existing advantages of the south-east region. The very success of the Common Market has been to create a natural economic core region which is in marked contrast to some of the peripheral regions such as Iberia, Greece, the Mezzogiorno, Ireland and south-western France.

Community policy-making has gradually changed to take account of this fact. Sectoral policies such as the ECSC and the CAP have increasingly been supplemented by policies and funds which have as their aim the strengthening of the peripheral regions. The proportion of EIB funds devoted to regional development schemes is over eighty per cent of the total. The European Social Fund devotes some three-quarters of its total aid quotas to the regions, and the ERDF is specifically designed for regional development and restructuring. The imbalance between the interdependent core areas and the peripheral regions must be corrected if further progress towards economic union is to be made.

Employment change has become the most serious problem affecting the Community. The recession has highlighted the new technological revolution with the rapidly increasing use of electronics, computers and robotisation in industrial production, and a corresponding reduction in levels of manpower. The changing international division of labour with the cheap labour low-cost competition from the industrialising Third World is associated with the structural decline of industries such as textiles, footwear, shipbuilding and steel. This combination of factors has led to the re-appearance in the industrialised world of unemployment on a large scale. Assistance to declining industries, retraining schemes and development grants for new technology, and to small specialised industries are all required on a large scale to reduce unemployment. The amount of Community funds going to agriculture is entirely out of proportion, and the Community's budget requires urgent restructuring to deal effectively with industrial and urban unemployment. The Social Policy may be one effective vehicle for this particular purpose as it has increasingly become an industrial restructuring and training policy.

Geographical extent and coherence has increased, with the addition in 1973 of the northern arc countries of the United Kingdom, Ireland and Denmark, and the southern arc countries of Greece in 1981, and Spain and Portugal in 1986. Territorial enlargement has not however led to greater economic coherence. The UK has special relationships with other parts of the world, and often takes different viewpoints on regional and agricultural issues. The three Mediterranean countries have major problems of integrating their economies inside the Community, they add to agricultural surpluses, and are major recipients of aid from the Community's resources. The core and periphery in geographical terms has led to 'discussion' of a politically and economically integrated core with an associated, more slowly integrating periphery, a two-tier Community.

A major structural improvement was the change in the composition of the European Parliament. The direct election first held in June 1979 gave the Community a new democratic base and political impetus. The parliamentarians have a more effective mandate from the population and are beginning to exercise greater control over the budget funds and economic policies, thus playing a larger part in the political development of the Community. The creation of the European Monetary System (EMS) during 1979 as a zone of monetary stability and cooperation was a major step forward in the consolidation of the Community, and an important stage in economic and monetary union.

The Community has made a significant change of direction from internal policies to external affairs. Political and diplomatic cooperation (POCO) have increased dramatically since the mid 1970s. The negotiations to construct a common fisheries policy involved considerable difficulties because of the divergent national requirements, particularly those of the maritime states of the Community. It reflects the fact that the European Community has 'de facto' jurisdiction over a large tract of sea which covers most of the continental shelf of Western Europe. The Common Fisheries Policy (CFP) successfully completed in January 1983, extends the principle of sovereignty beyond the traditional twelve miles to an 'economic zone' two hundred miles around the Community coastline. It corresponds to a Convention of the Law of the Sea. The Community speaks with one voice in some fields of external policy. The negotiations with the USSR, Canada and Norway over fishing rights, food aid, the energy conferences, GATT, the Conference on International Economic Cooperation are examples of the steps by which the European Community is emerging as a political entity on the international scene. The rapid growth in trade and aid to developing countries is reflected in the Lomé Convention (ACP Treaty). Trade agreements have also been concluded with many Mediterranean and North African countries, including Israel and the 'Maghred' countries of Algeria, Tunisia and Morocco. The Mediterranean is rapidly becoming part of the Community's economic sphere of interest. It is a major supplier of primary products and, lying as it does on the southern flank of the Community, is clearly an area of strategic interest to Western Europe (Chapter 7).

The targets set for European integration by the Treaty of Rome in 1958 had been only partially achieved by the early 1980s. The Community had expanded territorially to twelve members, which in itself brought major problems of adjustment. A whole series of internal problems included recession, regional imbalance, deindustrialisation, inflation and monetary instability, and environmental pollution. The market of 320 million people was only partially in place and needed much development. A period of internal consolidation was necessary, to change the CAP radically, to make the regional policy more effective and to develop policies on energy, research and development, aerospace, telecommunications, transport and the environment.

In a global context too, the superpowers now no longer have the dominating superiority in all aspects of life that they all possessed, and the European Community, a potential super-power in its own right, has substantial relationships with many other parts of the world. As the 1990s begin, political developments in Eastern Europe, with their potential social and economic implications, look set to continue. The emergence of Europe as a world power and the current economic disparities within the European Community may be significantly affected by developments in the Eastern European countries as the millennium draws to a close.

The painful economic adjustments of 1975–1986 were associated with a period of 'Eurosclerosis' during which few important decisions were taken and the Community was damaged by acrimonious discussions about the budget contributions of the UK and by the increasing problems of agricultural surpluses. The survival of the Community through this period seemed to focus thoughts, and to concentrate minds, upon possibilities for the future, and has formed the basis for a great psychological, economic and political 'leap forward'. The European Council, at Summit meetings in 1984 and 1985 drew up the **Single European Act**, which came into force in July 1987. Its principal objective was the creation by 1992 of the **Single European Market**.

The Single European Act

This came into force on 1 July 1987 and amended the Treaty of Rome in a number of ways. Most important of all, it extended the use of majority voting in the Council, as opposed to the old unanimity rule and veto, which was very restrictive. Secondly it enables the European Parliament, through a new 'cooperation procedure' with the Council, to play a more active part in decision-making. The work of the Council, Commission and Parliament is now much more clearly coordinated. The principal provisions of the Act in detail are as follows:-

1. **Institutional provisions**
    (a) Unanimity is replaced by qualified majority voting in the Council.
    (b) The European Council meets twice per year, bringing together the heads of government and the President of the Commission.
    (c) The Council acting on proposals from the Commission and in cooperation with the European Parliament issues directives for action by member states.

2. **Economic and Monetary Policy**
    (a) The Community adopts measures with the aim of progressively stabilising the internal market by December 1992.
    (b) The Commission takes into account the problems which certain member states at different levels of development, will have during the establishment of the single market, and may propose appropriate assistance, which nust be of a temporary nature.
    (c) The Council adopts provision for the harmonisation of legislation, including taxation and excise duties.
    (d) If member states feel it necessary to apply national provisions they must notify the Commission, who may take the matter to the European Court of Justice for a decision.
    (e) Member states should take account of the valuable experience of the EMS and the development of the ECU as part of the covergence of the economic and monetary policies.

3. **Social Policy**

Member states pay particular attention to improving the working environment and the health and safety of workers. The Court is empowered to adopt minimum requirements for implementation.

4. **Regional Policy (Economic and Social cohesion)**

The Community aims to reduce disparities between regions and takes action through the financial instruments and structural funds, including EAGGF (Guidance), ESF, EIB and ERDF. These will be amended and supplemented as necessary.

5. **Research and Technology**
    (a) The Community will strengthen the scientific and technological basis of European industry to encourage it to become more competitive at international level.
    (b) The internal market potential will be exploited fully by opening up

national public contracts, defining common standards, and removing legal and fiscal barriers.
  (c) The Community will implement cooperation in research, development and demonstration programmes and stimulate training and mobility of research workers.
6. **Environment**
The Community will preserve, protect and improve the quality of the environment, protect human health and ensure a prudent utilisation of natural resources.
7. **Cooperation in Foreign Policy**
  (a) The Community will formulate and implement a European foreign policy.
  (b) The President of the Commission initiates action and coordinates and represents the positions of the member states in relations with third countries. A secretariat prepares and implements the activities of European political cooperation.

1992: The Single Market

Despite the elimination of customs duties, tariffs and quota restrictions between the member states, the full common market which was the objective of the Treaty of Rome is not yet a reality. There remain a whole range of **physical**, **technical** and **fiscal** barriers, and restrictions upon **public purchases** and **service transactions** which must be removed. The free movement of goods for example, is impeded by technical barriers such as different product standards. A free market in services is blocked by national restrictions and the policy of national public purchasing distorts competition on a European scale.

The Commission White Paper, submitted to the Heads of Government at the Milan European Council in June 1985 outlined the programme to remove the remaining barriers to trade between the member states by 1992. The overall aim is that - 'the internal market shall comprise an area without internal frontiers in which the free movement of goods, persons, services and capital is ensured in accordance with the provision of the Treaty' (Article 13). There are some 300 measures which are likely to be implemented by this date, and their implementation may be much easier and faster because of the improvements to the Community decision-making process embodied in the Single European Act.

(a) **The removal of physical barriers**
  Although customs duties and quantitative restrictions in trade have been abolished, frontier delays for immigration administrative purposes impose further costs upon trade. The introduction of the single administrative document (SAD) in January 1988 for despatch and entry declarations on goods, has replaced over one hundred national trade forms, and is a considerable simplification of procedures. Changes in customs procedure such as immigration controls also need to be simplified. The Burgundy

coloured Euro-passport and green identification disc for drivers are two concrete proposals in the White Paper. Safeguards will still be required against terrorists, dangerous diseases and drugs, and common rules concerning extradition policies need to be formulated. Internal frontier posts are often used for making veterinary and plant health checks and these measures will have to depend upon mutual acceptance among member states of each other's arrangements. Goods transport hauliers will benefit from the end of time-consuming administrative delays at frontiers. It is estimated that the journey-time from London to Paris could be cut from 5½ hours to 3 hours, and from London to Amsterdam from 10 to 6 hours. These calculations, however, also take into account the likely effect of the Channel Tunnel.

(b) **The removal of technical barriers**

Technical regulations are rated by industrialists as the single most important category of trade barrier. More important, technical barriers are often greatest in hi-tech sectors, where market fragmentation places Europe at a competitive disadvantage compared to the United States and Japan.

National standards are a serious barrier to trade where there are different standards in each member state, or where member states do not recognise each other's arrangements for testing and certifying products. These practices add to costs and fragment the market. These barriers are being tackled in three ways:

(i) *the prevention of any new technical barriers*

Member states are required to notify the Commission in advance of any proposed new technical regulations or specifications and the Commission has power to intervene at this point.

(ii) *harmonised European standards*

Two standards bodies, CEN (European Standardisation Committee) and CENELEC (European Standardisation Committee for Electrical Products) are drawing up European standards for a wide range of products.

(iii) *testing and certification*

This involves the mutual recognition of test results and certificates. Consumer protection in the field of food products and pharmaceuticals is a good example of the value of harmonised standards.

(c) **The removal of fiscal barriers**

It is the existence of differing national rates of indirect taxation and excise duty which require the maintenance of expensive and time-consuming border controls.

Harmonisation of Community fiscal policies involves two Commission proposals:

(i) the introduction of two VAT bands, a standard rate of between fourteen and twenty per cent for most products, and a reduced rate of between four and nine per cent for socially sensitive items.

(ii) harmonisation of excise duties on cigarettes, tobacco, spirits, wine, beer and mineral oils.
(d) **Public purchasing**
Public procurement is still marked by the practice whereby governments and public bodies keep purchases and contracts within their own country. This activity accounts for up to fifteen per cent of the Community's GDP. Economic sectors such as defence, energy, water transport and telecommunications are included in this nationally reserved category. It is intended that the Supplies Directive (1 January 1989) will ensure that contracts and purchases by governments and public bodies will reflect European competition, not national indentity, thus reducing purchasing costs. In telecommunications, which are generally state-owned monopolies, full application of the competition policy by the introduction of open purchasing procedures, and greater emphasis on Europe-wide networks and services, will greatly enhance clarity and efficiency.
(e) **The freedom to engage in service transactions and activities in other Community countries**
Financial services are estimated to provide seven per cent of Community GDP, and the liberalisation of capital movements, insurance services, banking and investment involves the establishment of a Community-wide basic regulatory framework so that for example, insurance companies and building societies will be able to trade in any of the twelve Community members. Transport hauliers, at present restricted by quotas, will be able to deliver and return goods anywhere in the Community. Freedom of establishment for the professions involves the mutual recognition of training diplomas and degree qualifications, so that professionals such as doctors and lawyers may practice freely in other member states.

The Challenge of 1992

The European Community has come a long way, from the Treaty of Rome in 1958. With 35 per cent of world trade, it is the world's largest trading bloc. However, there is no European economy and no coherent European market. Instead, there are twelve countries, each with their own sovereignty, institutions, traditions and currencies. This heterogeneity may well be compounded in the future, as Austria is poised to make application for membership, and Norway is reconsidering its decision in 1973, not to join. Sweden and Switzerland may well consider membership at some stage.

The 1992 slogans which have been coined by politicians 'Europe without frontiers' and 'Europe open for business', genuinely represent a geographical picture of an enlarged European single domestic market of 320 million people, akin to that of the United States, but considerably larger, without physical, technical, fiscal or human barriers. Language differences will remain a constraint, and also the key spatial concept of accessibility, which continues to favour core areas as against peripheral ones.

Nevertheless, the economic gains will include better exploitation of the economies of scale, improved technical/economic/research efficiency, a stimulated flow of innovation, new processes and products, and rationalisation based upon the enhanced law of comparative advantage. It has been estimated by Commission research teams that the cost of 'non-Europe' (the damage inflicted by present market distortions) is about £135 billion per year. The overall effect of the single market is estimated to give potential for the raising of total GDP by between 2.5 and 6.5 per cent, with an extra two million jobs at least. Overall, it would seem possible to enhance the Community's annual growth rate by one per cent into the 1990s.

The political changes embodied in the Single European Act, and the economic integration envisaged in the 1992 Single Market, should together ensure continuing wealth creation and the ability to meet world-class economic competition. The programme implies interdependent converging economies with a common macro-economic policy and currencies moving freely across national borders. This further implies stable and competitive currencies within the EMS, the extended use of a parallel currency, the ECU, and ultimately the creation of a European Central Bank. The evolution of high level policy-making involves monetary union, a European currency, common fiscal policies and a common budget. Some visionaries see this as the first stages of the process in which the fragmented pattern of nation states in Western Europe is changed into the United States of Europe. This would be a state with resources, population and economic power superior to any other in the world.

Appendix

Key dates in European integration

September 19, 1946	Winston Churchill, in Zurich, urges Franco-German re-conciliation within 'a kind of United States of Europe'.
October 29, 1947	Creation of Benelux-economic union of Belgium, Luxembourg and the Netherlands.
April 18, 1951	The Treaty setting up the European Coal and Steel Community (ECSC) is signed in Paris.
February 10, 1953	ECSC common market for coal, iron-ore, and scrap is opened.
May 1, 1953	Opening of the ECSC common market for steel.
March 25, 1957	Signature of the Rome Treaties setting up the Common Market and Euratom.
January 1, 1959	First tariff reductions and quota enlargements in the Common Market.
July 18, 1961	The six Community countries issue Bonn Declaration aiming at political union.
November 8, 1961	Negotiations with Britain open in Brussels.
January 14, 1962	Community fixes basic features of Common Agricultural Policy.
January 14, 1963	President de Gaulle declares that Britain is not ready for Community membership. British negotiations broken off.
January 22, 1963	Franco-German Treaty of Cooperation signed in Paris.
June 1, 1964	Yaoundé Convention with 18 African countries (ex-colonies) as associated states, comes into operation.
March 31, 1965	Commission proposes that, as from 1st July 1967, all Community countries' import duties and levies be paid into Community budget and that powers of European Parliament be increased.
July 1, 1965	Council fails to reach agreement by deadline fixed on financing common farm policy; French boycott of Community Institutions begins seven-month crisis.
January, 1966	Crisis resolved by the Luxembourg Accords.
May 11, 1966	Council agrees that on 1st July, all tariffs on trade between the member states shall be removed and that the common external tariff shall come into effect, thus completing the Community's customs union.

May 10–11, 1967	Britain, Ireland and Denmark submit formal applications for membership of the Community.
July 1, 1967	Merger of Community executives–ECSC High Authority and EEC and Euratom Commissions.
November 27, 1967	General de Gaulle, in a press conference, objects to UK entry.
July 1, 1968	Customs union completed 18 months ahead of schedule; Common Agricultural Policy also complete.
July 18–19, 1968	Six adopt basic regulations for common transport policy.
July 28, 1968	Single market introduced for dairy and beef products.
July 29, 1968	Six decide to remove last remaining restrictions on free movement of workers.
December 10, 1968	Commission Vice-President Sicco Mansholt announces 'Agriculture 1980', Commissions radical ten-year plan to reform farming in the Six.
April 28, 1969	President de Gaulle resigns; succeeded in July by Georges Pompidou.
December 1–2, 1969	Heads of government of the Six, meeting at the Hague, agree to complete, enlarge and strengthen the Community.
December 19–22, 1969	Marathon Council session agrees on permanent arrangements for financing the common farm policy, providing the Community with its own resources from 1978 and strengthening the European Parliament's budgetary powers.
March 4, 1970	Commission submits a three-stage plan for full monetary and economic union by 1980.
June 30, 1970	Membership negotiations open in Luxembourg between the Six and Britain, Denmark, Ireland and Norway.
November 19, 1970	Foreign ministers of the Six meet for the first time in Munich to concert their views on foreign policy.
February 1, 1971	Common fisheries policy negotiations begin.
March 24, 1971	Six take first steps to carry out Mansholt Plan to modernise farming.
July 8, 1971	United Kingdom Government white paper recommending EEC entry issued upon successful completion of negotiations.
January 22, 1972	The Treaty of Accession signed by the United Kingdom, Ireland, Norway and Denmark (Norway later withdraws).
March, 1972	The 'Snake' (alignment of member states' currencies) introduced.

October 19, 1972	First Summit meeting of heads of Government in Paris. The summit meetings become part of the Community structure and are known as the 'European Council'.
January 1, 1973	The EEC is formally enlarged to nine members.
December, 1973	Copenhagen Summit meeting.
January, 1974	Oil crisis. Oil price quadrupled in three months.
December, 1974	Paris summit. Agreement to set up a Regional Development Fund.
March, 1975	Regional Development Fund in operation
March, 1975	European 'unit of account' (UA), to be used in the 'Snake' to relate each national currency to the others.
May, 1975	Lomé Convention signed by the EEC with 47 developing countries.
June 5, 1975	Referendum in the United Kingdom shows a two-thirds majority in favour of remaining a member of the EEC.
January, 1976	Tindemans Report on Economic and Political Union.
February, 1976	The Council of Ministers recommends that Greece be admitted to the Community over a phased period.
January, 1977	The 'new' EEC Commission takes office for a four-year period, headed by Mr Roy Jenkins of the United Kingdom.
July, 1977	Spain formally applied to become a member of the EEC.
April, 1978	The JET project (Nuclear Fusion) under the Euratom Treaty is established at Culham in Oxfordshire.
April, 1978	The EEC and the Peoples Republic of China sign first trade agreement.
July, 1978	The Heads of Governments at the European Council in Bremen, agree to set up the EMS (European Monetary System).
October, 1978	Formal negotiations opened for the accession of Portugal to the EEC.
February, 1979	Opening of formal negotiations with Spain for accession to the EEC.
March, 1979	EMS formally in operation.
June, 1979	First direct elections to the European Parliament.
October, 1979	The second ACP-EEC Convention signed at Lomé.
June, 1980	European Council and Western Economic Summit at Venice.
September, 1980	Co-operation agreement between the European Community and Brazil.

October, 1980	The Commission establishes production quotas and market regulations for the steel industry.
January 1, 1981	Greece becomes the tenth member of the European Community.
January 1, 1981	The second Lomé Convention comes into force.
March, 1981	Council adopts the resolution on a steel recovery policy.
June, 1981	European Investment Bank capital doubled.
March 25, 1982	Twenty-fifth anniversary of signing of the Treaties of Rome.
April, 1982	European Community support for the UK over the Falklands conflict, and the application by member states of sanctions and embargoes against Argentina.
July 6, 1982	The European Parliament passes a resolution giving guidelines for the reform of the Treaties and the achievement of European Union.
December, 1982	The European Council reaffirms its political commitment to the negotiations for the accession of Spain and Portugal.
January, 1983	Negotiations on the establishment of a Common Fisheries Policy (CFP) successfully completed.
June, 1983	European Council at Stuttgart: declaration on European Union.
March, 1983	European Council in Brussels: outline agreement on budgetary control, financing of the Community of twelve, and reform of the CAP.
June, 1983	Second elections to the European Parliament.
March, 1985	IMP's (the integrated Mediterranean programmes).
June, 1985	The Commission's White Paper on the single European Market.
July, 1985	The Commission's Green Paper on the agricultural situation in the Community.
December, 1985	The drawing up of the Single European Act at the Luxembourg European Council - the creation of a Single European market by the end of 1992.
January 1, 1986	The accession of Spain and Portugal to the European Community.
July, 1987	The Single European Act comes into force.
February, 1988	Summit Agreement where reform of the CAP, budget reform and the doubling of regional aid was agreed.
June, 1989	Third elections to the European Parliament.
1992	Single European Market comes into force.

Glossary

1. **CNABRL.** Compagnie nationale d'aménagement du Bas Rhône-Languedoc.
2. **European Currency Unit (ECU).** The monetary unit of account used in pricing agricultural commodities and in the Community budget. It is a 'basket unit', calculated according to the relative weighting of each Community currency. Formerly known as the EUA.
3. **FASASA.** Fonds d'adaptation sociale pour l'aménagement des structures agricoles.
4. **Federalism.** A more rapid approach to integration by which supranational political institutions are superimposed over national authorities.
5. **Footloose industry.** Manufacturing industry which is not based upon resource constraints such as coalfields, but which has the ability to choose a wide range of locations.
6. **Functionalism.** The step-by-step approach to integration, through agreements in economic sectors such as agriculture, tariffs, transport, etc.
7. **Geographical Inertia.** The tendency of older industrial regions to survive by the contraction and adaptation of old, heavy industries, and by the development of new light industry.
8. **Gross Domestic Product (GDP)** The total value of goods and services produced inside the country.
9. **Kennedy Round.** A major agreement in 1967 on tariff reductions amongst the industrial nations.
10. **Marshall Aid.** The aid and investment programme for the recovery of Europe provided by the United States after 1945 and named after its initiator, General Marshall.
11. **The 'Original Six' or 'EEC Six'.** Belgium, the Netherlands, Luxembourg, West Germany, France and Italy.
12. **PADOG.** Plan d'aménagement et d'organisation générale.
13. **Remembrement.** The French policy of rationalisation and enlargement of land-holdings to create a more efficient farming system.
14. **SAFER.** Société d'aménagement foncier et d'établissement rural.
15. **'Von Thunen Landscape'.** An idealised series of concentric agricultural zones around a city in which the most specialised farming is nearest the city market followed by less intensive farming types.
16. **DATAR** Délegation à l'aménagement du territoire et à l'action régionale.

References

Allen, K. and McClennan, M.C. *Regional Problems and Policies in Italy and France*. Allen and Unwin, 1970.
Bamford, C.G. and Robinson, H. *Geography of the EEC*. Macdonald and Evans, 1983.
Beckinsale, M. and R. *Southern Europe*. Hodder and Stoughton, 1979.
Blacksell, M. *A Political Geography of Post-War Europe*. Dawson, 1978
Broad, R. and Jarrett, R.H. *Community Europe Today*. O. Wolff, 1972.
Budd, S. *The EEC. A Guide to the Maze*. 2nd edn Kogan Page 1987.
Burke, L. *Greenheart Metropolis*. Macmillan, 1966.
Burtenshaw, D. *Saar-Lorraine*. Oxford University Press, 1976.
Burtenshaw, D. *Economic Geography of West Germany*. Macmillan, 1974.
Burtenshaw, D. *et al. The City in West Europe*. Wiley, 1981.
Cairncross, F. and McCrae, H. *The Second Great Crash*. Penguin, 1979.
Chapman, K. *People, Pattern and Process*. Arnold, 1982.
Charnley, A.H. *The EEC, A Study in Applied Economics*. Ginn, 1973.
Clark, C. *et al*. 'Industrial Location and Economic Potential in Western Europe'. *Regional Studies* 3, 1969.
Clout, H. *A Rural Policy for the EC?* Methuen, 1984.
Clout, H.D. 'French population growth 1962–68'. *Geography*, No. **56**. 1971.
Clout, H.D. *The Massif Central*. 2nd edn Oxford University Press, 1984.
Clout, H.D. *Regional Development in Western Europe*. 3rd edn Fulton, 1987.
Clout, H.D. *The Geography of Post-War France*. Pergamon, 1972.
Clout, H.D. *et al. Western Europe: Geographical Perspectives*, Longman, 1985.
Clout, H. *The Franco-Belgian Border Region*. Oxford University Press, 1979.
Clout, H. 'Nord coal miners prepare for 1983'. *Geographical Magazine*, March 1972.
Coffey, P. *The External Economic Relations of the EEC*. Macmillan, 1976.
Commission of the European Communities. *European Economy, 1992*. Belgium, 1988
Commission of the European Communities. *Making a Success of the Single Act*, Spokesmans Service, 1987.
Commission of the European Communities. *The Community Today*. Brussels-Luxembourg, 1979.
Commission of the European Communities. *The Origins and Growth of the European Community*. Luxembourg, 1986.
Commission of the European Communities. *Completing the Internal Market. COM (85) 310 Final*, June 1985.
Commission of the European Communities. *The European Community and the Mediterranean*, Luxembourg, 1985.
Commission of the European Communities. *Europe without Frontiers: Completing the Internal Market*, Luxembourg, 1987.
Council of the European Communities. *Single European Act*, Brussels, 1986.
De La Mahotiere, S. *Towards One Europe*. Pelican, 1970.
Department of Trade and Industry. *The Single Market: The Facts*. Central Office of Information, September 1988.
Economist. *Europe's Economies*. Economist, 1978.
Economist. *Offshore Ireland*. January 1975.
Economist. *Britain's Sunrise Strip*. January, 1982.
Economist. *The EEC in Transition*. Economist 1983.
Estall, R.C. and Buchanan, R.O. *Industrial Activity and Economic Geography*, 4th Edn, Hutchinson, 1983.
EEC Commission, *Report on the Regional Problems in the Enlarged Community. COM (73) 550*. Brussels, May 1973.
EEC Commission. *The Regions of Europe. COM (80) 816 Final*. 1981.
European Documentation. *The European Community's Industrial Strategy*. Luxembourg, 1983.
Eurostat. *Statistical Yearbook: Transport Communications. Tourism 1980*. Luxembourg, 1986.

Eurostat. *Review 1971-80*. Luxembourg, 1982.
Eurostat. *Europe in Figures: deadline 1992*. Statistical Office of the European Communities, Luxembourg, 1988.
Eurostat. *Energy and Industry*. Statistical Office of the European Communities, Luxembourg, 1988.
Eurostat. *Employment and Unemployment*. Statistical Office of the European Communities, Luxembourg, 1985.
Eurostat. *Yearbook of Regional Statistics*. Statistical Office of the European Communities, Luxembourg, 1984.
Eurostat. *Demographic Statistics*. Statistical Office of the European Communities, Luxembourg, 1988.
Eurostat. *Transport, Communication, Tourism (1970–1985)*. Statistical Office of the European Communities, Luxembourg, 1987.
Eurostat. *Agriculture: Statistical Yearbook*. Statistical Office of the European Communities, Luxembourg, 1988.
Eurostat. *Review 1976–1985*. Statistical Office of the European Communities, Luxembourg, 1987.
Eurostat. *Energy, 1985*. Statistical Office of the European Communities, Luxembourg, 1987.
Eurostat. *Basic Statistics*. (25th edition) Statistical Office of the European Communities, Luxembourg, 1987.
Evans, I. *Belgium and Luxembourg depend on Steel*. Geographical Magazine, September, 1982.
Evans, I. 'Europe's Steel Industry fights for survival in the 1980s.' *Geographical Magazine,* July 1982.
Fleming, D.K. 'Coastal steelworks in the Common Market countries.' *Geographical Review*, January 1967.
French Embassy Information Service. *Development of the Rhône*.
Gladwyn, Lord. *The European Idea*. Weidenfeld and Nicholson, 1979.
Hall, P. *Europe 2000*. Butterworth, 1976.
Hall, R. and Ogden, P. *Europe's Population in the 1970s and 1980s*. Q.M.C. University of London, 1983.
Hall, P. *The World Cities* 3rd edn, Weidenfeld and Nicholson, 1984.
Hall, P. and Hay, D. *Growth Centres in the European Urban System*. Heinemann, 1980.
Hellen, J.A. *North-Rhine Westphalia*. 2nd edn, Oxford University Press, 1984.
Hene, D. H. *Decision on Europe*. Jordan, 1970.
Heseltine, M. *The Challenge of Europe*. Weidenfeld, 1989.
Hill, B.E. *The CAP: Past, Present and Future*. Methuen, 1984.
Hodges, M. (ed.) *European Integration*. Weidenfeld and Nicholson, 1979.
House, J.W. *France*. Methuen, 1981.
Hudson, R. et al. *An Atlas of EEC Affairs*. Methuen, 1984.
Hull, A. et al. *Geographical Issues in Western Europe*. Longman, 1988.
Ilbery, B.W. *Western Europe: A Systematic Human Geography,* 2nd edn, Oxford, 1986.
Jordan, T.G. *The European Culture Area*. Harper and Row, 1973.
Keeble, D.E. *Core Periphery Disparities*. Geography, January, 1989.
Keeble, D.E. 'Models of Economic Development', in Chorley, R.J. and Haggett, P. *Models in Geography,* London, 1967.
King, R. 'From Six to Nine to Twelve'. Geographical Magazine April, 1979.
Kormoss, I.B.F. *Les Communautés Europeenes: Essai d'une Carte de Densité de Population*. Les Cahiers De Bruges, 1959.
Law, C. 'The British motor car assembly industry 1972-82'. *Geography*, January 1985.
Lee, R. and Ogden, P.E. *Economy and Society in the EEC*. Saxon House, 1976.
Mellor, R.E.H. *The Two Germanies*. Harper and Row. 1978.
Mellor, R.E.H. and Smith, A.E. *Europe: A Geographical Survey of the Continent*. Macmillan, 1979.
Monkhouse, F.J.A. *A Regional Geography of Western Europe,* Longman, 1974.
Mountjoy, A.B. 'Pressures and progress in southern Italy'. *Geographical Magazine,* November 1970.
Mountjoy, A.B. *The Mezzogiorno*. 2nd edn, Oxford University Press, 1982.
Mounfield, P.R. 'Nuclear Power'. *Geography*, October, 1985.
Myrdal, G. *Economic Theory and Underdeveloped Regions*. London, 1957.
Naylon, J. *Andalusia*. Oxford University Press, 1979.
Odell, P. 'Europe sits on its own energy.' *Geographical Magazine*, March 1974.

Odell, P. 'The Energy Economy of Western Europe.' *Geography*, January 1981.
Odell, P. *The Western European Energy Economy 1980–2000.* London, 1975.
Ogden, P.E. 'Counter-urbanisation in France'. *Geography*, January 1985.
Ogden, P.E. 'French Population Trends in the 1970s.' *Geography,* November 1981.
Open University Course Team. *The European Economic Community.* Open University Press, 1974.
   Unit 1–2 History and Institutions.
   Unit 3–4 National and International Impact
   Unit 5–6 Economics and Agriculture
   Unit 7–8 Work and Home
Oxford Regional Economic Atlas. *Western Europe,* Oxford University Press, 1971.
Parker, G. *The Countries of Community Europe.* Macmillan, 1979.
Parker, G. *The Logic of Unity.* 3rd edn., Longman, 1981.
Perroux, F. *Economic Policy for Development.* (translation by Livingstone, I). Penguin, 1971.
Pinder, D. *The Netherlands.* Dawson, 1982.
Postan, M. *An Economic History of Western Europe* 1945–64. Methuen, 1979.
Riley, R. and Ashworth, G.T. *Benelux.* Chatto and Windus, 1978.
Riley, R.C. *Belgium.* Dawson, 1976.
Scargill, D.I. 'Energy in France.' *Geography*, April 1973.
Scargill, D.I. *Economic Geography of France*, Macmillan, 1972.
Scargill, D.I. *Urban France.* Croom Helm, 1983.
Shackleton, M.R. *Europe.* 7th edn. Longman, 1974.
Sill, M. 'Coal in Western Europe'. *Geography,* January, 1984.
Spooner, D.J. 'Energy, 1973–83'. *Geography,* April, 1984.
Swann, D. *The Economics of the Common Market,* 4th edn. Penguin, 1980.
Thompson, I.B. *The Paris Basin,* Oxford University Press, 2nd edn. 1981.
Thompson, I.B. *Modern France.* Butterworth, 1970.
Times Review of Industry. *Times 1000.* 1985–86.
Tuppen, J.N. *France.* Dawson, 1980.
Tuppen, J.N. 'Fos - Europoort of the South?'. *Geography*, July 1975.
Vaughan, R. *Post-war Integration in Europe.* E. Arnold, 1976.
Wallace, H. *et al. Policy-Making in the European Communities.* Wiley, 1977.
Walker, D. *A Geography of Italy.* Methuen, 1967.
Warren, K. 'Iron and Steel'. *Geography,* April, 1984.
Warren, K. 'World Steel'. *Geography,* April, 1985.
White, P. *The West European City.* Longman. 1984.
Williams, A.M. *Southern Europe Transformed.* Harper and Row, 1984.
Williams, A.M. *Western European Economy.* Hutchinson, 1987.
Yuill, D. *et al. Regional Policy in the European Community.* Croom Helm, 1980.

Index

Abruzzi-Molise 277–80, 289
ACP countries 151
Agricultural labour force 130, 135
Agricultural regions 120–6, 250–3, 265–70, 294, 305–09
Air transport 297, 161–5
Alpine passes 264, 266, 281
Alps 123–4
Amsterdam 240–4
Andalusia 323, 324
Antwerp 227–9, 230, 238–9
Apennines 278
Apulia 278, 282–3, 289
Ardennes 227, 236–8
Artois 305, 308
Associated states 137, 151, 338
Athens-Piraeus 300–02
Automobile industry 99–105

Barcelona 108, 320, 327
Basic oxygen process 84
Basilicata 280, 288
Basque provinces 323, 328
Beauce 305, 308
Benelux Union 10
Bergstrasse 218, 223
Bilbao 28, 98, 318, 323
Bollenstreek 251
Bonn 178, 206, 213
Borinage coalfield 231–2, 234–5
Boskoop 251
Brie 305–06
British Empire 5
British Leyland 102
British Steel Corporation 84–5
Brittany 120, 125
Brussels 229–30
Budget 21

Calabria 187, 280, 287–8
Campania 280, 285
Cassa per il Mezzogiorno 281–90
Catalonia 108, 317, 323, 328
Cereals 119–21, 305–06
Champagne Humide 306–08
Champagne Pouilleuse 306–08
Channel Tunnel 156, 170, 330, 335
Charleroi 229, 233
Chemical industry 113–7
Cities, industry in 59–63
City regions 62, 179–81, 221–6, 309–12
Coal 49–55, 205–9, 231–5

Coastal industry 54–5, 92–3
Cologne 205, 214, 215
Colonial empires 5
Common Agricultural Policy (CAP) 131–43
Competition policy 72–3
Containerisation 248–50
Continental climate 120–4
Copenhagen 260–3
Corby 82
Core areas 188
Cork 293–4, 297
Cotton industry 107–11
Couronne 316
Customs Union 149–51

Dairy farming 120, 255–7
Danish islands 256–7
Donegal 294
Dortmund 205–6, 216
Dublin 293–4, 297
Duisburg 205–6
Düsseldorf 214, 215

Economic divergence 330
Economic growth rates 12–14
Economic potential 194–6
Economic regions 191–4
Economies of scale 8–9
Electricity 44
Emscher Valley 205, 211, 207, 209
Energy policies 44–6
'E' road system 153, 170
Essonne 313–4
Euratom 12, 41, 329
European Agricultural Guidance and Guarantee Fund (EAGGF) 21, 73, 135, 200, 333
European Coal and Steel Community (ECSC) 11–12, 28–30, 73, 90–1, 329
European Free Trade Association (EFTA) 10
European Investment Bank (EIB) 21, 201, 285, 333
European Monetary System (EMS) 23, 333, 337
European Parliament 19, 331–3
European Regional Development Fund (ERDF) 202–3, 333
European Social Fund (ESF) 202, 333
Europoort 246–50
Exports 148, 150–1

Farm structure   128–31
Fiat   102, 104, 270, 272
Flanders   105–6, 228–9, 238–9
Fontanili   268
Footloose industry   55–9
Fos industrial zone   56
Frankfurt   178, 217, 221–3
Free trade   147

Genoa   265, 269–70, 273–5
Geographical inertia   48, 233
Greater London industry   59–63
Greece   16–17, 18, 291, 298–303
Green pound   139
Gross Domestic Product   189–90
Growthpoles   198–9, 314–5

Hague, The   245
Heavy Industrial Triangle   26, 49, 50–1, 86–9, 95–6, 191
Hellweg   211–2, 214
Hercynian plateaux   123
Horizontal integration   63, 102
Horticulture   124, 250–3
Hydro-electric energy   42–4, 114

Ile de France   304–6, 312
Imperial Chemical Industries   64, 115, 117
Imports   54–5, 146, 147–51
Industrial corporations   64–7
Industrial location   47–8, 50–1, 54–9, 71
Industrial policy—EC   72–3
Industrial Revolution   5–6
Inland freight   155, 158, 161
Inland waterways   156–8
Institutions of the EC   19, 170, 333
Integrated steelworks   84–5, 92–3
Intensive farming   124–5, 250–3
International division of labour   69, 331
Intra-Community trade   147–9
IRI   65, 67, 70
Iron and steel   81–98
Iron ore   82, 86, 88, 90, 91

Joint European Torus (JET)   46
Jutland   255–7, 258

Kempenland   51, 229, 232, 239

'La Defense' redevelopment scheme   313–5
Lancashire   107–8
Land use   118–28
Languages   171, 227, 238–9
Latifundia   129, 279, 282
Liège   232–3
Liguria   264–5, 269–70
Lille   51, 53
Limburg   26
Limerick   293, 294, 297
Lisbon   174, 320, 325
Lombardy   264, 267, 268, 271

Lomé Convention   137, 147, 151
London, central area   59–63
London, industries   88–9
Lorraine   190–1
Lotharingian axis   190–1
Luxembourg   88–9, 173

Madrid   323–4, 326–7
Manchester   179–80, 181–2
Manchester to Milan axis   182, 191–2
Mannheim   219, 223–4
Marginal farming   122–3
Marseilles   55, 56–7
Mass production   63–4
Massif central   1, 54, 134, 175–7, 178, 184–6
MCAs   139
Mediterranean climate   120–4
Mediterranean region   6, 14, 16–17, 71, 77–80, 139–41, 151, 317–9, 320–1, 327–8
Meseta   319, 324
Metropoles d'Equilibre   198–9
Mezzogiorno   184, 276–90
Milan   59, 268, 270–1
Mineral resources   114, 145–6, 318–9
Minifundia   128–9, 280
Motorways   158–61, 162–3, 219, 225, 287

Nanterre   100, 310
Naples   278–9, 286
National Coal Board   26–8
Nationalised industries   70
NATO   329
Natural gas   36
Neckar valley   224–6
Nord coalfield   26, 51–3
Normandy   308
North-Rhine Westphalia   205–16
North Sea oil   36–9
North west Europe   120
Nuclear energy   40–2

OEEC   10
Oil   30, 35, 36–9

PADOG   199, 313–6
Paris   304–5, 309–16
Peripheral regions   193
Petrochemicals   114–7
Peugeot-Citroën   100, 102
Physical regions   1–4
Picardie   305, 308
Piedmont   264–5, 266, 270, 271
Population decline   175–7, 192–4, 292–3
Population density   171–3
Population growth   177–8, 312–3
Population migration   292–3, 302, 309, 311–2, 325–7
Ports   165–8, 245–50, 260–1
Port Talbot steelworks   83, 84, 85
Portugal   319–22, 324–8
Primeurs   125
Protection   131–5

Railways   155–6
Randstad   241–5, 254
Region Parisiènne   312–6
Regional imbalance   187, 192–4, 202, 322–3
Regional planning   253–4
Remembrement   134, 199, 307
Renault   100
Rheingau   217
Rhinelands   213–6, 217–26, 240
Rhône valley   43
Rice production   267–8
Riviera dei Fiori   269, 275
Road transport   158–61
Rotterdam   245–50
Ruhr   86, 205–12

Saarland   31, 88–9
Sambre-Meuse coalfield   86, 232–3, 238
Sardinia   282, 289
Sea Transport   165–8
Selby   31
Self-sufficiency   22, 126–8
Shannon airport   297–8
Shetland Basin   39–40
Shipping   165–8
Sicily   282, 285–7, 289
Siegerland   213
Single European Act   23, 332–4
Single European Market   23, 334–6
Spain   317–9, 320–2, 323–4, 327–8
Spain, steel   94–5
Spain, textiles   108
Steel—coastal   92–3
Steel—France   86, 88–9
Steel—Italy   91, 93–4
Steel production   82

Steel—United Kingdom   81–5
Stuttgart   224–6
Superpowers   7–8

Tachograph   169
Taranto   94, 285, 286, 287
Technology in industry   68–9
Tertiary sector   73–5
Textiles industry   105–13
TGV   156
Tourism   75–80, 303, 320–2
Trade barriers   151, 334–5
Trade creation and diversion   149–51
Transport policy   169–70
Turin   271–3

Underdeveloped regions   324
United Kingdom—coalfields   25–6, 31, 49–50
United Kingdom—industry   59–65

Val d'Aosta   266, 273
Vale of Evesham   125
Van Valkenburg   140
Viticulture   124, 217–8, 266–7, 307
Volkswagen   102–3
Von Thunen   126

Wallonia   238–9
Waterford   295
Westland   251
Wolfsburg   102–3

Yaounde agreement   13, 137, 147
Yorkshire textiles   107–8, 111
Yvelines   313–5